Computational Microelectronics

Edited by
S. Selberherr

Christoph Jungemann
Bernd Meinerzhagen

Hierarchical Device Simulation

The Monte-Carlo Perspective

Springer-Verlag Wien GmbH

PD Dr.-Ing. Christoph Jungemann
Prof. Dr. Bernd Meinerzhagen

Institut für Theoretische Elektrotechnik und Mikroelektronik
Universität Bremen
Bremen, Germany

Originally published by Springer-Verlag Wien New York in 2003
Softcover reprint of the hardcover 1st edition 2003

Printed on acid-free and chlorine-free bleached paper
SPIN 10719122

With 147 Figures

CIP data applied for

ISSN 0179-0307
ISBN 978-3-7091-7226-1 ISBN 978-3-7091-6086-2 (eBook)
DOI 10.1007/978-3-7091-6086-2

Preface

This book summarizes the research of more than a decade. Its early motivation dates back to the eighties and to the memorable talks Dr. C. Moglestue (FHG Freiburg) gave on his Monte-Carlo solutions of the Boltzmann transport equation at the NASECODE conferences in Ireland. At that time numerical semiconductor device modeling basically implied the application of the drift-diffusion model. On the one hand, those talks clearly showed the potential of the Monte-Carlo model for an accurate description of many important transport issues that cannot adequately be addressed by the drift-diffusion approximation. On the other hand, they also clearly demonstrated that at that time only very few experts were able to extract useful results from a Monte-Carlo simulator.

With this background, Monte-Carlo research activities were started in 1986 at the University of Aachen (RWTH Aachen), Germany. Different to many other Monte-Carlo research groups, the Monte-Carlo research in Aachen took place in an environment of active drift-diffusion and hydrodynamic model development. As a consequence, the algorithmic advantages of the classical models have never been underestimated. Therefore, the focus was not on a competition between both modeling approaches and a quick replacement of the classical models by the Monte-Carlo model. (Though we have to admit that this virus also popped up in our own group from time to time.) On the contrary, right from the beginning until today at the University of Bremen, where our research group is situated since 1995, our focus was and is on developing a modeling system for technology computer aided design applications (TCAD), where both the classical models and the Monte-Carlo model should coexist in the most meaningful and mutually benefitting manner. Not only the improvement of the Monte-Carlo model itself was important with respect to its physical accuracy and obvious deficiencies in terms of efficiency, stochastic error control and overall user-friendliness. In addition much effort was devoted to the improvement of both the Monte-Carlo model and the classical models by exchanging information between both modeling levels. This book tries to describe the state of the art of Monte-Carlo semiconductor device modeling with respect to both of these aspects.

This work would not have been possible without the support of many colleagues and institutions. We are deeply indebted to Prof. W. L. Engl (RWTH Aachen) and Prof. H. K. Dirks (University of Kiel) who initiated and supported this research in the early days. Special thanks go to Prof. C. Jacoboni and

Prof. R. Brunetti (University of Modena) for their generous support and patience at the beginning of this research, when our group had nothing but basic and probably boring questions and nothing to offer that could have been of interest for their own research. Moreover, we would especially like to thank Dr. R. Thoma (Motorola, Phoenix) for his continuous support and the development of the generalized hydrodynamic model, which cleared the path for the system integration of the Monte-Carlo model. We are indebted to Dr. M. Rieger and Prof. P. Vogl (Technical University of Munich) for discussions and provision of their nonlocal empirical pseudopotential code. For the productive and enjoyable two years at Fujitsu, Japan, where the first author developed several of the ideas described in this work, we like to thank Dr. H. Goto and S. Yamaguchi. We appreciate the discussions with Dr. B. Heinemann (IHP, Frankfurt/Oder) on transport and noise in bipolar transistors and the supply of profiles and measured data. Discussions about noise with Prof. F. Bonani (Politecnico di Torino) are gratefully acknowledged. We thank Drs. A. Emunds (Aachen), I. Bork (Synopsis, Munich), H.-J. Peifer (FH Aachen), F. Bufler (ETH Zürich), P. Graf (Micronas, Freiburg), S. Decker (Infineon, Munich), B. Neinhüs (University of Bremen) and all other present and former members of our device modeling group for their invaluable contributions to this work. Finally, we owe thanks to Prof. S. Selberherr (Technical University of Vienna) for his long lasting patience and continuous encouragement and to Prof. R. W. Dutton of Stanford University for his support during the final preparation of this manuscript.

The authors gratefully acknowledge the financial support of their work by the German Ministry for Education and Research (BMBF) and by Motorola, SPS.

The second author is deeply indebted to the first author for the tremendous effort of writing this book without much assistance.

Stanford and Bremen, February 2003 C. Jungemann and B. Meinerzhagen

Contents

Notation

Symbols

x	scalar
$\mathbf{x}$	vector
$\hat{x}$	matrix
$\underline{x}$	complex quantity
$\check{x}$	stochastic quantity
δx	first-order term of the linear response expansion or the fluctuation around the expected value
$\mathrm{E}\{x\}$	expectation
$\sigma\{x\}$	standard deviation
$r\{x\}$	relative error
$\dot{x}$	time derivative

A_l	surface of lth contact
$A_n(\varepsilon)$	equienergy surface in $\mathbf{k}$-space
$A_{n,w,i}^{\mathrm{tet}}(\varepsilon)$	area of intersection of tetrahedron and equienergy surface
a_0^{RSi}	lattice constant of relaxed silicon
$\mathbf{a}_i, \mathbf{a}_i^{\mathrm{RSi}}$	basis vectors of direct lattice (of relaxed silicon)
α^{NP}	nonparabolicity factor
α^{cost}	cost factor
α^{CT}	parameter of Caughey-Thomas expression
$\alpha_{n,w,i}^{\mathrm{tet}}$	tetrahedron parameter
α^{II}	impact ionization coefficient
$\alpha_{\infty}^{\mathrm{II}}$	parameter of Chynoweth's law
$\mathbf{b}_i, \mathbf{b}_i^{\mathrm{RSi}}$	basis vectors of reciprocal lattice (of relaxed silicon)
β^{sc}	inverse screening length
$\beta_{n,w,i}^{\mathrm{tet}}$	tetrahedron parameter
$C\{X\}(t)$	correlation coefficient of microscopic quantity X
$C_{i,j}$	capacitance coefficient
$\mathbf{c}_{n,w,i,j,k}^{\mathrm{tet}}(\varepsilon)$	intersection of tetrahedron edge with equienergy surface
$\chi_{1,2}$	eigenvalues of the amplification matrix

D, D_{mic}	diffusion constant (based on microscopic/macroscopic relaxation time)
D_ε	energy diffusion constant
$D_{n,w}(\varepsilon)$	density of states
$\overline{D}_{n,w}(\varepsilon)$	upper bound of density of states
$D_{n,w,i}^{\text{tet}}(\varepsilon)$	density of states for a single tetrahedron
$D_t K_\eta$	deformation potential constant for phonon mode η
δ	half width of confidence interval
dp_{scat}	differential probability density for scattering after collisionless propagation
$\mathbf{E}(\mathbf{r}, t)$	electric field
$\mathbf{E}_{i,j}$	electric field in grid cell (i, j)
$\mathbf{E}_{i,j,k}$	electric field in grid cell (i, j) for kth particle
$\mathbf{E}_{i,j,k}^{\text{par}}(x_k, y_k)$	electric field in grid cell (i, j) due to kth particle at position (x_k, y_k)
$E_{\text{c}}, E_{\text{v}}$	conduction/valence band edge
$E_{\text{c/v}}^{\text{Ge/Bgn}}$	conduction/valence band shift due to Ge content or band gap narrowing
E_{C}	parameter of Chynoweth's law
E_{conf}	confining field with respect to impact ionization
E_{eff}	effective channel field
eq	label for quantities calculated under equilibrium conditions
$\epsilon(\mathbf{r})$	dielectric constant
ϵ_0	dielectric constant of vacuum
ϵ^{si}	dielectric constant of silicon
$\varepsilon_n(\mathbf{k})$	band energy
ε^{gap}	band gap energy
$\varepsilon_n^{\text{II}}(\mathbf{k})$	final energy after impact ionization
ε_{TH}	threshold energy of impact ionization
$\varepsilon_{n',n}^{\text{trans}}(\mathbf{k}' \vert \mathbf{k})$	energy of transition
η	phonon mode
$\mathbf{F}(\boldsymbol{\zeta}, t)$	force
$F(x)$	probability distribution function
$F_\alpha, F_{\mathbf{j}}$	abbreviations for balance equations
$f(\boldsymbol{\zeta}, t)$	distribution function in phase space
$\check{f}_n(\mathbf{r}, \mathbf{k}, t)$	stochastic distribution function in phase space
$f_{\text{T}}, f_{\text{T}}^{\text{sat}}$	cutoff frequency (saturated velocity)
$G^{\text{II}}(x, y)$	generation rate by impact ionization
$\mathbf{G}_{i,j,k}$	reciprocal lattice vector
$\mathbf{G}_{i,j,m,n}$	Green's function of electric field due to box charge
$G_x, \mathbf{G}_x, \hat{G}_x, G_x^Y$	Green's function of quantity x
G_α^β	Green's function for quantity α and source β

$G_{l,k}$	ohmic conductance tensor
Γ	upper bound of scattering rate
Γ_i	upper bound of scattering rate in phase space primitive
$\gamma_{n,w,i}^{\mathrm{tet}}$	tetrahedron parameter
$h_{\max,n,i}^{*,\mathrm{tet}}$	maximum edge length of tetrahedron
$\hbar$	Planck's constant divided by 2π
$h_l(\mathbf{r})$	test function of lth contact
$h_l^{\mathrm{RS}}(\mathbf{r})$	Ramo-Shockley test function of lth contact
$h_l^{\mathrm{Y}}(\mathbf{r})$	Yoder test function of lth contact
$I_l, I_{\max}$	current of lth contact, maximum contact current
$I_{\mathrm{drain}}, I_{\mathrm{sub}}$	drain/substrate current
$I^{\mathrm{par}}, I^{\mathrm{dis}}$	particle/displacement current
$\mathbf{i}(\mathbf{r},t)$	conduction current density
$\mathbf{j}(\mathbf{r},t), \mathbf{j}_p, \mathbf{j}_n$	particle current density (electron,holes)
$\mathbf{j}_X(\mathbf{r},t)$	current density of microscopic quantity X
$\mathbf{k}, \mathbf{k}^{\mathrm{b,a}}, \mathbf{k}^*$	wave vector, before/after reflection, in the first irreducible wedge of relaxed silicon
k_{B}	Boltzmann's constant
$L_{\mathrm{gate}}, L_{\mathrm{met}}$	gate/metallurgical channel length
Λ_α	Jacobian of momentum-based models
λ_{D}	Debye length
λ	mean free path length
$\lambda_{1,2}$	nonuniformly distributed random numbers
$M_{n',n}(\mathbf{k}'\|\mathbf{k})$	matrix element
$\hat{m}_n(\mathbf{k})$	mass tensor
$1/m^*$	trace of inverse mass tensor for equilibrium
m_n^{DOS}	density of states mass
m_0	free electron mass
$\mu, \mu_{\mathrm{mic}}, \mu_{\mathrm{mac}}$	mobility (based on microscopic/macroscopic relaxation time)
μ_{eff}	effective channel mobility
μ_ε	energy mobility
$\mu_{\min}^{\mathrm{CT}}, \mu_{\max}^{\mathrm{CT}}$	parameters of Caughey-Thomas expression
N_{par}	real particle number
N_{sim}	number of simulation particles
$N^{\mathrm{b}}, N^{\mathrm{a}}$	particle number before/after refresh
N_x, N_y	number of real space grid nodes in x/y-direction
N_n^{nod}	number of grid nodes in $\mathbf{k}$-space
N_n^{tet}	number of tetrahedra
$N_n^{\mathrm{reject}}(\varepsilon)$	number of rejection steps
N_i^{scat}	number of scattering events
$N_{\mathrm{acc}}, N_{\mathrm{don}}$	acceptor, donor concentration

N_{cont}	number of contacts
N_{dop}	net doping concentration
$N_{\mathrm{c}}, N_{\mathrm{v}}$	effective density of conduction/valence bands
$N_{\mathrm{inv}}, N_{\mathrm{depl}}$	particle sheet densities (inversion/depletion layer)
$N_{\mathrm{ref}}^{\mathrm{CT}}$	parameter of Caughey-Thomas expression
$n(\mathbf{r}, t)$	particle/electron density
n_i	particle density at time $i\delta t$ time $i\delta t$
n_{i}	intrinsic particle density
$n_i^{\mathrm{a,b}}$	number of new particles with old particle state after refresh
n	energy band index
$n(\varepsilon)$	phonon occupation number
$\mathbf{n}_{n,w,i,j}^{\mathrm{tet}}$	normal surface area vector in $\mathbf{k}$-space
$\mathbf{n}$	vector of unit length normal to interface in $\mathbf{k}$-space
$\nu_n(\mathbf{k})$	valley index
Ω	system volume
$\Omega_{i,j}$	real space grid cell
$\omega, \omega_{1,2}$	angular frequencies
ω_η	phonon frequency
ω_{pl}	plasma frequency
$\hat{P}$	discretized Poisson operator
P_X	intrinsic simulation-time factor
$P_{\mathrm{conf}}(X, \delta)$	confidence level
P_i	probability of ith event
$P_{n,w,i}^{\mathrm{tet}}(\varepsilon)$	after scattering probability for tetrahedron
$P\{X\}$	probability of event X
$P_i(t\|\boldsymbol{\zeta}_0, t_0)$	probability for i scattering events in $[t_0, t]$
P_i^{b}	probability for particle state before refresh
P_{diff}	probability for diffusive scattering at $\mathrm{Si/SiO_2}$-interface
$p(x), p(x, y)$	probability densities
$p(y\|x)$	conditional probability density
$\check{p}(\mathbf{k}, t)$	stochastic single-particle probability density in $\mathbf{k}$-space
$p(\mathbf{r}, t)$	hole density
$p(\boldsymbol{\zeta}, t)$	single-particle probability density in phase space
$p(\boldsymbol{\zeta}, t\|\boldsymbol{\zeta}_0, t_0)$	conditional probability density
$p_i(\boldsymbol{\zeta}, t\|\boldsymbol{\zeta}_0, t_0)$	conditional probability density for i scattering events
$p_i^{\mathrm{a}}(\boldsymbol{\zeta})$	after scattering probability density for ith scattering mechanism
$p(\boldsymbol{\zeta}_n, t_n; \cdots; \boldsymbol{\zeta}_1, t_1)$	n-time joint probability density
$\Phi(\mathbf{r}, t)$	quasifermi potential
ϕ	basis function of quasifermi potential
Φ_{MS}	metal semiconductor workfunction difference
$\varphi_{\xi\xi}, \varphi_{\xi'\xi'}$	Correlation function of (single-particle) Langevin force

$\Psi(\mathbf{r},t), \Psi_{\mathrm{sc}}, \Psi_{\mathrm{cb}}$	electrostatic potential (due to space charge/contact bias)
$\Psi_{i,j}$	node potential
$\boldsymbol{\Psi}$	vector of node potentials
Ψ_h	workfunction difference relative to relaxed silicon
$\mathbf{Q}$	vector of box charge
$Q_{i,j}^{\mathrm{box}}$	box charge
$Q^{\mathrm{par,box}}$	particle charge in box
Q_k^{sim}	charge of simulation particle
$Q_{n,i}^{*,\mathrm{tet}}$	volume quality of tetrahedron
q	positive electron charge
$\overline{q}$	interface charge density
R_l	contact resistance of lth terminal
$R_{\mathrm{B},n}$	equivalent noise resistance of the base
$\underline{R}_\alpha$	abbreviation for the right-hand side of momentum-based noise models
$\mathbf{r}$	real space vector
r, r_1, r_2	uniformly distributed random numbers
r	relative error
$r(\eta, \nu(\mathbf{k}'), \nu(\mathbf{k}))$	selection rule
$r_{\mathrm{p}}, r_{\mathrm{s}}$	Multiple Refresh parameters
ρ	mass density
ϱ	space charge density
$\mathbf{S}_{n,w,i,j}^{\mathrm{tet}}$	surface area vector for tetrahedron face
$S(\boldsymbol{\zeta})$	scattering rate
$S(\boldsymbol{\zeta}\vert\boldsymbol{\zeta}')$	transition rate
$S_n^{\mathrm{vr,ie}}(\mathbf{r},\varepsilon)$	scattering rate of velocity-randomizing/isotropic-elastic scattering
$S^{\mathrm{vr,ie}}(\boldsymbol{\zeta}\vert\boldsymbol{\zeta}')$	transition rate of velocity-randomizing/isotropic-elastic scattering
$S_{\mathrm{self}}(\boldsymbol{\zeta}\vert\boldsymbol{\zeta}')$	transition rate of self-scattering
$S_{\mathrm{tot}}(\boldsymbol{\zeta})$	total scattering rate
$S_{\eta,n',n}^{\mathrm{op}}(\mathbf{k}'\vert\mathbf{k})$	transition rate of optical phonons
$S_{n',n}^{\mathrm{ac}}(\mathbf{k}'\vert\mathbf{k})$	transition rate of acoustic phonons
$S_{n',n}^{\mathrm{BH}}(\mathbf{k}'\vert\mathbf{k})$	transition rate of Brooks-Herring model
$S_{n',n}^{\mathrm{ABH}}(\mathbf{k}'\vert\mathbf{k})$	approximated transition rate of Brooks-Herring model
$S_{n',n}^{\mathrm{CBH}}(\mathbf{k}'\vert\mathbf{k})$	corrected transition rate of Brooks-Herring model
$S^{\mathrm{II}}(\varepsilon)$	impact ionization scattering rate
$S_{n',n}^{\mathrm{II}}(\mathbf{k}'\vert\mathbf{k})$	transition rate of impact ionization
$\hat{S}(x_{\mathrm{Ge}}, y_{\mathrm{Ge}})$	stress matrix
$S_{\xi'\xi'}$	spectral intensity of single-particle Langevin force
$S_{I_l I_k}^{\mathrm{II}}$	terminal current noise due to impact ionization
$\hat{S}$	amplification matrix

$S_{i,j}$	elements of amplification matrix
$s_\alpha, s_\beta, s_\gamma$	parameters of amplification matrix
$\underline{S}_{XY}$	spectral intensity of the cross-correlation of X and Y
$\mathbf{s}_p, \mathbf{s}_n$	hole/electron energy current density
T	temperature
T_0	lattice temperature
T^*	dynamic temperature
$\hat{T}_i$	transformation matrix in Brillouin zone
T_{CPU}	CPU time
T_{sim}	simulation time
$t, t_0, t_1, \ldots$	times
t_{scat}	time of scattering event
$t_{n,w,i,j}$	node indices of tetrahedron
$\tau, \tau^{\text{vr,ie}}$	microscopic relaxation time (velocity-randomizing/isotropic-elastic scattering)
τ_X, τ_X^*	macroscopic relaxation time of microscopic quantity X
τ_{r}	relaxation time
$\tau^{\text{ABH}}{}_n(\varepsilon)$	microscopic relaxation time of the approximated Brooks-Herring model
$\tau_{\text{c}}\{X\}$	correlation time of microscopic quantity X
$\tau(x_{\text{b}}, x_{\text{e}}), \tau_l$	transit time
u_{l}	sound velocity
$V^{\text{i}}, V^{\text{a}}, V^{\text{cb}}$	inner and applied voltage, contact bias
$V_{\text{gate}}, V_{\text{drain}}$	gate/drain voltage
V_{T}, V_{T^*}	thermal voltage
$V_{n,i}^{\text{tet}}$	tetrahedron volume
$\mathbf{v}_n(\mathbf{k}), v_{\text{sat}}$	velocity, saturated drift velocity
$\mathbf{v}_\perp$	velocity perpendicular to contact surface
$w(\boldsymbol{\zeta})$	statistical weight function
$w_i, w_i^{\text{b}}, w_i^{\text{a}}$	statistical particle weight (before/after refresh)
w_p, w_n	hole/electron energy density
$X(\boldsymbol{\zeta})$	microscopic quantity
$\check{X}$	stochastic variable
$\tilde{X}^{\text{a}}, \tilde{X}^{\text{b}}$	ensemble average of microscopic quantity X before/after refresh
$x(\mathbf{r}, t)$	macroscopic density corresponding to the microscopic quantity X
x_{Ge}	relative germanium content in the SiGe layer
x_i	grid coordinate in x-direction
Ξ	deformation potential
$\xi, \xi', \xi_{\check{X}}$	(single-particle) Langevin force (for the quantity X)
$\boldsymbol{\zeta}$	particle state vector
$\boldsymbol{\zeta}_{\text{New}}(t\|\boldsymbol{\zeta}_0, t_0)$	solution of equations of motion

$\underline{Y}_{k,l}$	Y-parameter
y_j	grid coordinate in y-direction
$\zeta(N_{\mathrm{acc}}, N_{\mathrm{don}})$	fit factor of Brooks-Herring model
$\langle\rangle$	expectation in phase space
$\langle\rangle_{\mathbf{k}}$	expectation in **k**-space
$\langle\rangle_{\mathrm{hom}}$	expectation under spatially homogeneous conditions
$\langle\rangle_{\mathrm{hom}}^{\mathrm{LF}}$	expectation of stochastic variable under spatially homogeneous conditions

Abbreviations

BJT	bipolar junction transistor
BTE	Boltzmann transport equation
BZ	Brillouin zone
CCD	charge coupled devices
DD	drift-diffusion
DIBL	drain induced barrier lowering
DOS	density of states
FB	full-band
FET	field effect transistor
HBT	heterojunction bipolar transistor
HD	hydrodynamic
II	impact ionization
ie	isotropic-elastic
KS	**k**-space
LF	local field
MC	Monte-Carlo
MOS	metal oxide semiconductor
MR	Multiple Refresh
NEC	nearest element center
NSC	nonself-consistent
PE	Poisson equation
PS	phase space
RF	radio frequency
RS	real space
RSi	relaxed silicon
SC	self-consistent
Si	silicon
SiGe	silicon-germanium
SOI	silicon on insulator
SSi	strained silicon
TCAD	technology computer aided design
vr	velocity-randomizing

1 Introduction

The ongoing advance of CMOS based microelectronics is mainly due to the continuous reduction of the device feature size which is expected to decrease for at least another decade [1.1–1.4]. State of the art are gate lengths of 0.13μm, volume production of devices with about 0.10μm gate length is now beginning, and technologies with much shorter gate lengths are under development [1.5]. Additional performance improvements beyond device scaling are obtained by using improved device structures, such as SOI (e.g. [1.6]) or FinFETs (e.g. [1.7]). The introduction of the SiGe technology has opened up new possibilities previously only available in expensive III-V technologies like band-gap engineering and enhancement of carrier mobility by strain [1.8–1.11]. By using strained Si layers pseudomorphically grown on relaxed SiGe layers in the channel region of CMOS devices the performance of MOSFETs has considerably been improved [1.12–1.15]. The performance of Si BJTs has been enhanced by fabricating the base with strained SiGe pseudomorphically grown on the Si bulk [1.16–1.18].

These technologies open up many new opportunities for the improvement of integrated circuits. Unfortunately, to obtain the maximum circuit performance it is necessary to use many different devices specifically designed for certain tasks. According to the International Technology Roadmap for Semiconductors future integrated circuits will contain more than ten different types of devices complicating the design process [1.1]. Development of such kind of circuits and devices is not more possible without the extensive use of technology computer aided design (TCAD) based on computer models for both the technological processing and the electric device characteristics [1.19–1.21]. On the one hand, TCAD reduces the cost for the development of a device technology by replacing expensive technological experiments by inexpensive computer simulations, and on the other hand, it is often the only possibility to gain insight into the complex processes of device fabrication and particle transport. The degree of success of TCAD depends on the reliability and efficiency of the utilized computer models. Due to the above mentioned rapid progress of the Si technology, the introduction of new device types and materials, the simulation models are again and again pushed to their limits entailing continuous improvement or new developments.

At the beginning analytical MOS models were sufficient for predictive device modeling [1.22, 1.23]. But when the gate length reached the order of 1μm short channel and hot-electron effects became a problem [1.22–1.24] and analytical modeling had to be backed by more accurate models and a hierarchy of simulation tools emerged. The lowest level in terms of physics comprises the semiempirical semianalytical models having the advantage of being numerically extremely efficient (e.g. [1.25, 1.26]). Without them analog simulation of the ever increasing integrated circuits would not be possible [1.27]. The next level consists of numerical device models based on balance equations derived for different moments of the semiclassical Boltzmann transport equation (BTE) [1.28–1.30]. The most widely used model of this level is the drift-diffusion (DD) model [1.31, 1.32]. It is the workhorse of today's TCAD and its shortcomings, especially its failure to reproduce hot-electron effects like the velocity overshoot or parasitic currents, have led to the development of hydrodynamic (HD) models [1.33]. But these models are not without their own problems. In the case of deep submicron devices problems like spurious velocity overshoot [1.34] or artificially enhanced particle diffusion [1.35] occur. These problems are avoided by solving the BTE directly, and these models constitute the third level in the modeling hierarchy. Depending on the method used to solve the BTE these models can be separated into two groups. One group consists of numerical solvers (e.g. [1.36–1.39]), and the other group comprises stochastic approaches based on the Monte-Carlo (MC) method [1.40–1.42] (e.g. [1.43–1.57]). An advantage of these models is their deep physical foundation allowing to include various effects at a very fundamental physical level [1.58]. Unfortunately, this is achieved only at the cost of a higher computational burden which is about three orders of magnitude larger than the CPU times for the models of the second level.

Although the MC models based on the semiclassical BTE contain quantum mechanical effects like the band structure, scattering rates calculated by Fermi's Golden Rule, or size quantization effects, the transport is basically classical [1.50]. This has led to the emergence of models based on quantum mechanical transport equations (e.g. [1.59–1.63]). But the numerical efficiency of these quantum transport models is two or more orders of magnitude lower than the one of the classical MC models. Application of the quantum transport models is today only possible for bulk systems or overly simplified one-dimensional device structures nevertheless entailing excessive CPU times. Therefore, these models are not yet feasible in device modeling and not considered in this work. Another effect not addressed in this work is electron-electron scattering. Although the self-consistent solution of the Poisson equation (PE) and BTE used in this work contains to a certain degree the long-range electron-electron interaction, its impact is reduced by performing the simulations in the collisionless plasma approximation [1.64] and by the coarseness of the discretization grid in the real space (RS) [1.65]. The short-range electron-electron interaction, which can be modeled as a scattering process [1.66], is, for example, of importance for impact ionization (II) or gate currents at low drain voltages [1.67,1.68]. Evaluation of these phenomena is not attempted in this

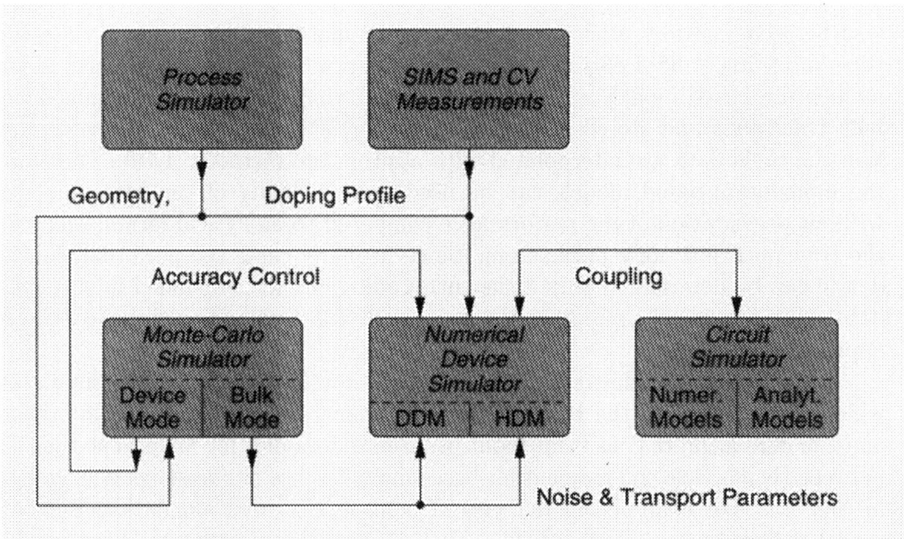

Figure 1.1: Hierarchical device simulation

work, because inclusion of electron-electron scattering increases the CPU times by two orders of magnitude [1.47]. On the other hand, it has been shown that the impact of electron-electron scattering on many quantities important for industrial TCAD is small (e.g. [1.69]).

Even without the above mentioned effects the CPU times of MC simulations are often prohibitive. These long CPU times have confined the MC models to the investigation of fundamental hot-carrier problems (e.g. [1.67, 1.70–1.75]). For the full characterization of the DC, AC, and noise properties of devices more efficient models are still mandatory, and the most widely used numerical device simulation tools are the momentum-based models [1.20, 1.76, 1.77]. On the other hand, these models contain many transport and noise parameters, which have not been determined by experiments. This problem was aggravated by the introduction of SiGe, which due to the higher degree of freedom (Ge content and strain level) requires many more parameters than unstrained Si. This led to the development of momentum-based models, which are derived from the BTE based on the same physical models as the MC model. This has the advantage of maximum consistency of the momentum and particle-based models, and all transport and noise parameters of the momentum-based models can be evaluated by the MC model [1.78–1.82]. This approach has been called hierarchical device simulation. Besides the evaluation of the transport and noise parameters by the more fundamental MC model, this approach has the advantage, that the accuracy of the momentum-based models for devices can be assessed by comparison with MC device simulations (e.g. [1.34, 1.35, 1.81, 1.83, 1.84]). The hierarchical device simulation approach is depicted in Fig. 1.1. The input of the device simulators is either generated by process simulation or by matching experimental data (inverse modeling [1.85–1.88]), or both. The transport and noise parameters of the momentum-based models are generated by MC simulations under homogeneous bulk conditions and stored in lookup tables for later use. For a device simulation the table data are interpolated with appropriate methods [1.79]. Therefore, the momentum-based models can be used independently of the MC simulator without degradation of the performance compared to analytical formulations for the transport and noise parameters. Since in many cases the performance of a device can be determined only if the surrounding circuit is included in the simulation, the numerical simulators can be operated in a mixed level device/circuit approach [1.89–1.92].

The main focus of this monograph is on the development of an MC simulator, which can be used as the basis of the hierarchical device simulation approach, and the derivation of consistent momentum-based models for SiGe. The relation between the particle and momentum-based models is not as one-sided as suggested in Fig. 1.1. For example, in the case of nonself-consistent simulations the MC model is based on an electric field evaluated with the momentum-based models [1.49, 1.69, 1.93, 1.94]. The CPU time necessary for certain MC simulations can be estimated in advance by the momentum-based noise models, and in the case of noise simulations for devices it is advisable to investigate the noise first with a momentum-based model in order to find out, whether an MC simulation is possible within the available finite CPU time or not [1.84, 1.95]. Thus, also MC simulation benefits from consistent momentum-based models. This synergy of the hierarchical approach improves the usefulness of each individual transport model of the hierarchy for TCAD.

In Chap. 2 a brief introduction into semiclassical transport physics based on the BTE [1.96] and the statistical description of the particle ensemble via the distribution function and conditional probability density [1.97, 1.98] is given. The derivation of balance equations from the BTE is outlined together with the macroscopic relaxation time approximation for the scattering integral, which form the basis for the derivation of the momentum-based models in Chap. 7 [1.28,1.33,1.78]. The microscopic relaxation time [1.99] is discussed in some detail, because it allows the construction of alternative estimators for important transport parameters, which are efficient in the warm-electron regime where the standard estimators are notoriously inefficient [1.45, 1.100]. Without these improved estimators generation of the lookup tables for the transport parameters containing over 50000 data points would be extremely CPU intensive. Furthermore, in the case of linear transport conditions, under which many of the fundamental physical experiments are performed, which are used to calibrate and verify the MC model, the respective transport parameters can be numerically calculated with the microscopic relaxation time avoiding lengthly MC simulations [1.99, 1.101]. In the last section of Chap. 2 fluctuations in the steady-state are discussed in the framework of the Langevin-type BTE [1.102]. Langevin-type balance equations are derived, which form the basis of the models for electronic noise given in Chap. 7 and of the analysis of the stochastic error of MC device simulations in Chap. 8. Again the microscopic relaxation time plays a fundamental role in the derivation of these models [1.103].

The MC method is introduced in Chap. 3. Basic MC methods for the generation of nonuniformly distributed random numbers are given [1.45, 1.104]. The MC algorithm for the BTE is developed with the corresponding Neumann series [1.98, 1.105] including self-scattering in order to improve the efficiency of the MC method [1.36, 1.49, 1.106]. As an important example the autocorrelation function of the velocity for a bulk system in equilibrium is analytically evaluated based on the Neumann series for the conditional probability density and the microscopic relaxation time. Basic statistics of MC simulations are discussed and a detailed analysis of convergence estimation based on confidence levels and intervals is given facilitating automatic error control during an MC simulation. The problem of bias in the estimated relative error due to the correlation in time of the sampled data is addressed [1.50]. A reliable automatic error control is a prerequisite for the efficient use of MC simulations in a TCAD framework, because otherwise efficient batch processing of MC runs is impossible.

In Chap. 4 the scattering mechanisms for SiGe are presented which are kept as simple as possible in order to allow the application of very efficient scattering algorithms but without sacrificing accuracy too much. The developed phonon model, for example, is based on the simple model of Refs. [1.104, 1.107]. Nevertheless, it yields results similar to the ones of much more complex models while being at the same time very CPU efficient. Impurity scattering is modeled in the same fashion including an empirical modification to obtain the correct low-field mobilities. The scattering models are carefully calibrated and verified, and the II model reproduces the crucial experimental data used in Ref. [1.108].

The full-band (FB) structure calculated by an appropriate method (e.g. [1.109]) is an indispensable ingredient of hot-electron modeling and hole transport in general. Since a closed analytical form of the FB structure does not exist, it has to be discretized over the entire Brillouin zone (BZ). The necessary algorithms for handling a discrete band structure are described in Chap. 5. In analogy to Refs. [1.110, 1.111] a nonuniform adaptive tetrahedral grid for the BZ is developed tailored to the special requirements of efficient FB-MC simulation of SiGe. Formulas for the calculation of the inverse mass tensor consistent with the nonuniform tetrahedral grid are given, and a highly efficient scheme for the calculation of the state after scattering is presented which exploits the special properties of the scattering processes. Together with the other procedures outlined in Chap. 5 this results in an FB-MC simulator being as efficient as an MC simulator based on analytical bands.

Chap. 6 contains the details of the device related parts of the MC simulator. The RS is discretized with a tensor-product grid for pseudomorphically grown structures. The self-consistent solution of the PE and the MC model with a forward Euler scheme is explained. The stability of the algorithm [1.112] and the reduction of artificial self-forces with a variant of the NEC scheme [1.113] is investigated. The nonself-consistent device mode for very efficient hot-electron simulations is described next [1.93]. Statistics of rare events is enhanced by a population-control method based on an ensemble MC method with nonuniformly weighted test particles. It is shown how the variance of selected quantities is reduced by importance sampling resulting in a variance redistribution without biasing the estimators for the expected values [1.114]. The terminal currents are estimated by a generalized Ramo-Shockley theorem [1.115] including contact resistances.

The momentum-based models are derived in Chap. 7, where the derivation of the HD model for a position-dependent band structure is discussed in detail. The noise analysis is based on a small-signal expansion of the HD model, where the Joule term leads to a distinctively different formulation of the Green's functions compared to the DD model. Modified Langevin forces are introduced, which yield a better agreement with fluctuations on the bulk level than exact ones. Since the DD model is a straight forward approximation of the HD model, its derivation is kept brief. Based on the DD model a proof of the Nyquist theorem [1.102] is given for multi-terminal and multi-dimensional devices [1.103]. Furthermore, the extraction of transport and noise parameters from MC bulk simulations is discussed.

The stochastic properties of MC device simulations are analyzed in Chap. 8. Based on a thorough noise analysis it is shown that the efficiency of terminal current estimators can not be improved by optimized test functions as proposed in Ref. [1.116]. In addition, the bias of the variance of terminal current estimators is investigated and shown to be more complicated than the bias in the case of bulk quantities easily leading to wrong results. A method for the in-advance estimation of the CPU time of MC simulations is developed next, which makes use of the momentum-based noise models. In addition, analytical expressions are derived.

In Chap. 9 the different aspects of the developed simulation hierarchy are demonstrated by application to three different device types, 1D N^+NN^+ and P^+PP^+ structures, MOSFETs, and SiGe HBTs. The 1D structures are investigated, because their simple structure makes very CPU efficient MC simulations possible, and noise is simulated with all three models in the GHz-range. The impact of the velocity overshoot in deep sub-micron NMOSFETs is investigated by a comparison of consistent MC and DD simulations, where the latter model does not contain any velocity overshoot effects at all. The differences between both models are quantified for the drain current and cutoff frequency of NMOSFETs with gate lengths from 80 to 250nm. The MC model is used to investigate the transport in MOSFETs with strained Si channels and PMOSFETs with strained SiGe channels. II and the resultant substrate currents are investigated by nonself-consistent simulations of which it is shown that they yield the same results as self-consistent simulations while they are orders of magnitude more efficient. The problem of the so-called surface II in MOSFETs [1.117] is addressed and it is found that there is no difference between bulk and surface II. The accuracy of the MC simulator is demonstrated by comparison with experimental data for NMOSFETs of technology generations from 256KBit to 1GBit. In the case of SiGe HBTs the DC and AC behavior is simulated and the accuracy of the momentum-based models is assessed by comparison with the MC model. It is shown that the velocity overshoot neglected in the DD model leads to a significant reduction of the base transit time. A detailed analysis of the noise in HBTs is performed with the HD model and some short-comings of compact models are addressed.

References

[1.1] SEMATECH, "The international technology roadmap for semiconductors", *http://public.itrs.net/Home.htm*, 2001.

[1.2] A. W. Wieder, "Status, trends and challenges in microelectronics for the next 10 to 15 years", *Electrical Engineering*, vol. 79, pp. 79–84, 1996.

[1.3] Y. Taur and E. J. Nowak, "CMOS devices below 0.1μm: How high will performance go?", *IEDM Tech. Dig.*, pp. 215–218, 1997.

[1.4] Y. Taur, C. H. Wann, and D. J. Frank, "25 nm CMOS design considerations", *IEDM Tech. Dig.*, pp. 789–792, 1998.

[1.5] IEEE, "International Electron Devices meeeting", *Technical Digest*, 2002.

[1.6] H.-S. P. Wong, D. J. Frank, and P. M. Solomon, "Device design considerations for double-gate, ground-plane, and single-gated ultra-thin SOI MOSFET's at the 25 nm channel length generation", *IEDM Tech. Dig.*, pp. 407–410, 1998.

[1.7] D. Hisamoto, W.-C. Lee, J. Kedzierski, H. Takeuchi, K. Asano, C. Kuo, and E. Anderson T.-J. King J. Bokor C. Hu, "FinFET–a self-aligned double-gate MOSFET scalable to 20 nm", *IEEE Trans. Electron Devices*, vol. 47, no. 12, pp. 2320–2325, 2000.

[1.8] U. König, "Electronic Si/SiGe devices: Basics, technology, performance", in *Advances in Solid State Physics (Festkörperprobleme)*, vol. 32, pp. 199–220. Vieweg, Braunschweig, 1992.

[1.9] J. D. Cressler, "Re-engineering silicon: Si-Ge heterojunction bipolar technology", *IEEE Spectrum*, vol. 3, pp. 49–55, 1995.

[1.10] F. Schäffler, "High-mobility Si and Ge structures", *Semicond. Sci. Technol.*, vol. 12, pp. 1515–1549, 1997.

[1.11] C. K. Maiti and G. A. Armstrong, *Applications of Silicon-Germanium Heterostructure Devices*, Series in Optics and Optoelectronics. Institute of Physics Publishing, Bristol, Philadelphia, 2001.

[1.12] U. König, M. Glück, and G. Höck, "Si/SiGe field-effect transistors", *J. Vac. Sci. Tech. B*, vol. 16, no. 5, pp. 2609–2614, 1998.

[1.13] J. L. Hoyt, H. M. Nayfeh, S. Eguchi, I. Aberg, G. Xia, T. Drake, E. A. Fitzgerald, and D. A. Antoniadis, "Strained silicon MOSFET technology", *IEDM Tech. Dig.*, pp. 23–26, 2002.

[1.14] K. Rim, S. Koester, M. Hargrove, J. Chu, P. M. Mooney, J. Ott, T. Kanarsky, P. Ronsheim, M. Ieong, A. Grill, and H.-S. P. Wong, "Strained Si NMOSFETs for high performance CMOS technology", *Symposium on VLSI Technology Digest of Technical Papers*, pp. 59–60, 2001.

[1.15] S. Thompson, N. Anand, M. Armstrong, C. Auth, B. Arcot, M. Alavi, P. Bai, J. Bielefeld, R. Bigwood, J. Brandenburg, M. Buehler, S. Cea, V. Chikarmane,

C. Choi, R. Frankovic, T. Ghani, G. Glass, W. Han, T. Hoffmann, M. Hussein, P. Jacob A. Jain, C. Jan, S. Joshi, C. Kenyon, J. Klaus, S. Klopic, J. Luce, Z. Ma, B. Mcintyre, K. Mistry, A. Murthy, P. Nguyen, H. Pearson, T. Sandford, R. Schweinfurth, R. Shaheed, S. Sivakumar, M. Taylor, B. Tufts, C. Wallace, P. Wang, C. Weber, and M. Bohr, "A 90nm logic technology featuring 50nm strained silicon channel transistors, 7 layers of Cu interconnects, low k ILD, and 1um^2 SRAM cell", *IEDM Tech. Dig.*, pp. 61–64, 2002.

[1.16] S. S. Iyer, G. L. Patton, J. M. C. Stork, B. S. Meyerson, and D. L. Harame, "Heterojunction bipolar transistors using Si-Ge alloys", *IEEE Trans. Electron Devices*, vol. 36, no. 10, pp. 2043–2064, 1989.

[1.17] A. Schüppen, "SiGe–HBTs for mobile communication", *Solid–State Electron.*, vol. 43, pp. 1373–1381, 1999.

[1.18] I. M. Anteney, G. Lippert, P. Ashburn, H. J. Osten, B. Heinemann, G. J. Parker, and D. Knoll, "Characterization of the effectiveness of carbon incorporation in SiGe for the elimination of parasitic energy barriers in SiGe HBTs", *IEEE Electron Device Lett.*, vol. 20, no. 3, pp. 116–118, 1999.

[1.19] W. L. Engl, H. K. Dirks, and B. Meinerzhagen, "Device modeling", *Proc. IEEE*, vol. 71, pp. 10–33, 1983.

[1.20] S. Selberherr, *Analysis and Simulation of Semiconductor Devices*, Springer, Wien, 1984.

[1.21] R. Dutton, "TCAD-Yesterday, today and tomorrow", *IEICE Trans. on Electronics*, vol. E82-C, no. 6, pp. 791–799, 1999.

[1.22] S. M. Sze, *Physics of Semiconductors Devices*, Wiley, New York, 1981.

[1.23] C.-T. Sah, *Fundamentals of Solid-State Electronics*, World Scientific, Singapore, 1991.

[1.24] T. H. Ning, P. W. Cook, R. H. Dennard, C. M. Osburn, S. E. Schuster, and H.-N. Yu, "1 μm MOSFET VLSI technology: Part iv—Hot-electron design constrains", *IEEE Trans. Electron Devices*, vol. 26, no. 4, pp. 346–353, 1979.

[1.25] N. Arora, *MOSFET Models for VLSI Circuit Simulation. Theory and Practice*, Springer, Wien, 1993.

[1.26] H. C. de Graaff and F. M. Klaassen, *Compact Transistor Modelling for Circuit Design*, Springer, Wien, 1990.

[1.27] W. L. Engl, R. Laur, and H. K. Dirks, "Medusa—A simulator for modular circuits", *IEEE Trans. Computer-Aided Des.*, vol. 1, pp. 85–93, 1982.

[1.28] M. Lundstrom, *Fundamentals of Carrier Transport*, vol. 10 of *Modular Series on Solid State Devices*, Addison-Wesley, New York, 1990.

[1.29] W. Hänsch, *The Drift Diffusion Equation and Its Application in MOSFET Modeling*, Springer, Wien, 1991.

[1.30] A. Schenk, *Advanced Physical Models for Silicon Device Simulation*, Springer, Wien, 1998.

[1.31] W. V. Roosbroeck, "Theory of the flow of electrons and holes in germanium and other semiconductors", *Bell System Technical Journal*, pp. 561–607, 1950.

[1.32] W. Shockley, *Electrons and Holes in Semiconductors*, van Nostrand, Princeton, New Jersey, 1950.

[1.33] K. Bløtekjær, "Transport equations for electrons in two-valley semiconductors", *IEEE Trans. Electron Devices*, vol. 17, no. 1, pp. 38–47, 1970.

[1.34] B. Meinerzhagen, R. Thoma, H. J. Peifer, and W. L. Engl, "On the consistency of the hydrodynamic and the Monte Carlo models", in *Proc. IWCE*, Illinois, 1992, Urbana–Champaign, pp. 7–12.

[1.35] I. Bork, C. Jungemann, B. Meinerzhagen, and W. L. Engl, "Influence of heat flux on the accuracy of hydrodynamic models for ultrashort Si MOSFETs", in *NUPAD Tech. Dig.*, Honolulu, 1994, vol. 5.

[1.36] H. D. Rees, "Calculation of steady state distribution functions by exploiting stabiblity", *Phys. Lett. A*, vol. 26, pp. 416–417, 1968.

[1.37] K. A. Hennacy and Neil Goldsman, "A generalized Legendre polynimial/sparse matrix approach for determining the distribution function in non-polar semiconductors", *Solid-State Electron.*, vol. 36, pp. 869–877, 1993.

[1.38] C. Jungemann, P. Graf, G. Zylka, R. Thoma, and W. L. Engl, "New highly efficient method for the analysis of correlation functions based on a spherical harmonics expansion of the BTE's Green's function", in *Proc. IWCE*, Portland, Oregon, May 1994, pp. 45–48.

[1.39] M. C. Vecchi and M. Rudan, "Modeling electron and hole transport with full-band structure effects by means of the spherical-harmonics expansion of the BTE", *IEEE Trans. Electron Devices*, vol. 45, no. 1, pp. 230–238, 1998.

[1.40] J. M. Hammarsley and D. C. Handscomb, *Monte Carlo Methods*, Methuen/Chapman and Hall, London, 1964.

[1.41] S. M. Ermakow, *Die Monte-Carlo-Methode und verwandte Fragen*, R. Oldenbourg, München, Wien, 1975.

[1.42] R. Y. Rubinstein, *Simulation and the Monte Carlo method*, Wiley series in probability and mathematical statistics. John Wiley & Sons, New York, 1981.

[1.43] T. Kurosawa, "Monte Carlo calculation of hot electron problems", *J. Phys. Soc. Jap.*, vol. 21, pp. 424–426, 1966.

[1.44] W. Fawcett, A. D. Boardman, and S. Swain, "Monte Carlo determination of electron transport properties in gallium arsenide", *J. Phys. Chem. Solids*, vol. 31, pp. 1963–1990, 1970.

[1.45] P. J. Price, "Monte Carlo calculation of electron transport in solids", *Semiconductors and Semimetals*, vol. 14, pp. 249–309, 1979.

[1.46] J. Y. Tang, H. Shichijo, K. Hess, and G. J. Iafrate, "Band-structure dependent impact ionization in silicon and gallium arsenide", *Journal de Physique*, vol. 42, pp. 63–69, 1981.

[1.47] M. V. Fischetti and S. E. Laux, "Monte Carlo analysis of electron transport in small semiconductor devices including band–structure and space–charge effects", *Phys. Rev. B*, vol. 38, pp. 9721–9745, 1988.

[1.48] H. J. Peifer, R. Thoma, A. Emunds, and W. L. Engl, "Hot carriers in a LDD–MOSFET investigated with a Monte Carlo simulator", in *VLSI Process/Device Modeling Workshop*, Tokyo, Aug. 1988.

[1.49] E. Sangiorgi, B. Riccò, and F. Venturi, "MOS^2: An efficient Monte Carlo simulator for MOS devices", *IEEE Trans. Computer–Aided Des.*, vol. 7, no. 2, pp. 259–271, 1988.

[1.50] C. Jacoboni and P. Lugli, *The Monte Carlo Method for Semiconductor Device Simulation*, Springer, Wien, 1989.

[1.51] K. Kometer, G. Zandler, and P. Vogl, "Lattice-gas cellular-automaton method for semiclassical transport in semiconductors", *Phys. Rev. B*, vol. 46, pp. 1382–1394, 1992.

[1.52] C. Moglestue, *Monte Carlo Simulation of Semiconductor Devices*, Chapman & Hall, London, 1993.
[1.53] K. Tomizawa, *Numerical Simulation of Submicron Semiconductor Devices*, Artech House, Boston, 1993.
[1.54] A. Abramo, L. Baudry, R. Brunetti, R. Castagne, M. Charef, F. Dessenne, P. Dollfus, R. Dutton, W. L. Engl, R. Fauquembergue, C. Fiegna, M. V. Fischetti, S. Galdin, N. Goldsman, M. Hackel, C. Hamaguchi, K. Hess, K. Hennacy, P. Hesto, J. M. Higman, T. Iizuka, C. Jungemann, Y.Kamakura, H. Kosina, T. Kunukiyo, S. E. Laux, H. Lin, C. Maziar, H. Mizuno, H. J. Peifer, S. Ramaswamy, N. Sano, P. G. Scrobohaci, S. Selberherr, M. Takenaka, T. Tang, J. L. Thobel, R. Thoma, K. Tomizawa, M. Tomizawa, T. Vogelsang, S. Wang, X. Wang, C.Yao, P. D. Yoder, and A. Yoshii, "A comparison of numerical solutions of the Boltzmann transport equation for high-energy electron transport silicon", *IEEE Trans. Electron Devices*, vol. 41, no. 9, pp. 1646–1654, 1994.
[1.55] J. J. Ellis-Monaghan, K. W. Kim, and M. A. Littlejohn, "A Monte Carlo study of hot electron injection and interface state generation model for silicon metal-oxide-semiconductor field-effect transistors", *J. Appl. Phys.*, vol. 75, pp. 5087–5094, 1994.
[1.56] T. Kunikiyo, M. Takenaka, Y. Kamakura, M. Yamaji, H. Mizuno, M. Morifuji, K. Taniguchi, and C. Hamaguchi, "A Monte Carlo simulation of anisotropic electron transport in silicon including full band structure and anisotropic impact-ionization model", *J. Appl. Phys.*, vol. 75, pp. 297–312, 1994.
[1.57] P. D. Yoder and K. Hess, "First-principles Monte Carlo simulation of transport in Si ", *Semicond. Sci. Technol.*, vol. 9, pp. 852–854, 1994.
[1.58] K. Hess, Ed., *Monte Carlo Device Simulation: Full Band and Beyond*, Kluwer, Boston, 1991.
[1.59] J. R. Barker and D. K. Ferry, "On the physics and modeling of small semiconductor devices -I", *Solid–State Electron.*, vol. 23, pp. 519–530, 1980.
[1.60] F. Rossi, C. Jacoboni, and M. Nedjalkov, "A Monte Carlo solution of the Wigner transport equation", *Semicond. Sci. Technol.*, vol. 9, pp. 934–936, 1994.
[1.61] C. Schulz-Mirbach, "The path integral Monte Carlo method for quantum transport on a parallel computer", in *Proc. SISDEP*, Erlangen, 1995, vol. 6, pp. 392–395.
[1.62] M. V. Fischetti, "Theory of electron transport in small semiconductor devices using the Pauli master equation", *J. Appl. Phys.*, vol. 83 (1), pp. 270–291, 1998.
[1.63] S. Datta, "The non-equilibrium green's function (NEGF) formalism: An elementary introduction", *IEDM Tech. Dig.*, pp. 701–704, 2002.
[1.64] A. Pacelli and U. Ravaioli, "Analysis of variance-reduction schemes for ensemble Monte-Carlo simulation of semiconductor devices", *Solid–State Electron.*, vol. 41, pp. 599–605, 1997.
[1.65] R.W. Hockney and J.W. Eastwood, *Computer Simulation Using Particles*, Institute of Physics Publishing, Bristol, Philadelphia, 1988.
[1.66] P. Lugli and D. K. Ferry, "Electron-electron interaction and high field transport in Si", *Appl. Phys. Lett.*, vol. 46, pp. 594–596, 1985.
[1.67] M. V. Fischetti, S. E. Laux, and E. Crabbe, "Understanding hot-electron transport in silicon devices: Is there a shortcut?", *J. Appl. Phys.*, vol. 78, pp. 1058–1087, 1995.

[1.68] M. Y. Chang, D. W. Dyke, C. C. C. Leung, and P. A. Childs, "Modelling of gate currents in MOSFETs operating at low drain voltages", in *Proc. ESSDERC*, Bologna, 1996, vol. 26, pp. 263–266.

[1.69] J. D. Bude and M. Mastrapasqua, "Impact ionization and distribution functions in sub-micron nMOSFET technologies", *IEEE Electron Device Lett.*, vol. 16, no. 10, pp. 439–441, 1995.

[1.70] R. Brunetti and C. Jacoboni, "Transient and stationary properties of hot-carrier diffusivity in semiconductors", in *Semiconductors Probed by Ultrafast Laser Spectroscopy*, vol. 1, pp. 367–412. Academic Press, New York, 1984.

[1.71] S. E. Laux and M. V. Fischetti, "Monte-Carlo Simulation of Submicrometer Si n-MOSFET's at 77 and 300 K", *IEEE Electron Device Lett.*, vol. 9, no. 9, pp. 467–469, 1988.

[1.72] J. Bude and K. Hess, "Impact ionization in semiconductors: Effects of high electric fields and high scattering rates", *Phys. Rev. B*, vol. 45, pp. 10958–10964, 1992.

[1.73] J. D. Bude, "Gate current by impact ionization feedback in sub-micron MOSFET technologies", in *1995 Symposium on VLSI technology*, AT & T Bell Laboratories, 600 Mountain Avenue, Murray Hill, NJ 07974, 1995.

[1.74] M. V. Fischetti and S. E. Laux, "Monte Carlo study of sub-band-gap impact ionization in small silicon field-effect transistors", in *IEDM*, 1995, pp. 305–308.

[1.75] R. B. Hulfachor, K. W. Kim, M. A. Littlejohn, and C. M. Osburn, "Comparative analysis of hot electron injection and induced device degradation in scaled 0.1 μm SOI n-MOSFET's using Monte Carlo simulation", *IEEE Electron Device Lett.*, vol. 17, no. 2, pp. 53–55, 1996.

[1.76] S. E. Laux, "Techniques for small-signal analysis of semiconductor devices", *IEEE Trans. Electron Devices*, vol. 32, no. 10, pp. 2028–2037, Oct. 1985.

[1.77] F. Bonani and G. Ghione, *Noise in Semiconductor Devices, Modeling and Simulation*, Advanced Microelectronics. Springer, Berlin, Heidelberg, New York, 2001.

[1.78] R. Thoma, A. Emunds, B. Meinerzhagen, H. J. Peifer, and W. L. Engl, "Hydrodynamic equations for semiconductors with nonparabolic bandstructures", *IEEE Trans. Electron Devices*, vol. 38, no. 6, pp. 1343–1352, 1991.

[1.79] B. Neinhüs, S. Decker, P. Graf, F. M. Bufler, and B. Meinerzhagen, "Consistent hydrodynamic and Monte-Carlo simulation of SiGe HBTs based on table models for the relaxation times", *VLSI Design*, vol. 8, pp. 387–391, 1998.

[1.80] P. Shiktorov, E. Starikov, V. Gruzinskis, T. Gonzalez, J. Mateos, D. Pardo, L. Reggiani, L. Varani, and J. C. Vaissere, "Langevin forces and generalized transfer fields for noise modeling in deep submicron devices", *IEEE Trans. Electron Devices*, vol. 47, no. 10, pp. 1992–1998, 2000.

[1.81] C. Jungemann, B. Neinhüs, and B. Meinerzhagen, "Comparative study of electron transit times evaluated by DD, HD, and MC device simulation for a SiGe HBT", *IEEE Trans. Electron Devices*, vol. 48, no. 10, pp. 2216–2220, 2001.

[1.82] C. Jungemann, B. Neinhüs, and B. Meinerzhagen, "Hierarchical 2–D DD and HD noise simulations of Si and SiGe devices: Part I—Theory", *IEEE Trans. Electron Devices*, vol. 49, no. 7, pp. 1250–1257, 2002.

[1.83] H. J. Peifer, B. Meinerzhagen, R. Thoma, and W. L. Engl, "Evaluation of impact ionization modeling in the framework of hydrodynamic equations", in

IEDM Tech. Dig., 1991, pp. 131–134.
[1.84] C. Jungemann, B. Neinhüs, S. Decker, and B. Meinerzhagen, "Hierarchical 2–D DD and HD noise simulations of Si and SiGe devices: Part II—Results", *IEEE Trans. Electron Devices*, vol. 49, no. 7, pp. 1258–1264, 2002.
[1.85] G. J. L. Ouwerling, "Physical parameter extraction by inverse device modelling: Application to one- and two-dimensional doping profiling", *Solid–State Electron.*, vol. 33, pp. 757–771, 1990.
[1.86] N. Khalil, J. Faricelli, D. Bell, and S. Selberherr, "The extraction of two-dimensional MOS transistor doping via inverse modeling", *IEEE Electron Device Lett.*, vol. 16, no. 1, pp. 17–19, 1995.
[1.87] H. Goto, S. Yamaguchi, and C. Jungemann, "Inverse modeling as a basis for predictive device simulation of deep submicron metal-oxide-semiconductor field effect transistors", *Jpn. J. Appl. Phys.*, vol. 37, pp. 5437–5443, 1998.
[1.88] Z. K. Lee, M. B. McIlrath, and D. A. Antoniadis, "Two-dimensional doping profile characterization of MOSFET's by inverse modeling using I-V characteristics in the subthreshold region", *IEEE Trans. Electron Devices*, vol. 46, no. 8, pp. 1640–1649, 1999.
[1.89] B. Meinerzhagen, J. M. J. Krücken, K. H. Bach, F. M. Stecher, and W. L. Engl, "A modular approach to parallel mixed level device/circuit simulation", in *VLSI Process/Device Modeling Workshop*, Kawasaki, Aug. 1990, pp. 170–172.
[1.90] T. Kobori and T. Wada, "Efficient device simulation for small scale circuit level analysis", *IEICE Trans. on Electronics*, vol. 74, pp. 1634–1640, 1991.
[1.91] K. Mayaram and D. O. Pederson, "Coupling algorithms for mixed-level circuit and device simulation", *IEEE Trans. Computer–Aided Des.*, vol. 11, pp. 1003–1012, 1992.
[1.92] T. Grasser and S. Selberherr, "Fully coupled electrothermal mixed–mode device simulation of SiGe HBT circuits", *IEEE Trans. Electron Devices*, vol. 48, no. 7, pp. 1421–1427, 2001.
[1.93] J. M. Higman, K. Hess, C. G. Hwang, and R. W. Dutton, "Coupled Monte Carlo-drift diffusion analysis of hot-electron effects in MOSFET's", *IEEE Trans. Electron Devices*, vol. 36, no. 5, pp. 930–937, 1989.
[1.94] C. Jungemann and B. Meinerzhagen, "On the applicability of nonself-consistent Monte Carlo device simulations", *IEEE Trans. Electron Devices*, vol. 49, no. 6, pp. 1072–1074, 2002.
[1.95] C. Jungemann and B. Meinerzhagen, "In-advance CPU time analysis for Monte Carlo device simulations", *Proc. SISPAD*, Kobe (Japan), 2002.
[1.96] O. Madelung, *Introduction to Solid State Theory*, Springer, Berlin, 1978.
[1.97] A. Reklaitis, "The calculation of electron transient response in semiconductors by the Monte Carlo technique", *Phys. Lett.*, vol. 13, pp. 367–370, 1982.
[1.98] H.-J. Peifer, "Monte–Carlo Simulation des Hochenergietransports von Elektronen in submikron MOS–Strukturen", Doctor thesis, RWTH Aachen, Aachen, 1992, Augustinus Buchhandlung.
[1.99] W. Brauer and H. W. Streitwolf, *Theoretische Grundlagen der Halbleiterphysik*, Vieweg, Braunschweig, 2nd edition, 1977.
[1.100] R. Stratton, "Diffusion of hot and cold electrons in semiconductor barriers", *Phys. Rev.*, vol. 126, pp. 2002–2013, 1962.
[1.101] H. Nakagawa and S. Zukotynski, "Drift mobility and Hall coefficient factor of

holes in germanium and silicon", *Can. J. Phys.*, vol. 56, pp. 364–372, 1978.

[1.102] Sh. Kogan, *Electronic Noise and Fluctuations in Solids*, Cambridge University Press, Cambridge, New York, Melbourne, 1996.

[1.103] E. V. Sukhorukov and D. Loss, "Noise in multiterminal diffuse conductors: Universality, nonlocality, and exchange effects", *Phys. Rev. B*, vol. 59, pp. 13054–13066, 1999.

[1.104] C. Jacoboni and L. Reggiani, "The Monte Carlo method for the solution of charge transport in semiconductors with application to covalent materials", *Rev. Mod. Phys.*, vol. 55, pp. 645–705, 1983.

[1.105] M. Nedjalkov and P. Vitanov, "Iteration approach for solving the Boltzmann equation with the Monte Carlo method", *Solid-State Electron.*, vol. 32, pp. 893–896, 1989.

[1.106] R. Brunetti, C. Jacoboni, A. Matulionis, and V. Dienys, "Effect of interparticle collisions on energy relaxation of carriers in semiconductors", *Physica*, vol. 134B, pp. 369–373, 1985.

[1.107] R. Brunetti, C. Jacoboni, F. Nava, L. Reggiani, G. Bosman, and R. J. J. Zijlstra, "Diffusion coefficient of electrons in silicon", *J. Appl. Phys.*, vol. 52, pp. 6713–6722, 1981.

[1.108] E. Cartier, M. V. Fischetti, E. A. Eklund, and F. R. McFeely, "Impact ionization in silicon", *Appl. Phys. Lett.*, vol. 62, pp. 3339–3341, 1993.

[1.109] M. L. Cohen and J. R. Chelikowsky, *Electronic Structure and Optical Properties of Semiconductors*, Springer, New York, 2nd edition, 1989.

[1.110] R. K. Smith and J. Bude, "Highly efficient full band Monte Carlo simulations", in *Proceedings of the International Workshop on Computational Electronics*, Leeds, Aug. 1993, pp. 224–230.

[1.111] E. X. Wang, M. D. Giles, S. Yu, F. A. Leon, A. Hiroki, and S. Odanaka, "Recursive M–tree method for 3–D adaptive tetrahedral mesh refinement and its application to Brillouin zone discretization", in *Proc. SISPAD*, Tokyo, Sept. 1996, pp. 67–68.

[1.112] P. W. Rambo and J. Denavit, "Time stability of Monte Carlo device simulation", *IEEE Trans. Computer-Aided Des.*, vol. 12, pp. 1734–1741, 1993.

[1.113] S. E. Laux, "On particle-mesh coupling in Monte Carlo semiconductor device simulation", *IEEE Trans. Computer-Aided Des.*, vol. 15, pp. 1266–1277, 1996.

[1.114] M. G. Gray, T. E. Booth, T. J. T. Kwan, and C. M. Snell, "A multi-comb variance reduction scheme for Monte Carlo semiconductor simulators", *IEEE Trans. Electron Devices*, vol. 45, no. 4, pp. 918–924, 1998.

[1.115] H. Kim, H. S. Min, T. W. Tang, and Y. J. Park, "An extended proof of the Ramo-Shockley theorem", *Solid-State Electron.*, vol. 34, pp. 1251–1253, 1991.

[1.116] P. D. Yoder, K. Gärtner, U. Krumbein, and W. Fichtner, "Optimized terminal current calculation for Monte Carlo device simulation", *IEEE Trans. Computer-Aided Des.*, vol. 16, pp. 1082–1087, 1997.

[1.117] J. W. Slotboom, G. Streutker, G. J. T. Davids, and P. B. Hartog, "Surface impact ionization in Silicon devices", in *IEDM*, 1987, pp. 494–497.

2 Semiclassical Transport Theory

Here, the classical theory of a kinetic gas is applied to the electron and hole ensembles in semiconductors with two quantum mechanical extensions. The particle kinetics are based on a position-dependent band structure calculated with the nonlocal empirical pseudopotential method [2.1–2.3] and scattering rates determined by Fermi's Golden Rule [2.4, 2.5]. In this theoretical framework the particle motion consists of a series of scattering events and accelerations by external forces, which is described by the semiclassical Boltzmann transport equation (BTE) [2.4–2.10].

In the first part of this section the distribution function, which characterizes the state of the particle gas, the evolution equation for the distribution function, the BTE, and the corresponding probability densities of the underlying stochastic process are introduced. In the next part balance equations for moments of the distribution function are analyzed, which are the basis of momentum-based models. In the third part the microscopic relaxation time is discussed, which allows to exactly solve the scattering integral of certain balance equations avoiding the macroscopic relaxation time approximation. In the last part of this section the theoretical background of fluctuations in the steady-state is given based on the Langevin-type Boltzmann transport equation, and corresponding balance equations are derived. The impact of the macroscopic relaxation time approximation is analyzed by comparison with exact solutions of the BTE for the velocity fluctuations in bulk systems.

2.1 The Boltzmann Transport Equation

The semiclassical Hamilton function of an electron in a conduction band is given by

$$H_n(\mathbf{r}, \mathbf{k}, t) = E_c(\mathbf{r}) + \varepsilon_n(\mathbf{r}, \mathbf{k}) - q\Psi(\mathbf{r}, t) \ , \tag{2.1}$$

where E_c is the conduction band edge, which is the minimum energy of all conduction bands, ε_n the energy of the nth band relative to the conduction band edge, Ψ the electrostatic potential, $\mathbf{k}$ the wave vector, $\mathbf{r}$ a vector of the real space (RS), t the time, and q the positive electron charge. In the case of holes it reads

$$H_n(\mathbf{r},\mathbf{k},t) = -E_v(\mathbf{r}) + \varepsilon_n(\mathbf{r},\mathbf{k}) + q\Psi(\mathbf{r},t) \ , \tag{2.2}$$

where E_v is the valence band edge, which is the maximum energy of the valence bands in the electron picture. The relative band energy ε is positive in the valence bands (hole picture). Electrons and holes are distinguished by the band index n.

The collisionless motion of the particles is described in the framework of semiclassical mechanics (Newton's laws) by

$$\hbar\frac{\partial \mathbf{k}}{\partial t} = -\nabla_\mathbf{r} H = \mathbf{F} \tag{2.3}$$

and

$$\frac{\partial \mathbf{r}}{\partial t} = \frac{1}{\hbar}\nabla_\mathbf{k} H = \mathbf{v} \ , \tag{2.4}$$

where $\hbar$ denotes Planck's constant divided by 2π, $\mathbf{v}$ the velocity, and $\mathbf{F}$ the force

$$\mathbf{F}_n(\mathbf{r},\mathbf{k},t) = \begin{cases} \nabla_\mathbf{r}\left(E_v(\mathbf{r}) - q\Psi(\mathbf{r},t) - \varepsilon_n(\mathbf{r},\mathbf{k})\right) & \text{for holes} \\ \nabla_\mathbf{r}\left(-E_c(\mathbf{r}) + q\Psi(\mathbf{r},t) - \varepsilon_n(\mathbf{r},\mathbf{k})\right) & \text{for electrons} \end{cases} . \tag{2.5}$$

The first and last terms on the right-hand side are due to the position-dependent band structure [2.11]. In the semiclassical theory the particle velocity is given by the group velocity of the particle's wave packet defined as [2.4]

$$\mathbf{v}_n(\mathbf{r},\mathbf{k}) = \frac{1}{\hbar}\nabla_\mathbf{k}\varepsilon_n(\mathbf{r},\mathbf{k}) \ . \tag{2.6}$$

The inverse mass tensor reads

$$\hat{m}_n^{-1}(\mathbf{r},\mathbf{k}) = \frac{1}{\hbar}\nabla_\mathbf{k}\mathbf{v}_n^\mathrm{T}(\mathbf{r},\mathbf{k}) = \frac{1}{\hbar^2}\nabla_\mathbf{k}\nabla_\mathbf{k}^\mathrm{T}\varepsilon_n(\mathbf{r},\mathbf{k}) \ . \tag{2.7}$$

The particles are scattered by various mechanisms (cf. Chap. 4) [2.9]. The scattering events are treated as local and instantaneous and they are characterized by a transition rate $S_{n',n}(\mathbf{k}'|\mathbf{k})(\mathbf{r},t)$, which is the rate of particles that are scattered from the occupied state $(n,\mathbf{r},\mathbf{k})$ into the empty state $(n',\mathbf{r},\mathbf{k}')$. Since the final state is not uniquely defined, the particle scattering contains a certain degree of randomness making it impossible to calculate the exact particle trajectory in the phase space (PS). Instead, stochastic methods are used and the particle gas is described by a distribution function $f_n(\mathbf{r},\mathbf{k},t)$ being the particle density in the PS [2.4,2.9,2.12,2.13]. The number of particles in an infinitesimal volume of the six-dimensional PS is given by

$$dN = \frac{2}{(2\pi)^3} f_n(\mathbf{r},\mathbf{k},t)d^3kd^3r \ . \tag{2.8}$$

$1/(2\pi)^3$ is the minimum PS volume of a particle according to the Heisenberg uncertainty principle and the factor of two is the result of the assumption of spin degeneracy. The Pauli exclusion principle necessitates a distribution function smaller than one [2.4]. Thus, the distribution function itself is the probability that the state $(n, \mathbf{r}, \mathbf{k})$ is occupied [2.14].

The combined effect of acceleration and scattering is described by the semiclassical Boltzmann transport equation [2.4, 2.9, 2.12, 2.14]

$$\begin{aligned}\left\{\frac{\partial}{\partial t} + \mathbf{F}_n^{\mathrm{T}}(\mathbf{r},\mathbf{k},t)\frac{1}{\hbar}\nabla_{\mathbf{k}} + \mathbf{v}_n^{\mathrm{T}}(\mathbf{r},\mathbf{k})\nabla_{\mathbf{r}}\right\} f_n(\mathbf{r},\mathbf{k},t) \quad &= \\ \frac{\Omega}{(2\pi)^3}\sum_{n'}\int_{\mathrm{BZ}} (1 - f_n(\mathbf{r},\mathbf{k},t)) S_{n,n'}(\mathbf{k}|\mathbf{k}')(\mathbf{r},t) f_{n'}(\mathbf{r},\mathbf{k}',t)& \\ -(1 - f_{n'}(\mathbf{r},\mathbf{k}',t)) S_{n',n}(\mathbf{k}'|\mathbf{k})(\mathbf{r},t) f_n(\mathbf{r},\mathbf{k},t) d^3k'&\ , \quad (2.9)\end{aligned}$$

where Ω denotes the system volume. The left-hand side is equivalent to the equations of motion (2.3) and (2.4) and the scattering integral on the right-hand side describes the scattering from an arbitrary state $(n', \mathbf{r}, \mathbf{k}')$ into the state $(n, \mathbf{r}, \mathbf{k})$ and the reverse process, where it has been assumed that the scattering process does not flip the particle's spin. The probability density rate for a transition from the initial state $(n', \mathbf{r}, \mathbf{k}')$ into the final state $(n, \mathbf{r}, \mathbf{k})$ is proportional to the probability that the initial state is occupied $f_{n'}(\mathbf{r}, \mathbf{k}', t)$, proportional to the transition rate $S_{n,n'}(\mathbf{k}|\mathbf{k}')(\mathbf{r}, t)$, and proportional to the probability that the final state is not occupied $(1 - f_n(\mathbf{r}, \mathbf{k}, t))$. The last factor is the result of the Pauli exclusion principle. Inclusion of particle generation caused by impact ionization (II) is discussed in Chap. 4. Suitable boundary conditions for the BTE are given in Sec. 6.1 (see also [2.15]).

The BTE (2.9) is nonlinear in the distribution function for three reasons. First, the right-hand side contains the product of the distribution function with itself due to the Pauli exclusion principle. Second, the transition rate depends on the distribution function. For example, in the case of impurity scattering the interaction is screened by the particle density being a function of the distribution function (cf. Sec. 4.3). Third, the charged particles contribute to the electromagnetic force on the left-hand side of the BTE. The technique used to solve the nonlinear BTE is based on a self-consistent (SC) iteration scheme of a *linear* version of the BTE (cf. Sec. 6.4), and the following discussion is therefore limited to the case of the linear BTE which is the basis of the Monte-Carlo (MC) method. The linear BTE reads

$$\begin{aligned}\left\{\frac{\partial}{\partial t} + \mathbf{F}_n^{\mathrm{T}}(\mathbf{r},\mathbf{k})\frac{1}{\hbar}\nabla_{\mathbf{k}} + \mathbf{v}_n^{\mathrm{T}}(\mathbf{r},\mathbf{k})\nabla_{\mathbf{r}} + S_n(\mathbf{r},\mathbf{k})\right\} f_n(\mathbf{r},\mathbf{k},t) \quad &= \\ \frac{\Omega}{(2\pi)^3}\sum_{n'}\int_{\mathrm{BZ}} S_{n,n'}(\mathbf{k}|\mathbf{k}')(\mathbf{r}) f_{n'}(\mathbf{r},\mathbf{k}',t) d^3k'&\ , \quad (2.10)\end{aligned}$$

where it is assumed that the transition rate and the force do *not* depend on the distribution function or time and the Pauli exclusion principle is dropped. The scattering rate is defined as

$$S_n(\mathbf{r},\mathbf{k}) = \frac{\Omega}{(2\pi)^3} \sum_{n'} \int_{\mathrm{BZ}} S_{n',n}(\mathbf{k}'|\mathbf{k})(\mathbf{r}) d^3k' \ , \tag{2.11}$$

which is the rate at which particles are scattered out of the initial state $(n, \mathbf{k}, \mathbf{r})$.

The linear BTE is the master equation of a particle gas consisting of independent and noninteracting particles [2.12, 2.16]. The corresponding probability density in the PS $p_n(\mathbf{r},\mathbf{k},t)$ for a single particle is given by

$$p_n(\mathbf{r},\mathbf{k},t) = \frac{1}{N_{\mathrm{par}}} \frac{2}{(2\pi)^3} f_n(\mathbf{r},\mathbf{k},t) \ , \tag{2.12}$$

where N_{par} is the total number of particles in the system

$$N_{\mathrm{par}} = \frac{2}{(2\pi)^3} \sum_{n} \int \int_{\mathrm{BZ}} f_n(\mathbf{r},\mathbf{k},t) d^3k d^3r \ . \tag{2.13}$$

The sum extends over all conduction bands in the case electrons and over all valence bands in the case of holes. The integral in **k**-space (KS) extends over the first Brillouin zone (BZ) and the one in the RS over the whole device domain. In the case that the single-particle probability density is used the total particle number of the system must be constant in time.

To simplify the notation of the following expressions, the variables of the particle state are lumped together into one quantity [2.17]

$$\zeta = (n, \mathbf{r}^{\mathrm{T}}, \mathbf{k}^{\mathrm{T}})^{\mathrm{T}} \tag{2.14}$$

and integrals over the PS are abbreviated by

$$\int d\zeta = \sum_{n} \int \int_{\mathrm{BZ}} d^3k d^3r \ . \tag{2.15}$$

With the two definitions

$$\dot{\zeta}^{\mathrm{T}} \nabla_{\zeta} = \mathbf{F}_n^{\mathrm{T}}(\mathbf{r},\mathbf{k}) \frac{1}{\hbar} \nabla_{\mathbf{k}} + \mathbf{v}_n^{\mathrm{T}}(\mathbf{r},\mathbf{k}) \nabla_{\mathbf{r}} \tag{2.16}$$

and

$$S(\zeta|\zeta') = \frac{\Omega}{(2\pi)^3} S_{n,n'}(\mathbf{k}|\mathbf{k}')(\mathbf{r}) \delta(\mathbf{r} - \mathbf{r}') \tag{2.17}$$

the linear BTE for the probability density reads

$$\left\{ \frac{\partial}{\partial t} + \dot{\zeta}^{\mathrm{T}} \nabla_{\zeta} + S(\zeta) \right\} p(\zeta, t) = \int S(\zeta|\zeta') p(\zeta', t) d\zeta' \ . \tag{2.18}$$

Since this master equation is linear, only contains a first-order derivative in time, and the scattering is instantaneous, the described process is Markovian. Therefore, the n-time joint probability density can be factored ($t_n > t_{n-1} > \cdots > t_2 > t_1$)

$$p(\zeta_n, t_n; \zeta_{n-1}, t_{n-1}; \cdots; \zeta_2, t_2; \zeta_1, t_1) = \\ p(\zeta_n, t_n | \zeta_{n-1}, t_{n-1}) \cdots p(\zeta_2, t_2 | \zeta_1, t_1) p(\zeta_1, t_1) , \quad (2.19)$$

where $p(\zeta', t' | \zeta, t)$ is the conditional probability density that a particle being in the state ζ at the time t is found in the state ζ' at the time t' [2.12]. Thus, the Markovian process is fully determined by the knowledge of the probability density $p(\zeta, t)$ and the conditional probability density $p(\zeta', t' | \zeta, t)$ for $(t' > t)$. Both densities are normalized

$$\int p(\zeta, t) d\zeta = 1 , \qquad \int p(\zeta', t' | \zeta, t) d\zeta' = 1 , \quad (2.20)$$

and for $t' = t$ the conditional probability density reduces to

$$p(\zeta', t | \zeta, t) = \delta(\zeta' - \zeta) . \quad (2.21)$$

The conditional probability density must also satisfy the BTE

$$\left\{ \frac{\partial}{\partial t} + \dot{\zeta}^{\mathrm{T}} \nabla_\zeta + S(\zeta) \right\} p(\zeta, t | \zeta_0, t_0) = \int S(\zeta | \zeta') p(\zeta', t | \zeta_0, t_0) d\zeta' . \quad (2.22)$$

The evolution of the probability density now reads

$$p(\zeta, t) = \int p(\zeta, t | \zeta_0, t_0) p(\zeta_0, t_0) d\zeta_0 \quad (2.23)$$

with $t > t_0$ and the Chapman-Kolmogorov equation applies ($t > t' > t_0$) [2.12]

$$p(\zeta, t | \zeta_0, t_0) = \int p(\zeta, t | \zeta', t') p(\zeta', t' | \zeta_0, t_0) d\zeta' . \quad (2.24)$$

Consequently, the evolution of the probability density can be stopped at any intermediate time t' and continued without affecting the final probability density.

Based on the joint probability density (2.19) expected values of the particle ensemble can be calculated [2.18]. The expectation $\langle\rangle$ of a product of n microscopic quantities $\prod_{i=1}^{n} X_i(\zeta_i)$ evaluated at the times t_i reads

$$\langle X_n, \cdots, X_2, X_1 \rangle (t_n, \cdots, t_2, t_1) = \\ \int \cdots \int \int X_n(\zeta_n) \cdots X_2(\zeta_2) X_1(\zeta_1) p(\zeta_n, t_n; \cdots; \zeta_2, t_2; \zeta_1, t_1) d\zeta_n \cdots d\zeta_2 d\zeta_1 . \quad (2.25)$$

The average particle velocity, for example, is given by

$$\langle \mathbf{v}\rangle(t) = \int \mathbf{v}(\zeta)p(\zeta,t)d\zeta \tag{2.26}$$

and its autocorrelation function by

$$\langle \mathbf{v},\mathbf{v}^{\mathrm{T}}\rangle(t,t') = \int\int \mathbf{v}(\zeta)\mathbf{v}^{\mathrm{T}}(\zeta')p(\zeta,t;\zeta',t')d\zeta d\zeta' \ . \tag{2.27}$$

In the case that two time arguments are equal, the order of the expectation is reduced because of Eq. (2.21)

$$\langle X_2, X_1\rangle(t_1,t_1) = \langle X_2 X_1\rangle(t_1) \ . \tag{2.28}$$

In the case of a stationary system expected values with one time argument are independent of time [2.18]

$$\langle X\rangle(t) = \langle X\rangle(0) = \langle X\rangle \tag{2.29}$$

and correlation functions only depend on the difference of the time arguments

$$\langle X_2, X_1\rangle(t_2,t_1) = \langle X_2, X_1\rangle(t_2-t_1,0) = \langle X_2, X_1\rangle(t_2-t_1) \ . \tag{2.30}$$

2.2 Balance Equations

Balance equations govern the evolution of macroscopic quantities $x(\mathbf{r},t)$, which are averages over the KS like, for example, particle or current density [2.12,2.19–2.21]. They are expectations of the form

$$x(\mathbf{r},t) = n(\mathbf{r},t)\left\langle X\right\rangle_{\mathbf{k}}(\mathbf{r},t) = \frac{2}{(2\pi)^3}\sum_n \int_{\mathrm{BZ}} X_n(\mathbf{r},\mathbf{k})f_n(\mathbf{r},\mathbf{k},t)d^3k \ , \tag{2.31}$$

where $n(\mathbf{r},t)$ is the particle density and the subscript of the expectation $\langle\rangle_{\mathbf{k}}$ indicates that the expected value is only calculated over the KS (including the sum over the valence or conduction bands in the case of holes or electrons, respectively) and normalized ($\langle 1\rangle_{\mathbf{k}} = 1$). In the following the arguments n, $\mathbf{r}$, $\mathbf{k}$, and t will be dropped as far as possible. The current density $\mathbf{j}_X$ corresponding to the macroscopic quantity x reads

$$\mathbf{j}_X = n\left\langle X\mathbf{v}\right\rangle_{\mathbf{k}} \ . \tag{2.32}$$

The accordant balance equation for the scalar microscopic quantity is obtained by taking the expectation of the BTE (2.9) multiplied with microscopic quantity X. The first term on the left-hand side of the BTE (2.9) yields

$$\frac{2}{(2\pi)^3}\sum_n \int_{\mathrm{BZ}} X_n(\mathbf{r},\mathbf{k})\frac{\partial f_n(\mathbf{r},\mathbf{k},t)}{\partial t}d^3k = \frac{\partial x}{\partial t} \ . \tag{2.33}$$

In the second term the gradient is shifted with integration by parts

$$\frac{2}{(2\pi)^3}\sum_n \int_{BZ} X_n(\mathbf{r},\mathbf{k})\mathbf{F}_n^{\mathrm{T}}(\mathbf{r},\mathbf{k},t)\frac{1}{\hbar}\nabla_{\mathbf{k}} f_n(\mathbf{r},\mathbf{k},t)d^3k$$
$$= -n\left\langle \mathbf{F}^{\mathrm{T}}\frac{1}{\hbar}\nabla_{\mathbf{k}}X\right\rangle_{\mathbf{k}} - n\left\langle X\frac{1}{\hbar}\nabla_{\mathbf{k}}^{\mathrm{T}}\mathbf{F}\right\rangle_{\mathbf{k}} . \quad (2.34)$$

The last term on the left-hand side evaluates to

$$\frac{2}{(2\pi)^3}\sum_n \int_{BZ} X_n(\mathbf{r},\mathbf{k})\mathbf{v}_n^{\mathrm{T}}(\mathbf{r},\mathbf{k})\nabla_{\mathbf{r}} f_n(\mathbf{r},\mathbf{k},t)d^3k$$
$$= \nabla_{\mathbf{r}}^{\mathrm{T}}\mathbf{j}_X - n\left\langle \mathbf{v}^{\mathrm{T}}\nabla_{\mathbf{r}}X\right\rangle_{\mathbf{k}} - n\left\langle X\nabla_{\mathbf{r}}^{\mathrm{T}}\mathbf{v}\right\rangle_{\mathbf{k}} . \quad (2.35)$$

The scattering integral is reformulated neglecting the Pauli exclusion principle

$$\frac{2}{(2\pi)^3}\sum_n \int_{BZ} X_n(\mathbf{r},\mathbf{k})\frac{\Omega}{(2\pi)^3}\sum_{n'} \int_{BZ} S_{n,n'}(\mathbf{k}|\mathbf{k}')(\mathbf{r},t)f_{n'}(\mathbf{r},\mathbf{k}',t)$$
$$- S_{n',n}(\mathbf{k}'|\mathbf{k})(\mathbf{r},t)f_n(\mathbf{r},\mathbf{k},t)d^3k'd^3k$$
$$= -n\left\langle \frac{\Omega}{(2\pi)^3}\sum_{n'} \int_{BZ} S_{n',n}(\mathbf{k}'|\mathbf{k})(\mathbf{r},t)\big(X_n(\mathbf{r},\mathbf{k}) - X_{n'}(\mathbf{r},\mathbf{k}')\big)d^3k'\right\rangle_{\mathbf{k}} . \quad (2.36)$$

With the abbreviation

$$S\{X\} = \frac{\Omega}{(2\pi)^3}\sum_{n'} \int_{BZ} S_{n',n}(\mathbf{k}'|\mathbf{k})(\mathbf{r},t)\big(X_n(\mathbf{r},\mathbf{k}) - X_{n'}(\mathbf{r},\mathbf{k}')\big)d^3k' \quad (2.37)$$

and

$$\left\langle X\nabla_{\mathbf{r}}^{\mathrm{T}}\mathbf{v}\right\rangle_{\mathbf{k}} = -\left\langle X\frac{1}{\hbar}\nabla_{\mathbf{k}}^{\mathrm{T}}\mathbf{F}\right\rangle_{\mathbf{k}} \quad (2.38)$$

the final balance equation is obtained

$$\frac{\partial x}{\partial t} - n\left\langle \mathbf{F}^{\mathrm{T}}\frac{1}{\hbar}\nabla_{\mathbf{k}}X\right\rangle_{\mathbf{k}} + \nabla_{\mathbf{r}}^{\mathrm{T}}\mathbf{j}_X - n\left\langle \mathbf{v}^{\mathrm{T}}\nabla_{\mathbf{r}}X\right\rangle_{\mathbf{k}} = -n\left\langle S\{X\}\right\rangle_{\mathbf{k}} . \quad (2.39)$$

The particle continuity equation is derived for $X = 1$

$$\frac{\partial n}{\partial t} + \nabla_{\mathbf{r}}^{\mathrm{T}}\mathbf{j} = 0 , \quad (2.40)$$

where $\mathbf{j} = \mathbf{j}_1$ is the particle current density. In the case of generation or recombination phenomena a nonzero generation/recombination rate appears on the right-hand side of Eq. (2.40). The balance equation for the vector-valued particle current density with $\mathbf{X} = \mathbf{v}$ reads

$$\frac{\partial \mathbf{j}}{\partial t} - n\left\langle \hat{m}^{-1}\mathbf{F}\right\rangle_{\mathbf{k}} + \left(\nabla_{\mathbf{r}}^{\mathrm{T}}\left[n\left\langle \mathbf{v}\mathbf{v}^{\mathrm{T}}\right\rangle_{\mathbf{k}}\right]\right)^{\mathrm{T}} - n\left\langle \mathbf{v}^{\mathrm{T}}\nabla_{\mathbf{r}}\mathbf{v}^{\mathrm{T}}\right\rangle_{\mathbf{k}}^{\mathrm{T}} = -n\left\langle S\{\mathbf{v}\}\right\rangle_{\mathbf{k}} , \quad (2.41)$$

where this equation was derived by applying Eq. (2.39) to each component of the vector-valued microscopic quantity.

Since each balance equation of the type (2.39) not only contains the density $x(\mathbf{r}, t)$ but also the corresponding current density $\mathbf{j}_X(\mathbf{r}, t)$ and other quantities (cf. Eq. (2.41)), it is not possible to derive a finite set of exact balance equations which is closed. In addition, it is not possible to calculate the expectation of the collision integral with a finite number of microscopic densities, because a finite number of microscopic densities does not contain the full information of the distribution function. In the case of certain balance equations the latter problem can be circumvented with the microscopic relaxation time (cf. Sec. 2.3) [2.19]. But in most cases the macroscopic relaxation time approximation has to be used

$$\langle S\{X\}\rangle_{\mathbf{k}} \approx \frac{\langle X\rangle_{\mathbf{k}} - \langle X\rangle_{\mathbf{k}}^{\mathrm{eq}}}{\tau_X} , \quad (2.42)$$

where $\langle\rangle_{\mathbf{k}}^{\mathrm{eq}}$ is the expectation in the case of thermal equilibrium. The approximation results from the fact that the macroscopic relaxation times are parameterized with a finite set of microscopic densities. In most cases this is only the average particle energy or temperature, and the macroscopic relaxation times are evaluated with the solutions of the corresponding balance equations under spatially homogeneous conditions [2.21].

2.3 The Microscopic Relaxation Time

The microscopic relaxation time approximation is based on a linear expansion of the stationary distribution function in the electric field in a homogeneous system

$$f_n(\mathbf{k}, \mathbf{E}) \approx f_n(\mathbf{k}, \mathbf{0}) + \mathbf{E}^{\mathrm{T}}\nabla_{\mathbf{E}} f_n(\mathbf{k}, \mathbf{E})|_{\mathbf{E}=\mathbf{0}} , \quad (2.43)$$

where, like in the following, the RS and time argument are dropped in order to improve the lucidity of the formulas. Under these conditions the solution of the BTE neglecting the Pauli exclusion principle for a zero electric field is given by the Maxwell-Boltzmann distribution [2.4]

$$f_n^{\mathrm{eq}}(\mathbf{k}) = c \exp\left(-\frac{\varepsilon_n(\mathbf{k})}{k_{\mathrm{B}}T_0}\right) , \quad (2.44)$$

where c is a normalization constant. Reformulating Eq. (2.43) by introducing the still unknown microscopic relaxation time tensor $\hat{\tau}_n(\mathbf{k})$ yields [2.8]

$$f_n(\mathbf{k}, \mathbf{E}) \approx f_n^{\mathrm{eq}}(\mathbf{k}) \pm \frac{q}{k_{\mathrm{B}}T_0}\mathbf{E}^{\mathrm{T}}\hat{\tau}_n(\mathbf{k})\mathbf{v}_n(\mathbf{k}) f_n^{\mathrm{eq}}(\mathbf{k}) . \quad (2.45)$$

The upper sign refers to holes and the lower one to electrons. Inserting this expansion into the BTE and retaining only terms up to the first-order in the electric field results in the following integral equation for the microscopic relaxation time tensor [2.4, 2.8, 2.22, 2.23]

$$\mathbf{v}_n(\mathbf{k}) = \frac{\Omega}{(2\pi)^3} \sum_{n'} \int_{\mathrm{BZ}} S_{n',n}(\mathbf{k}'|\mathbf{k}) \left[\hat{\tau}_n(\mathbf{k})\mathbf{v}_n(\mathbf{k}) - \hat{\tau}_{n'}(\mathbf{k}')\mathbf{v}_{n'}(\mathbf{k}')\right] d^3k' \ . \tag{2.46}$$

This integral equation is not limited to the case of linear transport conditions. The microscopic relaxation time is therefore more general than the microscopic relaxation time approximation, and below it will be shown how to solve exactly the scattering integral with the microscopic relaxation time under general transport conditions.

In the case that the transition rate is either velocity-randomizing (vr) or isotropic-elastic (ie) Eq. (2.46) can be solved [2.8]. In both cases the scattering rate (2.11) depends only on the energy and band index

$$S_n^{\mathrm{vr,ie}}(\varepsilon_n(\mathbf{k})) = \frac{\Omega}{(2\pi)^3} \sum_{n'} \int_{\mathrm{BZ}} S_{n',n}^{\mathrm{vr,ie}}(\mathbf{k}'|\mathbf{k}) d^3k' \tag{2.47}$$

Thus, the scattering rate must be isotropic on an equienergy surface in KS.

A velocity-randomizing scattering process is memoryless with respect to the velocity

$$\int_{\mathrm{BZ}} S_{n',n}^{\mathrm{vr}}(\mathbf{k}'|\mathbf{k})\mathbf{v}_{n'}(\mathbf{k}') d^3k' = \mathbf{0} \ , \tag{2.48}$$

where this relation must hold for every single final band n' and the energy transfer $\varepsilon_{n'}(\mathbf{k}') - \varepsilon_n(\mathbf{k})$ must be constant. The latter condition implies that the above relation holds on the equienergy surface in KS for the final energy. With the ansatz of a microscopic relaxation time being only a scalar energy dependent quantity ($\hat{\tau}_n^{\mathrm{vr}}(\mathbf{k}) = \tau_n^{\mathrm{vr}}(\varepsilon_n(\mathbf{k}))\hat{I}$) the second part of the integral on the right-hand side of Eq. (2.46) is zero

$$\int_{\mathrm{BZ}} S_{n',n}(\mathbf{k}'|\mathbf{k}) \tau_{n'}^{\mathrm{vr}}(\varepsilon_{n'}(\mathbf{k}'))\mathbf{v}_{n'}(\mathbf{k}') d^3k' = \mathbf{0} \ , \tag{2.49}$$

because the microscopic relaxation time $\tau_{n'}^{\mathrm{vr}}(\varepsilon_{n'}(\mathbf{k}'))$ is constant on the equienergy surface for the final energy $\varepsilon_{n'}(\mathbf{k}')$ and Eq. (2.48) applies. This allows to reduces Eq. (2.46) to

$$\tau_n^{\mathrm{vr}}(\varepsilon) = \frac{1}{S_n^{\mathrm{vr}}(\varepsilon)} \tag{2.50}$$

confirming the ansatz, because the scattering rate (2.47) was assumed to depend only on the energy and band index.

In the case of an isotropic-elastic scattering process it is not necessary that Eq. (2.48) holds, but the process must conserve the particle energy ($\varepsilon_{n'}(\mathbf{k}') = \varepsilon_n(\mathbf{k})$). Under the assumption of no interband transitions ($n' = n$) and an only energy and band index dependent microscopic relaxation time $\tau_n^{\mathrm{ie}}(\varepsilon)$ Eq. (2.46) reduces to

$$\frac{1}{\tau_n^{\mathrm{ie}}(\varepsilon_n(\mathbf{k}))} = \frac{\Omega}{(2\pi)^3} \int_{\mathrm{BZ}} S_{n,n}^{\mathrm{ie}}(\mathbf{k}'|\mathbf{k}) \left(1 - \frac{\mathbf{v}_n^{\mathrm{T}}(\mathbf{k})\mathbf{v}_n(\mathbf{k}')}{\mathbf{v}_n^{\mathrm{T}}(\mathbf{k})\mathbf{v}_n(\mathbf{k})}\right) d^3k' \,. \tag{2.51}$$

The ansatz holds only, if the right-hand side is isotropic on an equienergy surface.

For more than one scattering mechanism the total microscopic relaxation time is given by the Matthiessen rule [2.4]

$$\frac{1}{\tau_n(\varepsilon)} = \sum_i \frac{1}{\tau_{n,i}^{\mathrm{vr}}(\varepsilon)} + \sum_j \frac{1}{\tau_{n,j}^{\mathrm{ie}}(\varepsilon)} \,, \tag{2.52}$$

where the sums extend over all velocity-randomizing and isotropic-elastic processes.

In the case of isotropic-elastic interband transitions the Matthiessen rule does not hold, but if the microscopic relaxation time still is an energy and band index dependent scalar the total microscopic relaxation time is the solution of the following linear system of equations [2.24]

$$\begin{aligned} 1 = {} & \left(S_n^{\mathrm{vr}}(\varepsilon_n(\mathbf{k})) + S_n^{\mathrm{ie}}(\varepsilon_n(\mathbf{k}))\right) \tau_n(\varepsilon_n(\mathbf{k})) \\ & - \frac{\Omega}{(2\pi)^3} \sum_{n'} \int_{\mathrm{BZ}} S_{n',n}^{\mathrm{ie}}(\mathbf{k}'|\mathbf{k}) \frac{\mathbf{v}_n^{\mathrm{T}}(\mathbf{k})\mathbf{v}_{n'}(\mathbf{k}')}{\mathbf{v}_n^{\mathrm{T}}(\mathbf{k})\mathbf{v}_n(\mathbf{k})} \tau_{n'}(\varepsilon_{n'}(\mathbf{k}')) d^3k' \,. \end{aligned} \tag{2.53}$$

The rank of the system is given by the number of bands which contain the energy $\varepsilon_{n'}(\mathbf{k}') = \varepsilon_n(\mathbf{k}) = \varepsilon$.

With the microscopic relaxation time it is possible to circumvent the macroscopic relaxation time approximation (2.42) in the balance equation for the current density, because due to the definition of the microscopic relaxation time (2.46) the scattering integral (2.36) can be exactly evaluated for $\mathbf{X} = \tau\mathbf{v}$ [2.19]

$$\langle \mathbf{v} \rangle_{\mathbf{k}} = \langle S\{\tau\mathbf{v}\} \rangle_{\mathbf{k}} \,. \tag{2.54}$$

In the case that the microscopic relaxation time depends on the wave vector only via the energy the scattering integral can be solved for arbitrary microscopic quantities of the form $X = \tau Y \mathbf{v}$

$$\langle Y\mathbf{v} \rangle_{\mathbf{k}} = \langle S\{\tau Y\mathbf{v}\} \rangle_{\mathbf{k}} \,, \tag{2.55}$$

where Y must also depend on the wave vector only via the energy [2.22].

With $\mathbf{X} = \tau \mathbf{v}$ and Eq. (2.54) the current balance equation is obtained without any approximations

$$\frac{\partial n \left\langle \tau \mathbf{v} \right\rangle_{\mathbf{k}}}{\partial t} - n \left\langle \frac{\partial \tau}{\partial t} \mathbf{v} \right\rangle_{\mathbf{k}} - n \left\langle \frac{1}{\hbar} \nabla_{\mathbf{k}} \left[\tau \mathbf{v}^{\mathrm{T}} \right] \mathbf{F} \right\rangle_{\mathbf{k}} + \left(\nabla_{\mathbf{r}}^{\mathrm{T}} \left[n \left\langle \tau \mathbf{v} \mathbf{v}^{\mathrm{T}} \right\rangle_{\mathbf{k}} \right] \right)^{\mathrm{T}} - n \left\langle \mathbf{v}^{\mathrm{T}} \nabla_{\mathbf{r}} \tau \mathbf{v}^{\mathrm{T}} \right\rangle_{\mathbf{k}}^{\mathrm{T}} = -n \langle \mathbf{v} \rangle_{\mathbf{k}} = -\mathbf{j} \, . \quad (2.56)$$

The second term on the left-hand side is due to the time dependence of the microscopic relaxation time.

Under stationary and spatially homogeneous conditions the balance equation for the current density (2.56) reduces to a form similar to the standard DD formula (cf. (7.127))

$$\mathbf{j} = \pm n \hat{\mu} \mathbf{E} - \left(\nabla_{\mathbf{r}}^{\mathrm{T}} [n \hat{D}] \right)^{\mathrm{T}} , \quad (2.57)$$

where $\mathbf{E} = -\nabla_{\mathbf{r}} \Psi$ is the electric field. The mobility is defined as

$$\hat{\mu} = \left\langle \frac{q}{\hbar} \nabla_{\mathbf{k}} \left[\tau \mathbf{v}^{\mathrm{T}} \right] \right\rangle_{\mathbf{k}} = q \left\langle \tau \hat{m}^{-1} + \frac{d\tau}{d\varepsilon} \mathbf{v} \mathbf{v}^{\mathrm{T}} \right\rangle_{\mathbf{k}} . \quad (2.58)$$

The last equality holds for the important case of a microscopic relaxation time that depends on the wave vector only via the energy. The diffusion constant is given by

$$\hat{D} = \left\langle \tau \mathbf{v} \mathbf{v}^{\mathrm{T}} \right\rangle_{\mathbf{k}} . \quad (2.59)$$

Both tensor quantities are symmetric.

The equation corresponding to Eq. (2.57) for the energy current $\mathbf{s} = \mathbf{j}_{\varepsilon \mathbf{v}}$ is obtained with $\mathbf{X} = \tau \varepsilon \mathbf{v}$ and Eq. (2.55)

$$\mathbf{s} = \pm n \hat{\mu}_{\varepsilon} \mathbf{E} - \left(\nabla_{\mathbf{r}}^{\mathrm{T}} [n \hat{D}_{\varepsilon}] \right)^{\mathrm{T}} . \quad (2.60)$$

The energy current mobility and diffusion constant are given by

$$\hat{\mu}_{\varepsilon} = \left\langle \frac{q}{\hbar} \nabla_{\mathbf{k}} \left[\tau \varepsilon \mathbf{v}^{\mathrm{T}} \right] \right\rangle_{\mathbf{k}} = q \left\langle \tau \varepsilon \hat{m}^{-1} + \tau \mathbf{v} \mathbf{v}^{\mathrm{T}} + \frac{d\tau}{d\varepsilon} \varepsilon \mathbf{v} \mathbf{v}^{\mathrm{T}} \right\rangle_{\mathbf{k}} \quad (2.61)$$

and

$$\hat{D}_{\varepsilon} = \left\langle \tau \varepsilon \mathbf{v} \mathbf{v}^{\mathrm{T}} \right\rangle_{\mathbf{k}} . \quad (2.62)$$

Under equilibrium conditions the distribution function is given by the Maxwell-Boltzmann distribution function (2.44) and it is possible to evaluate the mobility and diffusion tensor by numerical integration. Furthermore, the Einstein relation holds for both microscopic quantities

$$\hat{\mu}^{\mathrm{eq}} V_{\mathrm{T}} = \hat{D}^{\mathrm{eq}} \quad \text{and} \quad \hat{\mu}_{\varepsilon}^{\mathrm{eq}} V_{\mathrm{T}} = \hat{D}_{\varepsilon}^{\mathrm{eq}} \quad (2.63)$$

as can easily be shown by integration by parts of Eqs. (2.58) and (2.61) and where the thermal voltage is given by

$$V_{\mathrm{T}} = \frac{k_{\mathrm{B}} T_0}{q} . \quad (2.64)$$

2.4 Fluctuations in the Steady-State

Realization of the stochastic process underlying the BTE (2.18) yields stochastic variables which fluctuate around their expected values [2.16, 2.25, 2.26]. Under stationary conditions the fluctuation of a stochastic variable $\check{X}$ is defined as

$$\delta X(t) = \check{X}(t) - \mathrm{E}\{X\} , \tag{2.65}$$

where E{} is the expectation with respect to the stochastic process. The fluctuations are characterized by spectral intensities. The cross-spectrum of two fluctuations δX and δY is given for a stationary process by Fourier transformation of the corresponding correlation function [2.18, 2.25]

$$\begin{aligned} \underline{S}_{XY}(\omega) &= \underline{S}^*_{YX}(\omega) \\ &= \lim_{T\to\infty} \mathrm{E}\left\{\frac{1}{T}\int_{-T}^{T} \delta X(t)\exp(-\mathrm{i}\omega t)dt \int_{-T}^{T} \delta Y(t)\exp(\mathrm{i}\omega t)dt\right\} \\ &= \lim_{T\to\infty} \mathrm{E}\left\{\frac{1}{T}\delta\underline{X}_T(\omega)\delta\underline{Y}^*_T(\omega)\right\} \\ &= 2\int_{-\infty}^{\infty} \mathrm{E}\{\delta X(t)\delta Y(0)\}\exp(-\mathrm{i}\omega t)dt , \end{aligned} \tag{2.66}$$

where the bar under a symbol denotes a Fourier transformed variable and ω is the angular frequency. In the case that the microscopic quantities are vector-valued the spectral intensity is matrix-valued

$$\hat{\underline{S}}_{\mathbf{XY}}(\omega) = \hat{\underline{S}}^{\dagger}_{\mathbf{YX}}(\omega) = 2\int_{-\infty}^{\infty} \mathrm{E}\left\{\delta\mathbf{X}(t)\delta\mathbf{Y}^{\mathrm{T}}(0)\right\}\exp(-\mathrm{i}\omega t)dt , \tag{2.67}$$

where † denotes the complex conjugate of the transpose.

Direct evaluation of the correlation function with the BTE (e.g. [2.27]) might be cumbersome (cf. Sec. 3.3). A different but mathematically equivalent approach for the evaluation of the spectral intensity is based on stochastic differential equations (Langevin equations) [2.16]. The Langevin-type Boltzmann transport equation (LBTE) is obtained by adding a Langevin force ξ to the BTE (2.9) [2.25, 2.28–2.31]

$$\begin{aligned} &\left\{\frac{\partial}{\partial t} + \mathbf{F}^{\mathrm{T}}_n(\mathbf{r},\mathbf{k},t)\frac{1}{\hbar}\nabla_{\mathbf{k}} + \mathbf{v}^{\mathrm{T}}_n(\mathbf{r},\mathbf{k})\nabla_{\mathbf{r}}\right\}\check{f}_n(\mathbf{r},\mathbf{k},t) \\ &= \frac{\Omega}{(2\pi)^3}\sum_{n'}\int_{\mathrm{BZ}} S_{n,n'}(\mathbf{k}|\mathbf{k}')(\mathbf{r})\check{f}_n(\mathbf{r},\mathbf{k}',t) - S_{n',n}(\mathbf{k}'|\mathbf{k})(\mathbf{r})\check{f}_n(\mathbf{r},\mathbf{k},t)d^3k' \\ &\qquad + \xi_n(\mathbf{r},\mathbf{k},t) , \end{aligned} \tag{2.68}$$

where the Pauli exclusion principle has been neglected and the transition rate does not depend on time or the distribution function. $\check{f}$ is a stochastic variable of which the fluctuations due to the Langevin force are abrupt and instantaneous. The expectation of the Langevin force vanishes ($\mathrm{E}\{\xi(\mathbf{r},\mathbf{k},t)\} = 0$) and the fluctuations induced by the Langevin forces are assumed to be so small that the corresponding response of the distribution function is linear and $\mathrm{E}\{\check{f}(\mathbf{r},\mathbf{k},t)\} = f(\mathbf{r},\mathbf{k})$ holds, where $f(\mathbf{r},\mathbf{k})$ is the solution of the corresponding stationary noiseless BTE.

The correlation function of the Langevin force reads for the steady-state [2.25, 2.28–2.31]

$$\begin{aligned}
&\varphi_{\xi\xi\; n,n'}(\mathbf{r},\mathbf{r}',\mathbf{k},\mathbf{k}',t,t') = \\
&\Omega\Bigg[\left(\int S_{n'',n}(\mathbf{k}''|\mathbf{k})(\mathbf{r})f_n(\mathbf{r},\mathbf{k}) + S_{n,n''}(\mathbf{k}|\mathbf{k}'')(\mathbf{r})f_{n''}(\mathbf{r},\mathbf{k}'')d^3k''\right)\delta_{n,n'}\delta(\mathbf{k}-\mathbf{k}') \\
&\qquad - S_{n,n'}(\mathbf{k}|\mathbf{k}')(\mathbf{r})f_{n'}(\mathbf{r},\mathbf{k}') - S_{n',n}(\mathbf{k}'|\mathbf{k})(\mathbf{r})f_n(\mathbf{r},\mathbf{k})\Bigg]\delta(\mathbf{r}-\mathbf{r}')\delta(t-t') \\
&\qquad\qquad = n(\mathbf{r})S_{\xi'\xi'\; n,n'}(\mathbf{r},\mathbf{k},\mathbf{k}')\delta(\mathbf{r}-\mathbf{r}')\delta(t-t')\;, \qquad (2.69)
\end{aligned}$$

where $S_{\xi'\xi'}$ is the spectral intensity of the single-particle Langevin force. Since the scattering events described by the Langevin force are independent, instantaneous, and do not change the position of a particle in the RS, the correlation function is proportional to a delta function in the RS and time and linear in the particle density.

With respect to the balance equation (2.39) the Langevin force yields an additional term

$$\frac{\partial \check{x}}{\partial t} - \check{n}\left\langle \mathbf{F}^{\mathrm{T}}\frac{1}{\hbar}\nabla_{\mathbf{k}}X\right\rangle_{\mathbf{k}}^{\mathrm{LF}} + \nabla_{\mathbf{r}}^{\mathrm{T}}\check{\mathbf{j}}_X - \check{n}\left\langle \mathbf{v}^{\mathrm{T}}\nabla_{\mathbf{r}}X\right\rangle_{\mathbf{k}}^{\mathrm{LF}} = -\check{n}\left\langle S\{X\}\right\rangle_{\mathbf{k}}^{\mathrm{LF}} + \xi_{\dot{X}} \qquad (2.70)$$

with

$$\check{n}(\mathbf{r},t)\langle X\rangle_{\mathbf{k}}^{\mathrm{LF}}(\mathbf{r},t) = \frac{2}{(2\pi)^3}\sum_n \int_{\mathrm{BZ}} X_n(\mathbf{r},\mathbf{k})\check{f}_n(\mathbf{r},\mathbf{k},t)d^3k \qquad (2.71)$$

and

$$\xi_{\dot{X}}(\mathbf{r},t) = \frac{2}{(2\pi)^3}\sum_n \int_{\mathrm{BZ}} X_n(\mathbf{r},\mathbf{k})\xi_n(\mathbf{r},\mathbf{k},t)d^3k\;. \qquad (2.72)$$

The correlation function of this Langevin force is given by

$$\varphi_{\xi_{\dot{X}}\xi_{\dot{X}}}(\mathbf{r},\mathbf{r}',t,t') = n(\mathbf{r})S_{\xi'_{\dot{X}}\xi'_{\dot{X}}}(\mathbf{r})\delta(\mathbf{r}-\mathbf{r}')\delta(t-t') \qquad (2.73)$$

with spectral intensity of the single-particle Langevin forces

$$S_{\xi'_{\dot{X}}\xi'_{\dot{X}}}(\mathbf{r}) = \frac{4}{(2\pi)^6}\sum_n\sum_{n'}\int_{\mathrm{BZ}}\int_{\mathrm{BZ}} X_n(\mathbf{k})S_{\xi'\xi'\; n,n'}(\mathbf{r},\mathbf{k},\mathbf{k}')X_{n'}(\mathbf{k}')d^3k'd^3k\;. \qquad (2.74)$$

The correlation function of the moments of the Langevin force (2.73) is therefore local in the RS, because the correlation function (2.69) of the Langevin force of the LBTE is local.

As an example for the application of the LBTE (2.68) the spectral intensity of the fluctuations of the average velocity (drift velocity) in a spatially homogeneous, stationary system is evaluated for holes. In this case the electric field, the transition rate, the spectral intensity of the single-particle Langevin force, and the distribution function do not depend on the position in the RS, and the RS argument can be dropped and it is sufficient to perform the calculations solely in the **k**-space. Assuming a constant particle number the distribution function can be replaced by the single-particle probability density $\check{p}_n(\mathbf{k}, t)$ which must be normalized in the KS

$$\frac{2}{(2\pi)^3} \sum_n \int_{\mathrm{BZ}} \check{p}_n(\mathbf{k}, t) d^3k = 1 \ . \tag{2.75}$$

The fluctuation of the single-particle probability density reads

$$\delta\check{p}_n(\mathbf{k}, t) = \check{p}_n(\mathbf{k}, t) - p_n(\mathbf{k}) \ , \tag{2.76}$$

where the stationary solution of the noiseless BTE is normalized to one

$$\frac{2}{(2\pi)^3} \sum_n \int_{\mathrm{BZ}} p_n(\mathbf{k}) d^3k = 1 \tag{2.77}$$

and the fluctuation to zero (particle number conservation)

$$\frac{2}{(2\pi)^3} \sum_n \int_{\mathrm{BZ}} \delta\check{p}_n(\mathbf{k}, t) d^3k = 0 \ . \tag{2.78}$$

The corresponding LBTE for the fluctuation of the single-particle probability density is obtained by subtracting the noiseless stationary BTE from the LBTE and it reads for holes with $\mathbf{F} = q\mathbf{E}$

$$\begin{aligned} &\frac{\partial \delta\check{p}_n(\mathbf{k}, t)}{\partial t} + \frac{q}{\hbar}\mathbf{E}^{\mathrm{T}} \nabla_{\mathbf{k}} \delta\check{p}_n(\mathbf{k}, t) \\ &= \frac{\Omega}{(2\pi)^3} \sum_{n'} \int_{\mathrm{BZ}} S_{n,n'}(\mathbf{k}|\mathbf{k}') \delta\check{p}_{n'}(\mathbf{k}', t) - S_{n',n}(\mathbf{k}'|\mathbf{k}) \delta\check{p}_n(\mathbf{k}, t) d^3k' \\ &\qquad + \xi'_n(\mathbf{k}, t) \ . \end{aligned} \tag{2.79}$$

Since the noiseless BTE is linear in the distribution function, the LBTE for the fluctuation is linear in the Langevin force. The correlation function of the single-particle Langevin force is given by

$$\varphi_{\xi'\xi'\,n,n'}(\mathbf{k},\mathbf{k}',t,t') =$$
$$\Omega\Big[\Big(\int S_{n'',n}(\mathbf{k}''|\mathbf{k})p_n(\mathbf{k}) + S_{n,n''}(\mathbf{k}|\mathbf{k}'')p_{n''}(\mathbf{k}'')d^3k''\Big)\,\delta_{n,n'}\delta(\mathbf{k}-\mathbf{k}')$$
$$- S_{n,n'}(\mathbf{k}|\mathbf{k}')p_{n'}(\mathbf{k}') - S_{n',n}(\mathbf{k}'|\mathbf{k})p_n(\mathbf{k})\Big]\,\delta(t-t')$$
$$= S_{\xi'\xi'\,n,n'}(\mathbf{k},\mathbf{k}')\delta(t-t')\;, \quad (2.80)$$

where the spectral intensity of the single-particle Langevin force for the bulk case (2.80) $S_{\xi'\xi'\,n,n'}(\mathbf{k},\mathbf{k}')$ is obtained from the general case (2.69) by replacing the single-particle probability density $f_n(\mathbf{r},\mathbf{k})/n(\mathbf{r})$ by $p_n(\mathbf{k})$ of the bulk. Under homogeneous conditions in the RS the general case therefore reduces to the bulk case.

The balance equation for the fluctuations of macroscopic quantities reads

$$\frac{\partial\langle\delta X\rangle^{\mathrm{LF}}_{\mathrm{hom}}}{\partial t} - q\mathbf{E}^{\mathrm{T}}\left\langle\frac{1}{\hbar}\nabla_{\mathbf{k}}\delta X\right\rangle^{\mathrm{LF}}_{\mathrm{hom}} = -\,\langle S\{\delta X\}\rangle^{\mathrm{LF}}_{\mathrm{hom}} + \xi'_{\dot{X}} \quad (2.81)$$

with

$$\langle\delta X\rangle^{\mathrm{LF}}_{\mathrm{hom}}(t) = \frac{2}{(2\pi)^3}\sum_n\int_{\mathrm{BZ}} X_n(\mathbf{k})\delta\breve{p}_n(\mathbf{k},t)d^3k \quad (2.82)$$

and

$$\xi'_{\dot{X}}(t) = \frac{2}{(2\pi)^3}\sum_n\int_{\mathrm{BZ}} X_n(\mathbf{k})\xi'_n(\mathbf{k},t)d^3k\;. \quad (2.83)$$

Since the process is stationary, it is more convenient to solve the problem in the frequency domain for which the balance equation reads

$$\mathrm{i}\omega\langle\delta X\rangle^{\mathrm{LF}}_{\mathrm{hom}} - q\mathbf{E}^{\mathrm{T}}\left\langle\frac{1}{\hbar}\nabla_{\mathbf{k}}\delta X\right\rangle^{\mathrm{LF}}_{\mathrm{hom}} = -\,\langle S\{\delta X\}\rangle^{\mathrm{LF}}_{\mathrm{hom}} + \underline{\xi}'_{\dot{X}}\;, \quad (2.84)$$

where the expectations are calculated with the Fourier transformed probability density $\delta\underline{\breve{p}}_n(\mathbf{k},\omega)$.

There are two ways to calculate the fluctuations of the drift velocity $\delta\mathbf{v}_d(t) = \langle\delta\mathbf{v}\rangle^{\mathrm{LF}}_{\mathrm{hom}}(t)$ depending on how the scattering integral is solved. In the case of the first approach, the microscopic relaxation time method, the scattering integral of the balance equation is solved exactly (cf. Sec. 2.3), whereas in the case of the second approach, the macroscopic relaxation time approximation, it is approximated (cf. Sec. 2.2).

In the case of the microscopic relaxation time method the balance equation (2.84) is calculated for the microscopic quantity

$$\underline{\mathbf{X}}_n(\mathbf{k},\omega) = \frac{\tau_n(\varepsilon_n(\mathbf{k}))\mathbf{v}_n(\mathbf{k})}{1+\mathrm{i}\omega\tau_n(\varepsilon_n(\mathbf{k}))}\;, \quad (2.85)$$

where it is assumed that the microscopic relaxation time is a scalar and depends on the wave vector only via the energy (cf. Sec. 2.3). Eq. (2.84) evaluates in this case without any approximations to (cf. Eq. (2.56))

$$\mathrm{i}\omega\left\langle\frac{\tau\delta\underline{\mathbf{v}}}{1+\mathrm{i}\omega\tau}\right\rangle_{\mathrm{hom}}^{\mathrm{LF}}-\left\langle\frac{q}{\hbar}\nabla_{\mathbf{k}}\frac{\tau\delta\underline{\mathbf{v}}^{\mathrm{T}}}{1+\mathrm{i}\omega\tau}\right\rangle_{\mathrm{hom}}^{\mathrm{LF}}\mathbf{E}=-\left\langle\frac{\delta\underline{\mathbf{v}}}{1+\mathrm{i}\omega\tau}\right\rangle_{\mathrm{hom}}^{\mathrm{LF}}+\underline{\boldsymbol{\xi}}'_{\frac{\tau\dot{\mathbf{v}}}{1+\mathrm{i}\omega\tau}}\ . \tag{2.86}$$

Solving for the velocity fluctuations yields

$$\delta\underline{\mathbf{v}}_d-\delta\underline{\hat{\mu}}_{\mathrm{mic}}\mathbf{E}=\underline{\boldsymbol{\xi}}'_{\frac{\tau\dot{\mathbf{v}}}{1+\mathrm{i}\omega\tau}} \tag{2.87}$$

with

$$\mathrm{i}\omega\left\langle\frac{\tau\delta\underline{\mathbf{v}}}{1+\mathrm{i}\omega\tau}\right\rangle_{\mathrm{hom}}^{\mathrm{LF}}+\left\langle\frac{\delta\underline{\mathbf{v}}}{1+\mathrm{i}\omega\tau}\right\rangle_{\mathrm{hom}}^{\mathrm{LF}}=\langle\delta\underline{\mathbf{v}}\rangle_{\mathrm{hom}}^{\mathrm{LF}}=\delta\underline{\mathbf{v}}_d \tag{2.88}$$

and

$$\delta\underline{\hat{\mu}}_{\mathrm{mic}}=\left\langle\frac{q}{\hbar}\nabla_{\mathbf{k}}\frac{\tau\delta\underline{\mathbf{v}}^{\mathrm{T}}}{1+\mathrm{i}\omega\tau}\right\rangle_{\mathrm{hom}}^{\mathrm{LF}}\ . \tag{2.89}$$

The spectral intensities corresponding to Eq. (2.87) read

$$\hat{S}_{\mathbf{v}_d-\hat{\mu}_{\mathrm{mic}}\mathbf{E}\ \mathbf{v}_d-\hat{\mu}_{\mathrm{mic}}\mathbf{E}}=\hat{S}_{\boldsymbol{\xi}'_{\frac{\tau\dot{\mathbf{v}}}{1+\mathrm{i}\omega\tau}}\boldsymbol{\xi}'_{\frac{\tau\dot{\mathbf{v}}}{1+\mathrm{i}\omega\tau}}}\ . \tag{2.90}$$

In the case of equilibrium ($\mathbf{E}=\mathbf{0}$) the left-hand side reduces to the spectral intensity of the drift velocity fluctuations

$$\hat{S}^{\mathrm{eq}}_{\mathbf{v}_d\mathbf{v}_d}=\hat{S}^{\mathrm{eq}}_{\boldsymbol{\xi}'_{\frac{\tau\dot{\mathbf{v}}}{1+\mathrm{i}\omega\tau}}\boldsymbol{\xi}'_{\frac{\tau\dot{\mathbf{v}}}{1+\mathrm{i}\omega\tau}}}\ . \tag{2.91}$$

Moreover, the right-hand side of Eq. (2.90) can be analytically evaluated under equilibrium conditions. With the principle of detailed balance holding in equilibrium [2.8]

$$S_{n,n'}(\mathbf{k}|\mathbf{k}')p^{\mathrm{eq}}_{n'}(\mathbf{k}')=S_{n',n}(\mathbf{k}'|\mathbf{k})p^{\mathrm{eq}}_{n}(\mathbf{k}) \tag{2.92}$$

where $p^{\mathrm{eq}}_n(\mathbf{k})$ is given by the Maxwell-Boltzmann distribution function (2.44), the spectral intensity of the Langevin force of the LBTE can be simplified

$$S^{\mathrm{eq}}_{\xi'\xi'\ n,n'}(\mathbf{k},\mathbf{k}')=16\pi^3\left[S_n(\mathbf{k})p^{\mathrm{eq}}_n(\mathbf{k})\delta_{n,n'}\delta(\mathbf{k}-\mathbf{k}')-\frac{\Omega}{(2\pi)^3}S_{n,n'}(\mathbf{k}|\mathbf{k}')p^{\mathrm{eq}}_{n'}(\mathbf{k}')\right]\ . \tag{2.93}$$

The spectral intensity of the Langevin force of Eq. (2.91) is obtained with Eqs. (2.74), (2.93) and (2.55) [2.29, 2.32]

$$\begin{aligned}&\hat{S}^{\mathrm{eq}}_{\boldsymbol{\xi}'_{\frac{\tau\dot{\mathbf{v}}}{1+\mathrm{i}\omega\tau}}\boldsymbol{\xi}'_{\frac{\tau\dot{\mathbf{v}}}{1+\mathrm{i}\omega\tau}}}=\\&\frac{4}{(2\pi)^6}\sum_n\sum_{n'}\int_{\mathrm{BZ}}\int_{\mathrm{BZ}}\frac{\tau_n(\varepsilon_n(\mathbf{k}))\mathbf{v}_n(\mathbf{k})}{1+\mathrm{i}\omega\tau_n(\varepsilon_n(\mathbf{k}))}S^{\mathrm{eq}}_{\xi'\xi'\ n,n'}(\mathbf{k},\mathbf{k}')\frac{\tau_{n'}(\varepsilon_{n'}(\mathbf{k}'))\mathbf{v}^{\mathrm{T}}_{n'}(\mathbf{k}')}{1-\mathrm{i}\omega\tau_{n'}(\varepsilon_{n'}(\mathbf{k}'))}d^3k'd^3k\\&=4\left\langle\frac{\tau\mathbf{v}\mathbf{v}^{\mathrm{T}}}{1+\omega^2\tau^2}\right\rangle_{\mathrm{hom}}^{\mathrm{eq}}=4\Re\left\{\left\langle\frac{\tau\mathbf{v}\mathbf{v}^{\mathrm{T}}}{1+\mathrm{i}\omega\tau}\right\rangle_{\mathrm{hom}}^{\mathrm{eq}}\right\}\ .\end{aligned} \tag{2.94}$$

The last term on the right-hand side is the real part of the diffusion constant for which the Einstein relation holds

$$\underline{\hat{D}}^{\mathrm{eq}}_{\mathrm{mic}}(\omega) = \left\langle \frac{\tau \mathbf{v}\mathbf{v}^{\mathrm{T}}}{1+\mathrm{i}\omega\tau} \right\rangle^{\mathrm{eq}}_{\mathrm{hom}} = V_{\mathrm{T}} \left\langle \frac{q}{\hbar} \nabla_{\mathbf{k}} \frac{\tau \mathbf{v}^{\mathrm{T}}}{1+\mathrm{i}\omega\tau} \right\rangle^{\mathrm{eq}}_{\mathrm{hom}} = V_{\mathrm{T}} \underline{\hat{\mu}}^{\mathrm{eq}}_{\mathrm{mic}}(\omega) \ . \tag{2.95}$$

Thus, under equilibrium conditions the spectral intensity of the drift velocity fluctuations is given by the real part of the diffusion constant [2.33]

$$\hat{S}^{\mathrm{eq}}_{\mathbf{v}_d \mathbf{v}_d} = 4\Re\left\{ \underline{\hat{D}}^{\mathrm{eq}}_{\mathrm{mic}} \right\} \ . \tag{2.96}$$

Since Eq. (2.87) is obtained from the LBTE without any approximations and the evaluation of the spectral intensity (2.96) does not involve any approximations, this spectral intensity is an exact result of the LBTE.

Inverse Fourier transformation yields the velocity autocorrelation function for a stationary and spatially homogeneous system in equilibrium [2.34]

$$\langle \delta\mathbf{v}, \delta\mathbf{v}^{\mathrm{T}} \rangle^{\mathrm{eq}}_{\mathrm{hom}}(t - t_0) = \left\langle \mathbf{v}\mathbf{v}^{\mathrm{T}} \exp\left(-\frac{|t-t_0|}{\tau} \right) \right\rangle^{\mathrm{eq}}_{\mathrm{hom}} \ . \tag{2.97}$$

This result can also be obtained by solving the BTE instead of the LBTE in the time domain (cf. Sec. 3.3). Integration over time yields the diffusion constant (2.59) [2.35]

$$\int_0^{\infty} \langle \delta\mathbf{v}, \delta\mathbf{v}^{\mathrm{T}} \rangle^{\mathrm{eq}}_{\mathrm{hom}}(t) dt = \langle \tau \mathbf{v}\mathbf{v}^{\mathrm{T}} \rangle^{\mathrm{eq}}_{\mathrm{hom}} = \hat{D}^{\mathrm{eq}} = \underline{\hat{D}}^{\mathrm{eq}}_{\mathrm{mic}}(\omega = 0) \ . \tag{2.98}$$

It is worth noting that this relation only holds under equilibrium conditions. Otherwise the integral over the velocity autocorrelation function (spreading diffusion constant) and the diffusion constant defined as a transport coefficient (2.59) differ.

The second approach for calculating the spectral intensity of the drift velocity fluctuations is based on the macroscopic relaxation time approximation (2.42) [2.36] and the balance equation (2.84). This yields with $\mathbf{X} = \mathbf{v}$

$$\mathrm{i}\omega\delta\underline{\mathbf{v}}_d - q\langle \delta\hat{m}^{-1} \rangle^{\mathrm{LF}}_{\mathrm{hom}} \mathbf{E} = -\frac{\delta\underline{\mathbf{v}}_d}{\tau_v} + \underline{\boldsymbol{\xi}}'_{\mathbf{v}} \ , \tag{2.99}$$

where τ_v is calculated in such a way that the correct longitudinal mobility is obtained under stationary noiseless conditions

$$\mathbf{E}^{\mathrm{T}} \mathbf{v}_d = \mathbf{E}^{\mathrm{T}} \hat{\mu}_{\mathrm{mac}} \mathbf{E} \tag{2.100}$$

with

$$\hat{\mu}_{\mathrm{mac}} = q\tau_v \langle \hat{m}^{-1} \rangle_{\mathrm{hom}} \ . \tag{2.101}$$

Solving Eq. (2.99) for the velocity fluctuations yields

$$\delta \underline{\mathbf{v}}_d - \delta \underline{\hat{\mu}}_{\mathrm{mac}} \mathbf{E} = \frac{\tau_v}{1 + \mathrm{i}\omega\tau_v} \underline{\boldsymbol{\xi}}'_{\dot{\mathbf{v}}} \tag{2.102}$$

with

$$\delta \underline{\hat{\mu}}_{\mathrm{mac}} = \frac{q\tau_v}{1 + \mathrm{i}\omega\tau_v} \langle \delta \hat{m}^{-1} \rangle_{\mathrm{hom}}^{\mathrm{LF}} \,. \tag{2.103}$$

The spectral intensities corresponding to Eq. (2.102) read

$$\hat{S}_{\mathbf{v}_d - \hat{\mu}_{\mathrm{mac}}\mathbf{E}\ \mathbf{v}_d - \hat{\mu}_{\mathrm{mac}}\mathbf{E}} = \frac{\tau_v^2}{1 + \omega^2\tau_v^2} \hat{S}_{\boldsymbol{\xi}'_{\dot{\mathbf{v}}}\boldsymbol{\xi}'_{\dot{\mathbf{v}}}} \tag{2.104}$$

With the same methods as in the case of the microscopic relaxation time approach the spectral intensity of the Langevin forces can be evaluated under equilibrium conditions without further approximations

$$\hat{S}_{\boldsymbol{\xi}'_{\dot{\mathbf{v}}}\boldsymbol{\xi}'_{\dot{\mathbf{v}}}} = \left\langle \frac{\mathbf{v}\mathbf{v}^{\mathrm{T}}}{\tau} \right\rangle_{\mathrm{hom}}^{\mathrm{eq}} \tag{2.105}$$

yielding

$$\hat{S}^{\mathrm{eq}}_{\mathbf{v}_d\mathbf{v}_d} = \frac{\tau_v^2}{1 + \omega^2\tau_v^2} \left\langle \frac{\mathbf{v}\mathbf{v}^{\mathrm{T}}}{\tau} \right\rangle_{\mathrm{hom}}^{\mathrm{eq}} , \tag{2.106}$$

and the autocorrelation function of the velocity fluctuations is given by

$$\langle \delta\mathbf{v}, \delta\mathbf{v}^{\mathrm{T}} \rangle_{\mathrm{hom}}^{\mathrm{eq}} (t - t_0) = \tau_v \left\langle \frac{\mathbf{v}\mathbf{v}^{\mathrm{T}}}{\tau} \right\rangle_{\mathrm{hom}}^{\mathrm{eq}} \exp\left(-\frac{|t - t_0|}{\tau_v} \right) \,. \tag{2.107}$$

Compared to the exact autocorrelation function (2.97) this autocorrelation function does not reduce to the variance of the velocity $\langle \mathbf{v}\mathbf{v}^{\mathrm{T}} \rangle_{\mathrm{hom}}^{\mathrm{eq}}$ for $t = t_0$, and the decay of the correlation is described by a simple relaxation time instead of the energy-dependent microscopic relaxation time. The spreading diffusion constant evaluates to

$$\int_0^{\infty} \langle \delta\mathbf{v}, \delta\mathbf{v}^{\mathrm{T}} \rangle_{\mathrm{hom}}^{\mathrm{eq}} (t) dt = \tau_v^2 \left\langle \frac{\mathbf{v}\mathbf{v}^{\mathrm{T}}}{\tau} \right\rangle_{\mathrm{hom}}^{\mathrm{eq}} \neq \tau_v \left\langle \mathbf{v}\mathbf{v}^{\mathrm{T}} \right\rangle_{\mathrm{hom}}^{\mathrm{eq}} = V_{\mathrm{T}} \hat{\mu}_{\mathrm{mac}} \,. \tag{2.108}$$

Thus, in the case of the macroscopic relaxation time approximation also the Einstein relation does not hold, and the spreading diffusion constant based on the fluctuations of the velocity does not equal the diffusion constant defined as a transport parameter based on the mobility.

References

[2.1] M. L. Cohen and J. R. Chelikowsky, *Electronic Structure and Optical Properties of Semiconductors*, Springer, New York, 2nd edition, 1989.

[2.2] M. M. Rieger and P. Vogl, "Electronic-band parameters in strained $Si_{1-x}Ge_x$ alloys on $Si_{1-y}Ge_y$ substrates", *Phys. Rev. B*, vol. 48, pp. 14276–14287, 1993.

[2.3] M. M. Rieger and P. Vogl, "Electronic-band parameters in strained $Si_{1-x}Ge_x$ alloys on $Si_{1-y}Ge_y$ substrates", *Phys. Rev. B*, vol. 50, pp. 8138, 1994, Erratum.

[2.4] O. Madelung, *Introduction to Solid State Theory*, Springer, Berlin, 1978.

[2.5] K. Hess, Ed., *Monte Carlo Device Simulation: Full Band and Beyond*, Kluwer, Boston, 1991.

[2.6] W. Shockley, *Electrons and Holes in Semiconductors*, van Nostrand, Princeton, New Jersey, 1950.

[2.7] J.M. Ziman, *Electrons and Phonons*, Clarendon Press, Oxford, 1960.

[2.8] W. Brauer and H. W. Streitwolf, *Theoretische Grundlagen der Halbleiterphysik*, Vieweg, Braunschweig, 2nd edition, 1977.

[2.9] C. Jacoboni and P. Lugli, *The Monte Carlo Method for Semiconductor Device Simulation*, Springer, Wien, 1989.

[2.10] M. Lundstrom, *Fundamentals of Carrier Transport*, vol. 10 of *Modular Series on Solid State Devices*, Addison-Wesley, New York, 1990.

[2.11] A. H. Marshak and K. M. van Vliet, "Electrical current in solids with position–dependent band structure", *Solid–State Electron.*, vol. 21, pp. 417–427, 1978.

[2.12] N. G. van Kampen, *Stochastic Process in Physics and Chemistry*, North-Holland Publishing, Amsterdam, 1981.

[2.13] G. Röpke, *Statistische Mechanik für das Nichtgleichgewicht*, Physik-Verlag, Weinheim, 1987.

[2.14] P. J. Price, "Monte Carlo calculation of electron transport in solids", *Semiconductors and Semimetals*, vol. 14, pp. 249–309, 1979.

[2.15] D. Schroeder, *Modelling of Interface Carrier Transport for Device Simulation*, Springer, Wien, 1994.

[2.16] C. W. Gardiner, *Handbook of Stochastic Methods*, Springer, Berlin, 1985.

[2.17] H.-J. Peifer, "Monte–Carlo Simulation des Hochenergietransports von Elektronen in submikron MOS–Strukturen", Doctor thesis, RWTH Aachen, Aachen, 1992, Augustinus Buchhandlung.

[2.18] A. Papoulis, *Probability,Random Variables and Stochastic Processes*, Mc Graw-Hill, 3rd edition, 1991.

[2.19] R. Stratton, "Diffusion of hot and cold electrons in semiconductor barriers", *Phys. Rev.*, vol. 126, pp. 2002–2013, 1962.

[2.20] K. Bløtekjær, "Transport equations for electrons in two-valley semiconductors", *IEEE Trans. Electron Devices*, vol. 17, no. 1, pp. 38–47, 1970.

[2.21] R. Thoma, "Entwicklung eines hydrodynamischen Gleichungssystems zur Beschreibung von Halbleiterbauelementen", Doktorarbeit, RWTH Aachen, 1991, Aachen.

[2.22] E. M. Lifshitz and L. P. Pitaevskii, *Physical Kinetics*, vol. 10 of *Courses of Theoretical Physics*, Butterworth–Heinemann, Oxford, UK, 1981.

[2.23] K. Hess, *Advanced Theory of Semiconductor Devices*, IEEE Press, Piscataway, NJ, USA, 2000.

[2.24] T. Ando, A. B. Fowler, and F. Stern, "Electronic properties of two-dimensional systems", *Rev. Mod. Phys.*, vol. 54, pp. 437–672, 1982.

[2.25] Sh. Kogan, *Electronic Noise and Fluctuations in Solids*, Cambridge University Press, Cambridge, New York, Melbourne, 1996.

[2.26] F. Bonani and G. Ghione, *Noise in Semiconductor Devices, Modeling and Simulation*, Advanced Microelectronics. Springer, Berlin, Heidelberg, New York, 2001.

[2.27] C. E. Korman and I. D. Mayergoyz, "Semiconductor noise in the framework of semiclassical transport", *Phys. Rev. B*, vol. 54, pp. 17620–17627, 1996.

[2.28] S. M. Kogan, "Equations for the correlation functions using a generalized Keldysh technique", *Phys. Rev. A*, vol. 44, pp. 8072–8082, 1991.

[2.29] H. S. Min and D. Ahn, "Langevin noise sources for the Boltzmann transport equations with the relaxation-time approximation in nondegenerate semiconductors", *J. Appl. Phys.*, vol. 58, pp. 2262–2265, 1985.

[2.30] H. S. Min, "A unified theory of noise in nondegenerate semiconductors", *J. Appl. Phys.*, vol. 61, pp. 4549–4565, 1987.

[2.31] H. S. Min, "Steady-state Nyquist theorem for nondegenerate semiconductors", *J. Appl. Phys.*, vol. 64, pp. 6339–6344, 1988.

[2.32] C. Jungemann and B. Meinerzhagen, "Analysis of the stochastic error of stationary Monte Carlo device simulations", *IEEE Trans. Electron Devices*, vol. 48, no. 5, pp. 985–992, 2001.

[2.33] J.-P. Nougier, "Fluctuations and noise of hot carriers in semiconducor materials and devices", *IEEE Trans. Electron Devices*, vol. ED-41, no. 11, pp. 2034–2049, 1994.

[2.34] W. Shockley, John A. Copeland, and R. P. James, "The impedance field method of noise calculation in active semiconductor devices", in *Quantum theory of atoms, molecules and solid state*, P. O. Lowdin, Ed., pp. 537–563. Academic Press, 1966.

[2.35] L. Reggiani, E. Starikov, P. Shiktorov, V. Gružinskis, and L. Varani, "Modelling of small-signal response and electronic noise in semiconductor high-field transport", *Semicond. Sci. Technol.*, vol. 12, pp. 141–156, 1997.

[2.36] P. Shiktorov, E. Starikov, V. Gruzinskis, T. Gonzalez, J. Mateos, D. Pardo, L. Reggiani, L. Varani, and J. C. Vaissere, "Langevin forces and generalized transfer fields for noise modeling in deep submicron devices", *IEEE Trans. Electron Devices*, vol. 47, no. 10, pp. 1992–1998, 2000.

3 The Monte-Carlo Method

The MC method is a stochastic method for the solution of integrals [3.1–3.4]. By formal integration the BTE is transformed into an integral equation, which can be solved with the MC method [3.5–3.12].

In the first section of this chapter basic MC methods are introduced, which allow the generation of nonuniformly distributed pseudorandom numbers which are the basis of the MC method. In the second part the integral equation corresponding to the BTE is presented and based on the Neumann series the MC algorithm is derived including self-scattering. Based on the integral representation of the BTE the velocity autocorrelation function for a bulk system in equilibrium is derived as an example in the next section. The fourth part covers basic statistics for the estimation of expectations (e.g. mean energy, terminal currents, etc.), and in the last section the stochastic error of MC estimation is analyzed.

3.1 Basic Monte-Carlo Methods

All MC methods discussed in this work require uniformly and identically distributed random numbers. Since the generation of real random numbers on a computer is difficult and the results are not reproducible, here, pseudorandom numbers are used [3.2, 3.13]. A sequence of pseudorandom numbers is generated with the combined linear congruential generator of Ref. [3.14] which produces a sequence of a length of about $2.3 \cdot 10^{18}$. The pseudorandom numbers are (pseudo-) independent and identically distributed with the properties

$$0 < r < 1 \ , \quad p(r) = 1 \ . \tag{3.1}$$

The generator is always initialized with the same seed, because in this case the results of the MC simulation are reproducible allowing for a simpler debugging of the program [3.15]. Of course, in the case of a series of MC simulations the final state of the random number generator of the last run is passed to the next simulation as the seed to avoid a reproduction of the pseudorandom sequence.

Random numbers with nonuniform distributions can be generated with the direct method [3.4,3.14,3.16]. For a given probability density $p(x) \geq 0$ the distribution function reads [3.17]

$$F(x) = \int_{-\infty}^{x} p(x')dx' \tag{3.2}$$

with

$$F(-\infty) = 0 \; , \quad 0 \leq F(x) \leq 1 \; , \text{ and } \quad F(\infty) = 1 \; . \tag{3.3}$$

A random number x with the probability density $p(x)$ is generated with the inverse distribution function

$$x = F^{-1}(r) \; , \tag{3.4}$$

where r is a random number of the type (3.1).

In the case of a two-dimensional probability density $p(x, y)$ the random numbers x and y are calculated successively [3.17]. First, x is determined with the marginal probability density

$$p(x) = \int_{-\infty}^{\infty} p(x, y)dy \tag{3.5}$$

using the method for one-dimensional probability densities. Then y is determined with the conditional probability density

$$p(y) = p(y|x) = \frac{p(x, y)}{p(x)} \; , \tag{3.6}$$

where x has already been determined. The extension to higher dimensional problems is straight forward.

In most cases the inverse distribution function (3.4) does not have a simple, closed form rendering the direct method useless. In the case that the probability density $p(x)$ is bounded the rejection method can be used [3.15,3.16]. The rejection method consists of two steps. In the first step a random number x_1 is generated with a probability density $\tilde{p}(x)$ with

$$M\tilde{p}(x) \geq p(x) \; , \tag{3.7}$$

where M is a positive, finite constant. The probability density $\tilde{p}(x)$ is modeled in such a way that it reproduces the original probability density $p(x)$ as good as possible. On the other hand, a closed form of the inverse distribution function corresponding to $\tilde{p}(x)$ must exist to allow the application of Eq. (3.4) to generate the random number x_1 with a uniformly distributed random number r_1. The random number x_1 is only accepted, if the following condition holds

$$p(x_1) > r_2 M\tilde{p}(x_1) \; , \tag{3.8}$$

where r_2 is a further uniformly distributed random number. If the random number x_1 is rejected, the whole procedure is repeated with two new random numbers r_1 and r_2. The expected value of the number of rejection steps is

$$\mathrm{E}\left\{N^{\text{reject}}\right\} = M \ . \tag{3.9}$$

Thus, M should be as small as possible.

In the case of a discrete distribution the integer random number i can be calculated with a discrete version of the direct method [3.15]

$$i \ : \ \sum_{j=1}^{i-1} P_j < r \leq \sum_{j=1}^{i} P_j \ , \tag{3.10}$$

where P_j is the probability of the jth event and r a uniformly distributed random number. The rejection method can be used as well.

3.2 The Monte-Carlo Solver for the Boltzmann Transport Equation

The formal integration of the linearized BTE (2.22) for the conditional probability density yields [3.6–3.11, 3.18, 3.19]

$$\begin{aligned} p(\zeta,t|\zeta_0,t_0) &= p_0(\zeta,t|\zeta_0,t_0) \\ &+ \int_{t_0}^{t} \int \int p_0(\zeta,t|\zeta_1',t_1) S(\zeta_1'|\zeta_1) p(\zeta_1,t_1|\zeta_0,t_0) d\zeta_1' d\zeta_1 dt_1 \ , \end{aligned} \tag{3.11}$$

where $p_0(\zeta,t|\zeta_0,t_0)$ is the conditional probability density that a particle found at the time t_0 in the state ζ_0 appears at t in the state ζ without being scattered

$$p_0(\zeta,t|\zeta_0,t_0) = \delta\left(\zeta - \zeta_{\text{New}}(t|\zeta_0,t_0)\right) \exp\left(-\int_{t_0}^{t} S\left(\zeta_{\text{New}}(\tau|\zeta_0,t_0)\right) d\tau\right) \ . \tag{3.12}$$

$\zeta_{\text{New}}(t|\zeta_0,t_0)$ is the solution of the equations of motion (2.3) and (2.4) at the time t for a particle that was at time t_0 in the state ζ_0. The usual MC algorithm is based on the conjugate[1] of Eq. (3.11) [3.10]

$$\begin{aligned} p(\zeta,t|\zeta_0,t_0) &= p_0(\zeta,t|\zeta_0,t_0) \\ &+ \int_{t_0}^{t} \int \int p(\zeta,t|\zeta_1',t_1) S(\zeta_1'|\zeta_1) p_0(\zeta_1,t_1|\zeta_0,t_0) d\zeta_1' d\zeta_1 dt_1 \ . \end{aligned} \tag{3.13}$$

[1] Direct application of Eq. (3.11) yields the so-called backward MC method [3.20].

Equations (3.11) and (3.13) are integral equations, because the conditional probability density appears also in the integral on the right-hand side. Inserting Eq. (3.13) into itself yields

$$\begin{aligned} p(\zeta,t|\zeta_0,t_0) &= p_0(\zeta,t|\zeta_0,t_0) \\ &+ \int_{t_0}^t \int\int p_0(\zeta,t|\zeta_1',t_1)S(\zeta_1'|\zeta_1)p_0(\zeta_1,t_1|\zeta_0,t_0)d\zeta_1'd\zeta_1dt_1 \\ &+ \int_{t_0}^t \int\int \int_{t_1}^t \int\int p(\zeta,t|\zeta_2',t_2)S(\zeta_2'|\zeta_2)p_0(\zeta_2,t_2|\zeta_1',t_1) \\ &\qquad S(\zeta_1'|\zeta_1)p_0(\zeta_1,t_1|\zeta_0,t_0)d\zeta_2'd\zeta_2dt_2d\zeta_1'd\zeta_1dt_1 \ . \end{aligned} \tag{3.14}$$

Repeating this step infinite times gives the Neumann series

$$\begin{aligned} p(\zeta,t|\zeta_0,t_0) &= p_0(\zeta,t|\zeta_0,t_0) \\ &+ \int_{t_0}^t \int\int p_0(\zeta,t|\zeta_1',t_1)S(\zeta_1'|\zeta_1)p_0(\zeta_1,t_1|\zeta_0,t_0)d\zeta_1'd\zeta_1dt_1 \\ &+ \int_{t_0}^t \int\int \int_{t_1}^t \int\int p_0(\zeta,t|\zeta_2',t_2)S(\zeta_2'|\zeta_2)p_0(\zeta_2,t_2|\zeta_1',t_1) \\ &\qquad S(\zeta_1'|\zeta_1)p_0(\zeta_1,t_1|\zeta_0,t_0)d\zeta_2'd\zeta_2dt_2d\zeta_1'd\zeta_1dt_1 \\ &+ \quad \ldots \\ &\vdots \end{aligned} \tag{3.15}$$

The first term on the right-hand side is the probability density that a particle is not scattered during its propagation from the initial state ζ_0 into the final state ζ. The second term gives the probability density that the particle is scattered exactly once between the initial and final state. The third term includes two scattering events and so forth. Thus, the conditional probability density $p(\zeta,t|\zeta_0,t_0)$ is the sum of all conditional probability densities with zero, one, two, ... scattering events

$$p(\zeta,t|\zeta_0,t_0) = \sum_{i=0}^{\infty} p_i(\zeta,t|\zeta_0,t_0) \tag{3.16}$$

with

$$p_i(\zeta,t|\zeta_0,t_0) = \int_{t_0}^{t}\int\int p_{i-1}(\zeta,t|\zeta_1',t_1)S(\zeta_1'|\zeta_1)p_0(\zeta_1,t_1|\zeta_0,t_0)d\zeta_1'd\zeta_1dt_1 \ . \tag{3.17}$$

The probability of i scattering events in the interval $[t_0,t]$ is given by the integral of the probability density for i scattering events (3.17) over all final states

$$P_i(t|\zeta_0,t_0) = \int p_i(\zeta,t|\zeta_0,t_0)d\zeta \ . \tag{3.18}$$

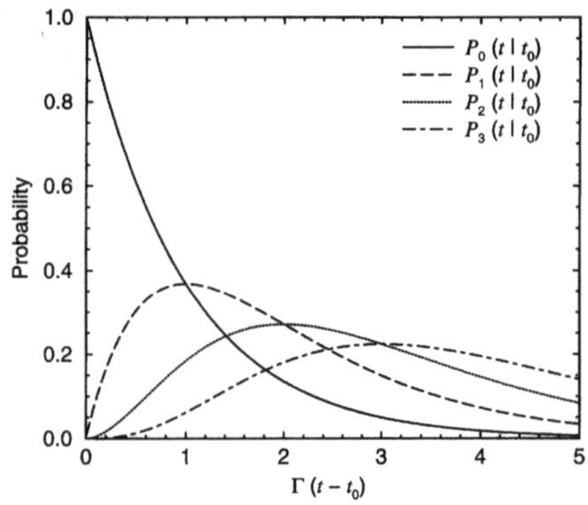

Figure 3.1: Probability $P_i(t|t_0)$ of i scattering events in the interval $[t_0, t]$ for a Poisson process with scattering rate Γ

The integral in Eq. (3.12) is difficult to evaluate, because the scattering rate depends in a complex way on the particle state $\boldsymbol{\zeta}$. To circumvent this problem self-scattering is included in the scattering integral [3.6, 3.9, 3.18], which has a transition rate of the following form

$$S_{\mathrm{self}}(\boldsymbol{\zeta}'|\boldsymbol{\zeta}) = (\Gamma - S(\boldsymbol{\zeta}))\delta(\boldsymbol{\zeta}' - \boldsymbol{\zeta}) \ . \tag{3.19}$$

The Dirac function on the right-hand side ensures that the particle state is not changed by this scattering process and the solution of the BTE is therefore unchanged. Nevertheless, the total scattering rate is changed

$$S_{\mathrm{tot}}(\boldsymbol{\zeta}) = \int S_{\mathrm{self}}(\boldsymbol{\zeta}'|\boldsymbol{\zeta})d\boldsymbol{\zeta}' + S(\boldsymbol{\zeta}) = \Gamma \ , \tag{3.20}$$

where Γ must be larger than the scattering rate $S(\boldsymbol{\zeta})$ for all possible states $\boldsymbol{\zeta}$. Now it is possible to evaluate the probability for i scattering events in the interval $[t_0, t]$ including self-scattering

$$P_i(t|t_0) = \frac{(\Gamma(t - t_0))^i}{i!} \exp(-\Gamma(t - t_0)) \ , \tag{3.21}$$

where the probability does not depend on the initial state $\boldsymbol{\zeta}_0$ anymore. Thus, the scattering events have a Poisson distribution. In Fig. 3.1 the probabilities for up to three scattering events are shown. The average number of scattering events in the interval $[t_0, t]$ is

$$\sum_{i=0}^{\infty} i P_i(t|t_0) = \Gamma(t - t_0) \tag{3.22}$$

and the maximum of the probability $P_i(t|t_0)$ is found for $\Gamma(t - t_0) = i$. The sum

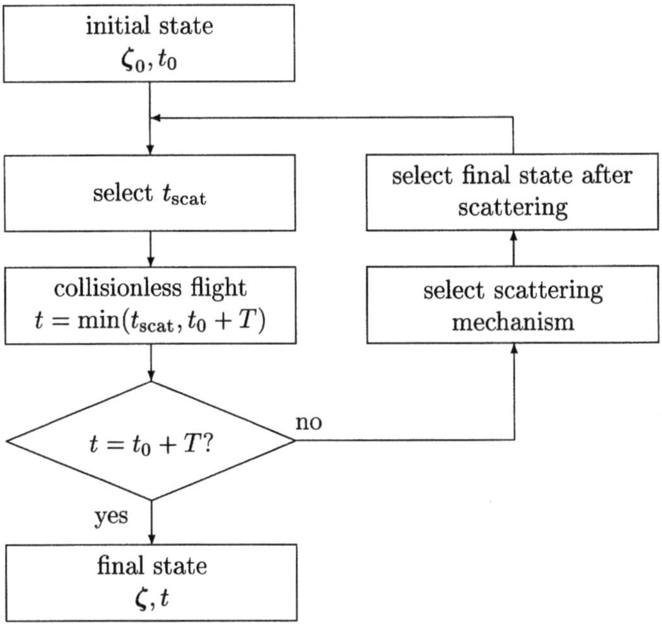

Figure 3.2: Flowchart of the MC method for a simulation lasting T

over all probabilities yields

$$\sum_{i=0}^{\infty} P_i(t|t_0) = \int p(\zeta, t|\zeta_0, t_0) d\zeta = 1 \tag{3.23}$$

in accordance with Eq. (2.20).

According to Eq. (3.15) the particle motion consists of a series of instantaneous scattering events and collisionless accelerations by external forces. The MC method exactly reproduces this sequence by simulating particles with stochastic methods [3.5–3.11, 3.21]. The flowchart of this procedure is shown in Fig. 3.2, where the particle is simulated for the time T. First, the duration of the collisionless propagation is determined after which the particle is scattered. The probability for a scattering event at time τ in an interval of width $d\tau$ after being propagated in the interval $[t_0, \tau]$ without suffering a collision is

$$dp_{\text{scat}} = \Gamma P_0(\tau|t_0) d\tau \ . \tag{3.24}$$

Using the direct method (3.4) the time t of the scattering event is generated

$$\int_0^r dp_{\text{scat}} = \int_{t_0}^{t_{\text{scat}}} \Gamma P_0(\tau|t_0) d\tau = 1 - \exp(-\Gamma(t_{\text{scat}} - t_0)) \ , \tag{3.25}$$

where r is a uniformly distributed random number between zero and one. Solving for t_{scat} yields

$$t_{\mathrm{scat}} = t_0 - \frac{1}{\Gamma}\ln(1-r) \ . \tag{3.26}$$

Second, the particle state ζ is determined just before the scattering event by integrating the equations of motion (2.3) and (2.4)

$$\zeta = \zeta_{\mathrm{New}}(t_{\mathrm{scat}}|\zeta_0, t_0) \ . \tag{3.27}$$

Third, one of the physical scattering processes (e.g. phonon scattering, impurity scattering; cf. Chap. 4) or self-scattering is selected. The ith scattering process is chosen with the direct method for discrete probabilities (3.10)

$$\sum_{j=1}^{i-1} \frac{S_j(\zeta)}{\Gamma} < r \leq \sum_{j=1}^{i} \frac{S_j(\zeta)}{\Gamma} \ , \tag{3.28}$$

where r is another uniformly distributed random number and $S_j(\zeta)$ is the scattering rate of the jth scattering process.

Fourth, the final state after scattering is generated with the probability density

$$p_i^{\mathrm{a}}(\zeta') = \frac{S_i(\zeta'|\zeta)}{S_i(\zeta)} \ , \tag{3.29}$$

where the final band and wedge are determined with the direct method (3.10) and the final wave vector with the procedure described in Sec. 5.7. These four steps are repeated until the particle has been simulated for the time T.

The constant Γ must be larger than the sum of the scattering rates of all physical scattering processes $S(\zeta)$ for all possible particle states ζ (cf. Fig. 3.3). This results in frequent self-scattering for particle states for which the sum of the physical scattering rates is small compared to the self-scattering rate and CPU time is wasted. The number of self-scattering events is reduced by the variable Γ scheme [3.22]. To this end, the PS is partitioned into small regions Λ_i (here the primitives of the PS grid) and for each region Λ_i a Γ_i is determined

$$\Gamma_i = \max_{\zeta\in\Lambda_i}[S(\zeta)] \ . \tag{3.30}$$

Since the local Γ_i is smaller than the global Γ, the number of self-scattering events is reduced. The duration of a collisionless flight is given by

$$\Gamma_{i(j)}(t_{\mathrm{scat}} - t_j) = -\ln(1-r) - \sum_{k=1}^{j} \Gamma_{i(k)}(t_k - t_{k-1}) \quad \text{for} \quad t_j \leq t_{\mathrm{scat}} < t_{j+1} \ , \tag{3.31}$$

where t_j is the time at which the particle passes from one primitive of the PS grid into another one during its collisionless flight. $i(j)$ is the index of the primitive

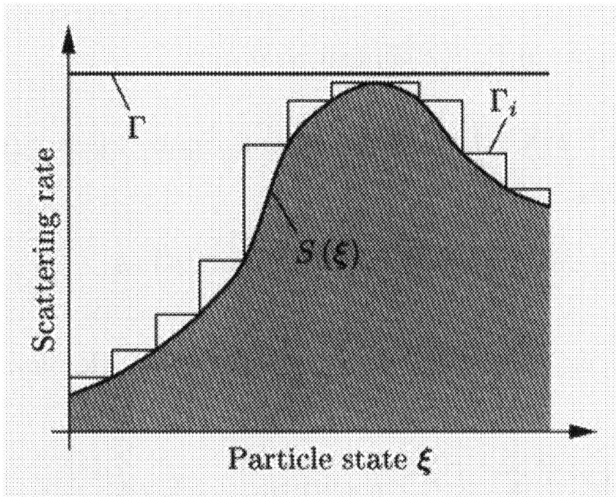

Figure 3.3: The scattering rate $S(\zeta)$, the variable Γ_i, and the global Γ

entered by the particle at time t_j. At the beginning the particle is in the primitive $i(0)$ at t_0. The sum on the right-hand side of equation (3.31) is evaluated during the collisionless flight. Whenever the particle enters a new primitive, t_{scat} is evaluated for this primitive with the local Γ. Is t_{scat} larger than the time of the next border crossing, the particle is moved into the next primitive and t_{scat} is evaluated for the new primitive with the new local Γ. When the end of the collisionless flight falls into the current primitive, the flight terminates there and the scattering event is performed. During the whole collisionless flight the same random number r is used.

In the case of a transition rate with a complicated dependence on the initial or final state the internal self-scattering method might be used to reduce the CPU time or to simplify the algorithms [3.23,3.24]. In analogy to the rejection method (3.7) an upper bound of the transition rate is used

$$\tilde{S}_i(\zeta'|\zeta) \geq S_i(\zeta'|\zeta) \tag{3.32}$$

for which the scattering rate and the final state are easy to calculate. After the final state has been selected, a test (3.8) is performed to decide whether the scattering event is fictitious or real

$$r\tilde{S}_i(\zeta'|\zeta) \leq S_i(\zeta'|\zeta) \ , \tag{3.33}$$

where r is a random number of the type (3.1). In the case that the relation (3.33) is satisfied the scattering event is accepted. Otherwise it is rejected and the particle is not scattered.

3.3 Velocity Autocorrelation Function

The autocorrelation function of the velocity (2.27) for a homogeneous system can be calculated with Eq. (3.11). The electric field is assumed to be zero. In this case the RS arguments in Eq. (2.27) can be dropped

$$\langle \mathbf{v}, \mathbf{v}^{\mathrm{T}} \rangle (t, t_0) = \sum_{n_0} \int_{\mathrm{BZ}} \sum_{n} \int_{\mathrm{BZ}} \mathbf{v}_n(\mathbf{k}) p_{n,n_0}(\mathbf{k}, t|\mathbf{k}_0, t_0) \mathbf{v}_{n_0}^{\mathrm{T}}(\mathbf{k}_0) p_{n_0}(\mathbf{k}_0, t_0) d^3k d^3k_0 \ . \tag{3.34}$$

With Eq. (3.11) the inner integral can be formulated as

$$\sum_{n} \int_{\mathrm{BZ}} \mathbf{v}_n(\mathbf{k}) p_{n,n_0}(\mathbf{k}, t|\mathbf{k}_0, t_0) d^3k = \sum_{n} \int_{\mathrm{BZ}} \mathbf{v}_n(\mathbf{k}) p_{0,n,n_0}(\mathbf{k}, t|\mathbf{k}_0, t_0) d^3k$$
$$+ \frac{\Omega}{(2\pi)^3} \int_{t_0}^{t} \sum_{n_1} \int_{\mathrm{BZ}} \sum_{n_1'} \int_{\mathrm{BZ}} \sum_{n} \int_{\mathrm{BZ}} \mathbf{v}_n(\mathbf{k}) p_{0,n,n_1'}(\mathbf{k}, t|\mathbf{k}_1', t_1) S_{n_1',n_1}(\mathbf{k}_1'|\mathbf{k}_1)$$
$$p_{n_1,n_0}(\mathbf{k}_1, t_1|\mathbf{k}_0, t_0) d^3k d^3k_1' d^3k_1 dt_1 \ . \tag{3.35}$$

For a zero force in a collisionless homogeneous system the band index and the wave vector of a particle are constant in time and the collisionless conditional probability density reduces to

$$p_{0,n,n_0}(\mathbf{k}, t|\mathbf{k}_0, t_0) = \delta_{n,n_0} \delta(\mathbf{k} - \mathbf{k}_0) \exp\left(-S_{n_0}(\mathbf{k}_0)(t - t_0)\right) \ , \tag{3.36}$$

and Eq. (3.35) reads

$$\sum_{n} \int_{\mathrm{BZ}} \mathbf{v}_n(\mathbf{k}) p_{n,n_0}(\mathbf{k}, t|\mathbf{k}_0, t_0) d^3k = \mathbf{v}_{n_0}(\mathbf{k}_0) \exp\left(-S_{n_0}(\mathbf{k}_0)(t - t_0)\right)$$
$$+ \frac{\Omega}{(2\pi)^3} \int_{t_0}^{t} \sum_{n_1} \int_{\mathrm{BZ}} \sum_{n_1'} \int_{\mathrm{BZ}} \exp\left(-S_{n_1'}(\mathbf{k}_1')(t - t_1)\right) \mathbf{v}_{n_1'}(\mathbf{k}_1') S_{n_1',n_1}(\mathbf{k}_1'|\mathbf{k}_1)$$
$$p_{n_1,n_0}(\mathbf{k}_1, t_1|\mathbf{k}_0, t_0) d^3k_1' d^3k_1 dt_1 \ . \tag{3.37}$$

In the case of a velocity-randomizing scattering event the inner most integral of the second term on the right-hand side of Eq. (3.37) is always zero

$$\int_{\mathrm{BZ}} \exp\left(-S_{n_1'}(\varepsilon_{n_1'}(\mathbf{k}_1'))(t - t_1)\right) \mathbf{v}_{n_1'}(\mathbf{k}_1') S_{n_1',n_1}^{\mathrm{VT}}(\mathbf{k}_1'|\mathbf{k}_1) d^3k_1' = 0 \tag{3.38}$$

due to Eq. (2.48) and the constant energy transfer $\varepsilon_{n_1'}(\mathbf{k}_1') - \varepsilon_{n_1}(\mathbf{k}_1) = \mathrm{const.}$, and the scattering rate $S_{n_1'}(\varepsilon_{n_1'}(\mathbf{k}_1'))$ must be constant on an equienergy surface (cf. Eq. (2.47)). Thereby, all contributions from velocity-randomizing scattering events to the right-hand side of Eq. (3.37) vanish.

In the case of intraband isotropic-elastic scattering events ($n_1' = n_1$) the innermost integral can be expressed with Eq. (2.46) as

$$\frac{\Omega}{(2\pi)^3}\sum_{n_1'}\int_{BZ} \exp\left(-S_{n_1'}(\varepsilon_{n_1'}(\mathbf{k}_1'))(t-t_1)\right) \mathbf{v}_{n_1'}(\mathbf{k}_1')S^{\text{ie}}_{n_1',n_1}(\mathbf{k}_1'|\mathbf{k}_1)d^3k_1' =$$
$$\frac{\Omega}{(2\pi)^3}\int_{BZ} \exp\left(-S_{n_1}(\varepsilon_{n_1}(\mathbf{k}_1'))(t-t_1)\right) \mathbf{v}_{n_1}(\mathbf{k}_1')S^{\text{ie}}_{n_1,n_1}(\mathbf{k}_1'|\mathbf{k}_1)d^3k_1' =$$
$$\exp\left(-S_{n_1}(\varepsilon_{n_1}(\mathbf{k}_1))(t-t_1)\right)\left(S^{\text{ie}}_{n_1}(\varepsilon_{n_1}(\mathbf{k}_1)) - \frac{1}{\tau^{\text{ie}}_{n_1}(\varepsilon_{n_1}(\mathbf{k}_1))}\right)\mathbf{v}_{n_1}(\mathbf{k}_1) \tag{3.39}$$

because the scattering process is isotropic-elastic ($S^{\text{ie}}_{n_1}(\varepsilon_{n_1}(\mathbf{k}_1')) = S^{\text{ie}}_{n_1}(\varepsilon_{n_1}(\mathbf{k}_1))$, $\tau^{\text{ie}}_{n_1}(\varepsilon_{n_1}(\mathbf{k}_1')) = \tau^{\text{ie}}_{n_1}(\varepsilon_{n_1}(\mathbf{k}_1))$), and $\varepsilon_{n_1}(\mathbf{k}_1') = \varepsilon_{n_1}(\mathbf{k}_1)$. Under the assumption that all scattering processes are either velocity-randomizing or intraband isotropic-elastic and inclusion of a vanishing term for the velocity-randomizing processes ($S^{\text{vr}}_{n_1'}(\varepsilon_{n_1'}(\mathbf{k}_1')) - 1/\tau^{\text{vr}}_{n_1'}(\varepsilon_{n_1'}(\mathbf{k}_1')) = S^{\text{vr}}_{n_1}(\varepsilon_{n_1}(\mathbf{k}_1)) - 1/\tau^{\text{vr}}_{n_1}(\varepsilon_{n_1}(\mathbf{k}_1)) = 0$) Eq. (3.37) reads

$$\sum_n\int_{BZ} \mathbf{v}_n(\mathbf{k})p_{n,n_0}(\mathbf{k},t|\mathbf{k}_0,t_0)d^3k = \mathbf{v}_{n_0}(\mathbf{k}_0)\exp\left(-S_{n_0}(\varepsilon_{n_0}(\mathbf{k}_0))(t-t_0)\right)$$
$$+\int_{t_0}^{t}\sum_{n_1}\int_{BZ} \exp\left(-S_{n_1}(\varepsilon_{n_1}(\mathbf{k}_1))(t-t_1)\right)\left(S_{n_1}(\varepsilon_{n_1}(\mathbf{k}_1)) - \frac{1}{\tau_{n_1}(\varepsilon_{n_1}(\mathbf{k}_1))}\right)$$
$$\mathbf{v}_{n_1}(\mathbf{k}_1)p_{n_1,n_0}(\mathbf{k}_1,t_1|\mathbf{k}_0,t_0)d^3k_1dt_1\ . \tag{3.40}$$

Inserting Eq. (3.11) into the second term on the right-hand side of the above equation and repeating the above steps infinite times yields

$$\sum_n\int_{BZ} \mathbf{v}_n(\mathbf{k})p_{n,n_0}(\mathbf{k},t|\mathbf{k}_0,t_0)d^3k$$
$$=\sum_{n=0}^{\infty}\frac{\left[\left(S_{n_0}(\varepsilon_{n_0}(\mathbf{k}_0)) - \frac{1}{\tau_{n_0}(\varepsilon_{n_0}(\mathbf{k}_0))}\right)(t-t_0)\right]^n}{n!}$$
$$\mathbf{v}_{n_0}(\mathbf{k}_0)\exp\left(-S_{n_0}(\varepsilon_{n_0}(\mathbf{k}_0))(t-t_0)\right)$$
$$= \mathbf{v}_{n_0}(\mathbf{k}_0)\exp\left(-\frac{t-t_0}{\tau_{n_0}(\varepsilon_{n_0}(\mathbf{k}_0))}\right) \tag{3.41}$$

and the autocorrelation function reads

$$\langle \mathbf{v}, \mathbf{v}^{\mathrm{T}} \rangle (t, t_0) =$$
$$\sum_{n_0} \int_{\mathrm{BZ}} \sum_{n} \int_{\mathrm{BZ}} \mathbf{v}_n(\mathbf{k}) p_{n,n_0}(\mathbf{k}, t|\mathbf{k}_0, t_0) \mathbf{v}_{n_0}^{\mathrm{T}}(\mathbf{k}_0) p_{n_0}(\mathbf{k}_0, t_0) d^3k d^3k_0 =$$
$$\sum_{n_0} \int_{\mathrm{BZ}} \mathbf{v}_{n_0}(\mathbf{k}_0) \mathbf{v}_{n_0}^{\mathrm{T}}(\mathbf{k}_0) \exp\left(-\frac{t-t_0}{\tau_{n_0}(\varepsilon_{n_0}(\mathbf{k}_0))}\right) p_{n_0}(\mathbf{k}_0, t_0) d^3k_0$$
$$= \left\langle \mathbf{v}\mathbf{v}^{\mathrm{T}} \exp\left(-\frac{t-t_0}{\tau}\right) \right\rangle (t_0) \ . \tag{3.42}$$

Thus, for a forceless system the velocity autocorrelation function of a given particle state is a simple exponential where the decay time is given by the corresponding microscopic relaxation time [3.25]. Integration of the velocity autocorrelation function over time yields the (spreading) diffusion constant

$$\int_{t_0}^{\infty} \langle \mathbf{v}, \mathbf{v}^{\mathrm{T}} \rangle (t, t_0) dt = \langle \tau \mathbf{v}\mathbf{v}^{\mathrm{T}} \rangle (t_0) = \hat{D}(t_0) \ . \tag{3.43}$$

This result is consistent with Eq. (2.98) in the case of a stationary system.

3.4 Basic Statistics

The MC method allows to estimate the expected values (2.25) of microscopic quantities (e.g. velocity, energy) [3.9]. To this end, an ensemble of N_{sim} independent and identically distributed test particles is simulated with the MC method described in Sec. 3.2. The single-particle probability density (2.12) is estimated by

$$\tilde{p}(\zeta, t) = \frac{1}{N_{\mathrm{sim}}} \sum_{i=1}^{N_{\mathrm{sim}}} w_i \delta(\zeta - \zeta_i(t)) \ , \tag{3.44}$$

where the tilde indicates an estimation, $\zeta_i(t)$ is the state of the ith particle and w_i is its statistical weight

$$0 \le w_i \le N_{\mathrm{sim}} \quad \wedge \quad \sum_{j=1}^{N_{\mathrm{sim}}} w_j = N_{\mathrm{sim}} \ . \tag{3.45}$$

Here, in contrast to Sec. 6.7 the statistical weight is not correlated with the particle state.

The estimator for an expected value based on an ensemble average reads

$$\tilde{X}(t) = \int X(\zeta) \tilde{p}(\zeta, t) d\zeta = \frac{1}{N_{\mathrm{sim}}} \sum_{i=1}^{N_{\mathrm{sim}}} w_i X(\zeta_i(t)) \ . \tag{3.46}$$

Under the given assumptions the estimator is unbiased [3.17], because the expected value of the estimator equals the expectation of the microscopic quantity (2.25) [3.26]

$$\mathrm{E}\left\{\tilde{X}(t)\right\} = \frac{1}{N_{\mathrm{sim}}} \sum_{i=1}^{N_{\mathrm{sim}}} w_i \mathrm{E}\{X(t)\} = \mathrm{E}\{X(t)\} \ , \tag{3.47}$$

where E{} denotes the expected value. The variance of the estimator is given by

$$\sigma^2\left\{\tilde{X}(t)\right\} = \mathrm{E}\left\{\left(\tilde{X}(t) - \mathrm{E}\left\{\tilde{X}(t)\right\}\right)^2\right\} = \frac{\sum_{i=1}^{N_{\mathrm{sim}}} w_i^2}{N_{\mathrm{sim}}^2} \sigma^2\{X(t)\} \ . \tag{3.48}$$

The estimator is consistent [3.17], if

$$\lim_{N_{\mathrm{sim}}\to\infty} \sigma^2\left\{\tilde{X}(t)\right\} = \sigma^2\{X(t)\} \lim_{N_{\mathrm{sim}}\to\infty} \frac{\sum_{i=1}^{N_{\mathrm{sim}}} w_i^2}{N_{\mathrm{sim}}^2} = 0 \ . \tag{3.49}$$

This condition necessitates that the variance $\sigma^2\{X(t)\}$ is finite and that the sum of the squares of the statistical weights increases with a sub-square dependence on the particle number. For the sum of the squares of the statistical weights the following relation holds

$$N_{\mathrm{sim}} \leq \sum_{j=1}^{N_{\mathrm{sim}}} w_j^2 \leq N_{\mathrm{sim}}^2 \ . \tag{3.50}$$

The minimum is obtained, if all particles have the same weights ($w_i = 1$) and the maximum, if one particle has the weight N_{sim} and all other particles have zero weight [3.26].

The most efficient simulation is the one which produces the smallest variance for a given computational cost. Since the computational cost of the simulation of a particle is independent of its statistical weight, the most efficient simulation is the one for which the sum of the squares of the statistical weight is minimal (cf. Eq. (3.48)). Thus, uniform statistical weights give the optimum result in the case that the statistical weight and the particle state are uncorrelated [3.26]. However, in Sec. 6.7 it will be shown that the variance of estimates for certain microscopic variables can be improved by using nonuniform statistical weights correlated with the particle state.

In the case of stationary MC simulations it is possible to average the microscopic quantities not only over the ensemble but also over time assuming an ergodic system [3.9, 3.17]

$$\overline{X} = \frac{1}{T_{\mathrm{sim}}} \int_0^{T_{\mathrm{sim}}} \tilde{X}(t) dt \ , \tag{3.51}$$

where the bar indicates the time average and T_{sim} is the length of the simulated time interval. This estimator is also unbiased and consistent, if the sampled stochastic process has a finite variance and correlation time (cf. Sec. 3.5) [3.17]. The exact evaluation of the integral in Eq. (3.51) involves integrals over the particles' paths which are in general cumbersome to evaluate. This problem is avoided by using before-scattering statistics or fixed frequency sampling [3.9,3.15]. In the latter case the time average is evaluated at fixed time intervals

$$\overline{X} = \frac{1}{N} \sum_{i=1}^{N} \tilde{X}(i\Delta t) \; . \tag{3.52}$$

In the case of before-scattering statistics the data are sampled before every scattering event including self-scattering, and the data are weighted with the inverse scattering rate [3.9,3.15]

$$\overline{X} = \frac{1}{N_{\rm sim}} \sum_{i=1}^{N_{\rm sim}} \frac{w_i}{\sum_{j=1}^{N_i^{\rm scat}} \frac{1}{\Gamma(\zeta_i(t_{i,j}^{\rm scat}))}} \sum_{j=1}^{N_i^{\rm scat}} \frac{X(\zeta_i(t_{i,j}^{\rm scat}))}{\Gamma(\zeta_i(t_{i,j}^{\rm scat}))} \; , \tag{3.53}$$

where $t_{i,j}^{\rm scat}$ is the time of the jth scattering event of the ith particle, $N_i^{\rm scat}$ the number of scattering events during the simulation period, $\zeta_i(t_{i,j}^{\rm scat})$ the particle state just before the jth scattering event, and $\Gamma(\zeta)$ the total scattering rate including self-scattering.

The latter two estimators (3.52) and (3.53) sample less information from the simulation than the estimator (3.51) and are therefore less efficient. But the degradation of the efficiency is small, as long as the sampling frequency is of the same order as the scattering rate and the integral over the absolute value of the correlation function of the sampled quantity is not much larger than the integral of the correlation function itself. The latter restriction does not hold, for example, in the case of plasma oscillations and the sampling frequency must be larger than the plasma frequency. This is obviously the case for fixed frequency sampling during a device simulation (Sec. 6.4), because the statistics are evaluated at the end of each time step, which is of the order of femtoseconds, and the time between two sampling events is therefore in most cases much smaller than the time between two scattering events or the period of plasma oscillations. The condition, that the sampling frequency must be of the same order as the scattering rate is obviously also satisfied for before-scattering statistics. But this does not necessarily imply, that the second condition, that the sampling frequency must be larger than the oscillation frequency of the correlation function, is fulfilled. In the case of plasma oscillation this is usually not the case, and a terminal current estimator based on before-scattering statistics has a much lower efficiency than an estimator based on exact integration over time.

3.5 Convergence Estimation

The MC method only estimates expected values of the distribution function and the results contain a stochastic error [3.1,3.9,3.14,3.26,3.27]. In general this error decreases with the simulation time and the MC simulations are performed until a certain degree of convergence is achieved. Thus, a reliable convergence criterion is necessary to avoid on the one hand a waste of CPU time in the case that the error is overestimated and on the other hand erroneous results in the case that the error is underestimated.

The convergence estimation is based on the concept of the confidence interval and confidence level [3.1, 3.14]. The confidence interval gives a range for the variation of the estimate

$$|\tilde{X} - \mathrm{E}\{\tilde{X}\}| \leq \delta \tag{3.54}$$

with δ being the half width of the interval. The confidence level P_{conf} is the probability that the estimate is inside the confidence interval

$$P_{\mathrm{conf}}(\tilde{X}, \delta) = P\{|\tilde{X} - \mathrm{E}\{\tilde{X}\}| \leq \delta\} , \tag{3.55}$$

where $P\{\}$ is the probability of the event enclosed in the curly brackets. The Tchebycheff inequality states [3.17]

$$1 - P_{\mathrm{conf}}(\tilde{X}, \delta) = P\{|\tilde{X} - \mathrm{E}\{\tilde{X}\}| \geq \delta\} \leq \frac{\sigma^2\{\tilde{X}\}}{\delta^2} . \tag{3.56}$$

For a half width, which is two times the standard deviation of the process ($\delta = 2\sigma\{\tilde{X}\}$), the confidence level is larger than 75%. Based on the distribution function of the estimate the confidence level can be exactly calculated. Although the distribution function of the estimate is in general not known, it can be approximated by a Gaussian distribution function, if the estimation, which is based on an arithmetical mean or time integration, is averaged over a sufficiently large number of particles or a sufficiently long simulation time, so that the central limit theorem applies [3.17]. In this case the distribution function of the estimate is given by

$$p(\tilde{X}) = \frac{1}{\sqrt{2\pi}\sigma\{\tilde{X}\}} \exp\left(-\frac{(\tilde{X} - \mathrm{E}\{\tilde{X}\})^2}{2\sigma^2\{\tilde{X}\}}\right) , \tag{3.57}$$

and the confidence level reads

$$P_{\mathrm{conf}}(\tilde{X}, \delta) = \int\limits_{\mathrm{E}\{\tilde{X}\}-\delta}^{\mathrm{E}\{\tilde{X}\}+\delta} p(\tilde{X}) d\tilde{X} = \frac{1}{\sqrt{2\pi}} \int\limits_{-\delta/\sigma\{\tilde{X}\}}^{\delta/\sigma\{\tilde{X}\}} \exp\left(-\frac{x^2}{2}\right) dx \tag{3.58}$$

(see Fig. 3.4). For the half width being two times the standard deviation the confidence level is 95.45% [3.1]. The corresponding relative error is given by

$$r\{\tilde{X}\} = \frac{\delta}{\mathrm{E}\{\tilde{X}\}} = \frac{2\sigma\{\tilde{X}\}}{\mathrm{E}\{\tilde{X}\}} . \tag{3.59}$$

To calculate the confidence level the variance of $\tilde{X}$ must be evaluated. This is in general difficult for MC simulations, because the particles are not independent

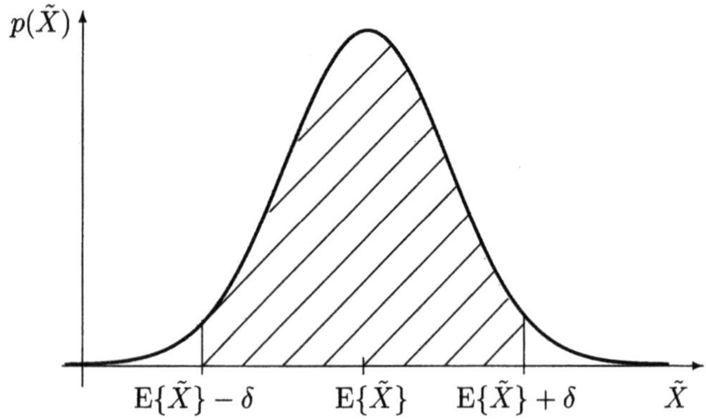

Figure 3.4: The distribution function of the estimated mean $\tilde{X}$, the confidence interval, and the confidence level given by the hatched area

as assumed in Sec. 3.4 and Eq. (3.48) does not apply. The particles are correlated among each other and in time due to the SC simulation of the MC method and the Poisson equation (PE), carrier-carrier scattering, screening, Pauli exclusion principle, statistical enhancement methods, and the finite scattering rate [3.9, 3.28–3.32]. In general, an evaluation of the variance is only possible by running independent simulations of the same problem and estimating the variance on the basis of these uncorrelated data [3.32]. In the important case of stationary simulations this can be avoided as discussed below.

The impact of a correlation in time is exemplified in Figs. 3.5 and 3.6 for a stationary and Markovian Gaussian process (Ornstein-Uhlenbeck process [3.33]) with and without correlation. The mean is sampled with Eq. (3.52). Although the variance of both processes is the same, the variance of the estimated mean strongly differs and is much larger in the case of the correlated process manifested in the stronger fluctuations of the estimated mean. This example clearly demonstrates that the correlation can not be neglected when evaluating the variance of the estimated mean. Unfortunately, the correlation is a priori unknown and difficult to estimate.

In the case of stationary simulations the problem of the unknown correlation can be circumvented by extracting the variance from the time-dependent fluctuations of the ensemble average $\tilde{X}(t)$ [3.9]. The variance of the estimator (3.51) is given by [3.17]

$$\begin{aligned}\sigma^2\{\overline{X}\} &= \mathrm{E}\{(\overline{X} - \mathrm{E}\{\overline{X}\})^2\} \\ &= \frac{1}{T_{\mathrm{sim}}^2} \int_0^{T_{\mathrm{sim}}} \int_0^{T_{\mathrm{sim}}} \mathrm{E}\{\tilde{X}(t)\tilde{X}(t')\} dt' dt - \mathrm{E}^2\{\tilde{X}\}\end{aligned}$$

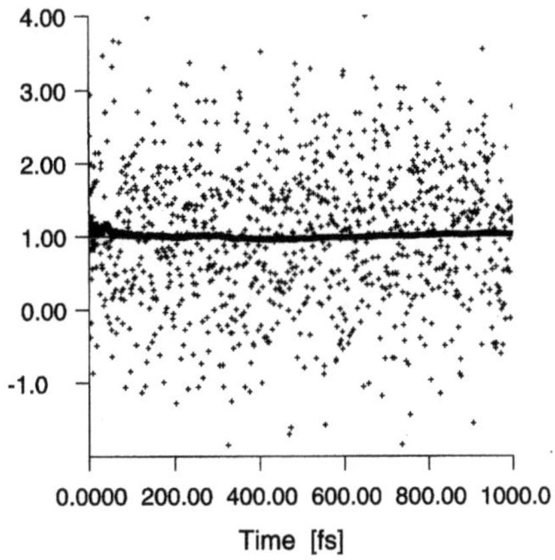

Figure 3.5: An uncorrelated, stationary and Markovian Gaussian process sampled every 1fs (filled symbols), the estimated mean (+) obtained by averaging over all previous samples and the mean value of the process (solid line), where the mean value and the standard deviation equal one

$$= \frac{\sigma^2\{\tilde{X}\}}{T_{\rm sim}^2} \int_0^{T_{\rm sim}} \int_0^{T_{\rm sim}} C\{\tilde{X}\}(t-t')dt'dt$$
$$= \frac{\sigma^2\{\tilde{X}\}}{T_{\rm sim}} \int_{-T_{\rm sim}}^{T_{\rm sim}} \left(1 - \frac{|\tau|}{T_{\rm sim}}\right) C\{\tilde{X}\}(\tau)d\tau \ , \tag{3.60}$$

where $\sigma^2\{\tilde{X}\}$ is the variance of the stationary ensemble average and $C\{\tilde{X}\}(\tau)$ its correlation coefficient

$$\mathrm{E}\{\tilde{X}(t)\tilde{X}(t')\} = \sigma^2\{\tilde{X}\}C\{\tilde{X}\}(t-t') + \mathrm{E}^2\{\tilde{X}\} \ . \tag{3.61}$$

Like the mean the standard deviation has to be estimated. An unbiased estimator for the variance of the estimated time average is given by

$$\overline{\sigma}^2\{\overline{X}\} = \frac{1}{\frac{T_{\rm sim}}{\int_{-T_{\rm sim}}^{T_{\rm sim}} \left(1-\frac{|\tau|}{T_{\rm sim}}\right)C\{\tilde{X}\}(\tau)d\tau} - 1} \left[\overline{X^2} - \overline{X}^2\right] \ . \tag{3.62}$$

with

$$\overline{X^2} = \frac{1}{T_{\rm sim}} \int_0^{T_{\rm sim}} \tilde{X}^2(t)dt \ . \tag{3.63}$$

The estimator (3.62) is unbiased because of $\mathrm{E}\{\overline{\sigma}^2\{\overline{X}\}\} = \sigma^2\{\overline{X}\}$ (cf. Eq. (3.60)).

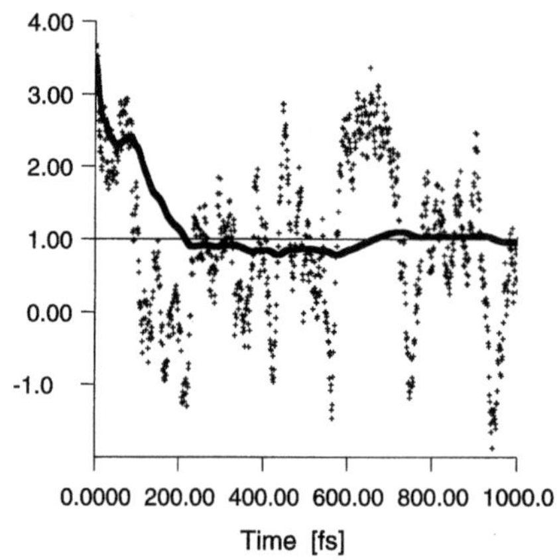

Figure 3.6: A correlated, stationary and Markovian Gaussian process sampled every 1fs (filled symbols), the estimated mean (+) obtained by averaging over all previous samples and the mean value of the process (solid line), where the mean value and the standard deviation equal one and the correlation time is 30fs

The evaluation of Eq. (3.62) requires the exact knowledge of the in most cases unknown correlation coefficient. In the cases that the simulated time T_{sim} is much larger than the correlation time defined as

$$\tau_{\mathrm{c}}\{\tilde{X}\} = \int_0^\infty C\{\tilde{X}\}(\tau)d\tau \tag{3.64}$$

the following approximation holds under quite general conditions [3.17]

$$\frac{1}{T_{\mathrm{sim}}} \int_{-T_{\mathrm{sim}}}^{T_{\mathrm{sim}}} \left(1 - \frac{|\tau|}{T_{\mathrm{sim}}}\right) C\{\tilde{X}\}(\tau)d\tau \approx \frac{2\tau_{\mathrm{c}}\{\tilde{X}\}}{T_{\mathrm{sim}}} . \tag{3.65}$$

With this approximation Eq. (3.62) reads

$$\overline{\sigma}^2\{\overline{X}\} = \frac{1}{\frac{T_{\mathrm{sim}}}{2\tau_{\mathrm{c}}\{\tilde{X}\}} - 1} \left[\overline{X^2} - \overline{X}^2\right] . \tag{3.66}$$

The evaluation of this formula still requires the precise value of the in general unknown correlation time. This problem is avoided by splitting the simulation history into N subhistories of the length T_{sub} being much longer than the correlation time [3.16]

$$X_i = \frac{1}{T_{\mathrm{sub}}} \int_{(i-1)T_{\mathrm{sub}}}^{iT_{\mathrm{sub}}} \tilde{X}(t)dt \quad \wedge \quad T_{\mathrm{sub}} \gg \tau_{\mathrm{c}}\{\tilde{X}\} \tag{3.67}$$

with $T_{\mathrm{sim}} = NT_{\mathrm{sub}}$. The estimator for the time average is now given by

$$\widehat{X} = \frac{1}{N}\sum_{i=1}^{N} X_i = \overline{X} \tag{3.68}$$

and exactly reproduces the estimator (3.51). The hat indicates the average over the different subhistories. The corresponding estimator for the variance of the estimated mean reads

$$\widehat{\sigma}^2\{\widehat{X}\} = \frac{1}{N-1}\left[\widehat{X^2} - \widehat{X}^2\right] \tag{3.69}$$

with

$$\widehat{X^2} = \frac{1}{N}\sum_{i=1}^{N} X_i^2 . \tag{3.70}$$

Equations (3.68) and (3.69) resemble the well-known formulas for sampling independent random events [3.17]. The expected value of the estimator (3.69) is

$$\begin{aligned} \mathrm{E}\{\widehat{\sigma}^2\{\widehat{X}\}\} &= \frac{\sigma^2\{\tilde{X}\}}{N-1}\left(\frac{1}{T_{\mathrm{sub}}}\int_{-T_{\mathrm{sub}}}^{T_{\mathrm{sub}}}\left(1-\frac{|\tau|}{T_{\mathrm{sub}}}\right)C\{\widehat{X}\}(\tau)d\tau\right. \\ &\quad \left. - \frac{1}{T_{\mathrm{sim}}}\int_{-T_{\mathrm{sim}}}^{T_{\mathrm{sim}}}\left(1-\frac{|\tau|}{T_{\mathrm{sim}}}\right)C(\{\widehat{X}\}\tau)d\tau\right) . \end{aligned} \tag{3.71}$$

If the approximation (3.65) holds for T_{sub}, the expected value of the estimator is

$$\mathrm{E}\{\widehat{\sigma}^2\{\widehat{X}\}\} \approx \frac{2\tau_{\mathrm{c}}\{\tilde{X}\}}{T_{\mathrm{sim}}}\sigma^2\{\tilde{X}\} \tag{3.72}$$

and equals the value of Eq. (3.60) calculated under the same approximation

$$\sigma^2\{\overline{X}\} \approx \frac{2\tau_{\mathrm{c}}\{\tilde{X}\}}{T_{\mathrm{sim}}}\sigma^2\{\tilde{X}\} . \tag{3.73}$$

Therefore, Eq. (3.69) is an unbiased estimator for the variance of the time average (3.51) as long as the length of the subhistories T_{sub} is much longer than the correlation time $\tau_{\mathrm{c}}\{\tilde{X}\}$. Thus, a rough estimate of the correlation time is sufficient to use this scheme. In addition, the scheme also works in the case that the time average is calculated with Eq. (3.52) or (3.53).

With Eq. (3.73) the relative error (3.59) of the time average reads

$$r\{\widehat{X}\} \approx \sqrt{\frac{2\tau_{\mathrm{c}}\{\tilde{X}\}}{T_{\mathrm{sim}}}}\,\frac{2\sigma\{\tilde{X}\}}{\mathrm{E}\{\tilde{X}\}} \tag{3.74}$$

and is inversely proportional to the square root of the simulation time. The simulation is stopped, when the error is below a given level (e.g. 10%). The results are then converged within ±10% with a probability of 95.45% assuming a Gaussian distribution of the samples. Since the evaluation of the mean and its variance is not possible in a MC simulation, the corresponding estimates are used

$$\hat{r}\{\widehat{X}\} \approx \frac{2\widehat{\sigma}\{\widehat{X}\}}{\widehat{X}} . \tag{3.75}$$

As explained above, the length of the subhistories must be much longer than the correlation time to eliminate the bias of the estimator of the variance. On the other hand, the subhistories should not be too long, to ensure that the number of subhistories is large enough for a reliable estimation of the variance [3.16]. Assuming that the samples X_i are distributed according to a Gaussian distribution function and that they are uncorrelated ($T_{\text{sub}} \gg \tau_{\text{c}}\{\tilde{X}\}$), the estimated variance $\widehat{\sigma}^2\{\widehat{X}\}$ has a χ^2 distribution [3.14]. The relative error of the estimated variance is [3.14]

$$r\{\widehat{\sigma}^2\{\widehat{X}\}\} \approx \sqrt{\frac{8}{N-1}} \tag{3.76}$$

for a confidence level of 95.45%. The relative error depends only on the number of subhistories N and not on the sampled quantity $\tilde{X}$ or on the length of the simulation T_{sim}. For example, the relative error of the estimated variance is ±10% for about 800 subhistories. Thus, the length of the subhistories should be as short as possible to achieve a sufficient large number of subhistories. On the other hand, the estimator of the variance becomes biased for short subhistories and the length of the subhistories must be much longer than the correlation time. This trade-off will be discussed next.

For many processes (e.g. velocity of a homogeneous system in equilibrium, Fig. 3.7, Eq. (2.97)) the correlation coefficient is approximately given by an exponential function

$$C\{\tilde{X}\}(\tau) = \exp\left(-\frac{|\tau|}{\tau_{\text{r}}}\right) , \tag{3.77}$$

where τ_{r} is the relaxation time of the processes. In this case the relaxation time and the correlation time are the same ($\tau_{\text{c}}\{\tilde{X}\} = \tau_{\text{r}}$). The relative error of the estimated mean as a function of the subhistory length for $T_{\text{sim}} \gg \tau_{\text{r}}$ is

$$\frac{r\{\widehat{X}\}(T_{\text{sub}})}{\frac{2\sigma\{\overline{X}\}}{\mathrm{E}\{\overline{X}\}}} = \frac{\sqrt{\mathrm{E}\{\widehat{\sigma}^2\{\widehat{X}\}\}}}{\sigma\{\overline{X}\}} \approx \sqrt{1 - \frac{\tau_{\text{r}}}{T_{\text{sub}}}\left[1 - \exp\left(-\frac{T_{\text{sub}}}{\tau_{\text{r}}}\right)\right]} , \tag{3.78}$$

where the relative error is normalized to the correct, unbiased value.

A realization of a process with an exponential correlation function is the velocity v of the Ornstein-Uhlenbeck process (Fig. 3.6) which is Gaussian, Markovian, and stationary [3.33]. In Fig. 3.8 the normalized relative error (3.78) of the velocity is shown together with the relative error of the estimated variance for $T_{\text{sim}} = 10^4 \cdot \tau_{\text{r}}$. The relative error of the estimated mean is always underestimated

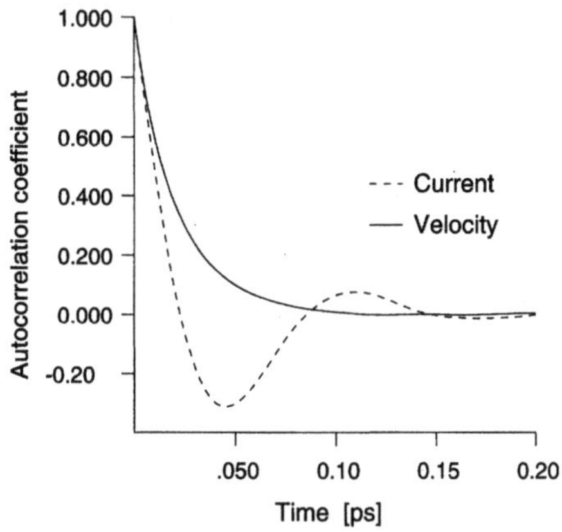

Figure 3.7: Autocorrelation coefficient of the single particle velocity of an inhomogeneous bulk system and of the terminal current of a device example under equilibrium conditions as resulting from MC simulations

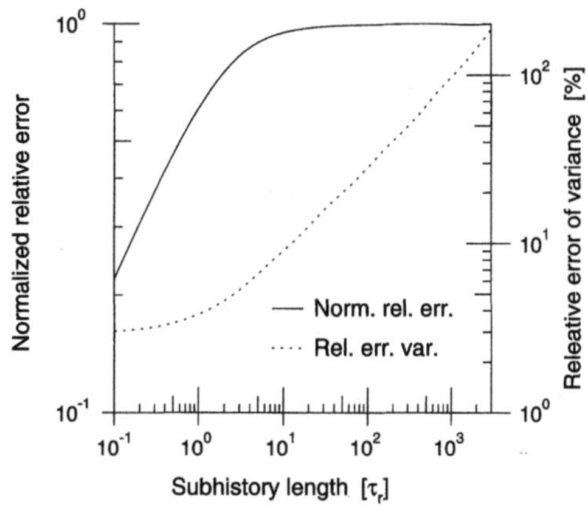

Figure 3.8: Normalized relative error of the estimated mean (3.78) and the relative error of the estimated variance $r\{\widehat{\sigma}^2\{\widehat{X}\}\}$ for an exponential correlation coefficient ($T_{\mathrm{sim}} = 10^4 \cdot \tau_{\mathrm{r}}$)

and approaches the correct value for large T_{sub} in a monotonic fashion. If the subhistory length is about ten times the relaxation time the deviation is only 5%. On the other hand, the relative error of the estimated variance increases with T_{sub} and is about 10% for $T_{\mathrm{sub}} = 10 \cdot \tau_{\mathrm{r}}$. Above this value the subhistories are sufficiently uncorrelated and the relative error of the estimated variance agrees well with the result of Eq. (3.76).

Another class of correlation coefficients (e.g. contact current, Fig. 3.7) is similar

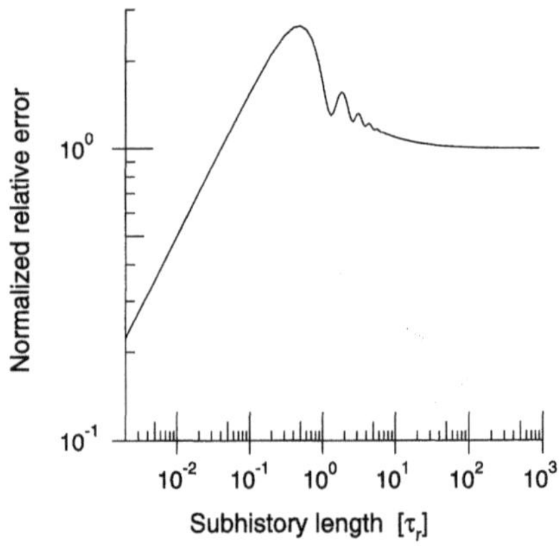

Figure 3.9: Normalized relative error of the estimated mean (3.81) for an oscillating correlation coefficient

to the following function

$$C\{\tilde{X}\}(\tau) = \exp\left(-\frac{|\tau|}{2\tau_{\mathrm{r}}}\right) \cos\left(\sqrt{\omega_{\mathrm{pl}}^2 - \frac{1}{4\tau_{\mathrm{r}}^2}}\,|\tau|\right) , \tag{3.79}$$

where τ_{r} is again the relaxation time and ω_{pl} the plasma frequency with $\omega_{\mathrm{pl}} > 1/(2\tau_{\mathrm{r}})$. The correlation time is given by

$$\tau_{\mathrm{c}}\{\tilde{X}\} = \frac{1}{2\omega_{\mathrm{pl}}^2 \tau_{\mathrm{r}}} . \tag{3.80}$$

The normalized relative error for $T_{\mathrm{sim}} \gg \tau_{\mathrm{r}}$ reads

$$\frac{r\{\widehat{X}\}(T_{\mathrm{sub}})}{\frac{2\sigma\{\tilde{X}\}}{\mathrm{E}\{\tilde{X}\}}} \approx \left(1 - \frac{2\tau_{\mathrm{r}}}{T_{\mathrm{sub}}}\left[\frac{2 - 4\omega_{\mathrm{pl}}^2\tau_{\mathrm{r}}^2}{4\omega_{\mathrm{pl}}^2\tau_{\mathrm{r}}^2}\left(1 - \exp\left(-\frac{T_{\mathrm{sub}}}{2\tau_{\mathrm{r}}}\right)\cos\left(\sqrt{\omega_{\mathrm{pl}}^2 - \frac{1}{4\tau_{\mathrm{r}}^2}}\,T_{\mathrm{sub}}\right)\right) + \frac{\sqrt{4\omega_{\mathrm{pl}}^2\tau_{\mathrm{r}}^2 - 1}}{2\omega_{\mathrm{pl}}^2\tau_{\mathrm{r}}^2}\exp\left(-\frac{T_{\mathrm{sub}}}{2\tau_{\mathrm{r}}}\right)\sin\left(\sqrt{\omega_{\mathrm{pl}}^2 - \frac{1}{4\tau_{\mathrm{r}}^2}}\,T_{\mathrm{sub}}\right)\right]\right)^{\frac{1}{2}} . \tag{3.81}$$

In Fig. 3.9 the normalized relative error is shown for $\omega_{\mathrm{pl}} = 5/\tau_{\mathrm{r}}$. In contrast to the case of the exponential correlation coefficient the relative error does not increase in a monotonic fashion and it can be overestimated. For $T_{\mathrm{sub}} = 10\tau_{\mathrm{r}} = 500\tau_{\mathrm{c}}\{\tilde{X}\}$ the relative error is overestimated by about 10%.

Both examples (Figs. 3.8 and 3.9) show that the length of the subhistories has to be chosen carefully. They should be long enough to avoid a biasing of the estimated error and, at the same time, they should be short enough to obtain a sufficient number of subhistories for a reliable estimation of the variance (cf. Eq. (3.76)). On the other hand, the estimated mean (3.51) is independent of the length of the subhistories and an error of about 30% in the estimated variance due to noise or bias is acceptable.

References

[3.1] J. M. Hammarsley and D. C. Handscomb, *Monte Carlo Methods*, Methuen/Chapman and Hall, London, 1964.

[3.2] F. James, "Monte Carlo theory and practice", *Rep. Prog. Phys.*, vol. 43, pp. 1144–1189, 1980.

[3.3] R. Y. Rubinstein, *Simulation and the Monte Carlo method*, Wiley series in probability and mathematical statistics. John Wiley & Sons, New York, 1981.

[3.4] M. H. Kalos and P. A. Whitlock, *Monte Carlo Methods, Basics*, vol. 1, John Wiley & Sons, New York, 1986.

[3.5] T. Kurosawa, "Monte Carlo calculation of hot electron problems", *J. Phys. Soc. Jap.*, vol. 21, pp. 424–426, 1966.

[3.6] W. Fawcett, A. D. Boardman, and S. Swain, "Monte Carlo determination of electron transport properties in gallium arsenide", *J. Phys. Chem. Solids*, vol. 31, pp. 1963–1990, 1970.

[3.7] A. Reklaitis, "The calculation of electron transient response in semiconductors by the Monte Carlo technique", *Phys. Lett.*, vol. 13, pp. 367–370, 1982.

[3.8] L. Reggiani, *Hot-Electron Transport in Semiconductors*, Springer, Berlin, 1985.

[3.9] C. Jacoboni and P. Lugli, *The Monte Carlo Method for Semiconductor Device Simulation*, Springer, Wien, 1989.

[3.10] M. Nedjalkov and P. Vitanov, "Iteration approach for solving the Boltzmann equation with the Monte Carlo method", *Solid–State Electron.*, vol. 32, pp. 893–896, 1989.

[3.11] H.-J. Peifer, "Monte–Carlo Simulation des Hochenergietransports von Elektronen in submikron MOS–Strukturen", Doctor thesis, RWTH Aachen, Aachen, 1992, Augustinus Buchhandlung.

[3.12] C. Moglestue, *Monte Carlo Simulation of Semiconductor Devices*, Chapman & Hall, London, 1993.

[3.13] S. K. Park and K. W. Miller, "Random number generators: Good ones are hard to find", *Transactions of the ACM*, Nov. 1988.

[3.14] S. Brandt, *Datenanalyse*, BI Wissenschaftsverlag, 1992.

[3.15] P. J. Price, "Monte Carlo calculation of electron transport in solids", *Semiconductors and Semimetals*, vol. 14, pp. 249–309, 1979.

[3.16] C. Jacoboni and L. Reggiani, "The Monte Carlo method for the solution of charge transport in semiconductors with application to covalent materials", *Rev. Mod. Phys.*, vol. 55, pp. 645–705, 1983.

[3.17] A. Papoulis, *Probability,Random Variables and Stochastic Processes*, Mc Graw-Hill, 3rd edition, 1991.

[3.18] H. D. Rees, "Calculation of steady state distribution functions by exploiting stabiblity", *Phys. Lett. A*, vol. 26, pp. 416–417, 1968.

[3.19] H. Kosina, M. Nedjalkov, and S. Selberherr, "Theory of the Monte Carlo method for semiconductor device simulation", *IEEE Trans. Electron Devices*, vol. 47, no. 10, pp. 1898–1908, 2000.

[3.20] C. Jacoboni, P. Poli, and L. Rota, "A new Monte Carlo technique for the solution of the Boltzmann transport equation", *Solid-State Electron.*, vol. 31, pp. 523–526, 1988.

[3.21] M. Nedjalkov and P. Vitanov, "Application of the iteration approach to the ensemble Monte Carlo technique", *Solid-State Electron.*, vol. 33, pp. 407–410, 1990.

[3.22] E. Sangiorgi, B. Riccò, and F. Venturi, "MOS^2: An efficient Monte Carlo simulator for MOS devices", *IEEE Trans. Computer-Aided Des.*, vol. 7, no. 2, pp. 259–271, 1988.

[3.23] R. Brunetti, C. Jacoboni, A. Matulionis, and V. Dienys, "Effect of interparticle collisions on energy relaxation of carriers in semiconductors", *Physica*, vol. 134B, pp. 369–373, 1985.

[3.24] P. Konijn, T. G. van de Roer, and F. P. Widdershoven, "An efficient Monte Carlo device simulation code based on rigorous application of internal scattering", *Solid-State Electron.*, vol. 36, pp. 1579–1581, 1993.

[3.25] W. Shockley, John A. Copeland, and R. P. James, "The impedance field method of noise calculation in active semiconductor devices", in *Quantum theory of atoms, molecules and solid state*, P. O. Lowdin, Ed., pp. 537–563. Academic Press, 1966.

[3.26] S. M. Ermakow, *Die Monte-Carlo-Methode und verwandte Fragen*, R. Oldenbourg, München, Wien, 1975.

[3.27] C. Jungemann, S. Yamaguchi, and H. Goto, "Convergence estimation for stationary ensemble Monte Carlo simulations", *IEEE J. Tech. Comp. Aided Design*, , no. 10, 1998.

[3.28] R.W. Hockney and J.W. Eastwood, *Computer Simulation Using Particles*, Institute of Physics Publishing, Bristol, Philadelphia, 1988.

[3.29] M. V. Fischetti and S. E. Laux, "Monte Carlo analysis of electron transport in small semiconductor devices including band–structure and space–charge effects", *Phys. Rev. B*, vol. 38, pp. 9721–9745, 1988.

[3.30] C. Jungemann, S. Decker, R. Thoma, W.-L. Engl, and H. Goto, "Phase space multiple refresh: A general purpose statistical enhancement technique for Monte Carlo device simulation", *IEEE J. Tech. Comp. Aided Design*, , no. 2, 1997.

[3.31] A. Pacelli and U. Ravaioli, "Analysis of variance-reduction schemes for ensemble Monte-Carlo simulation of semiconductor devices", *Solid-State Electron.*, vol. 41, pp. 599–605, 1997.

[3.32] M. G. Gray, T. E. Booth, T. J. T. Kwan, and C. M. Snell, "A multi-comb variance reduction scheme for Monte Carlo semiconductor simulators", *IEEE Trans. Electron Devices*, vol. 45, no. 4, pp. 918–924, 1998.

[3.33] N. G. van Kampen, *Stochastic Process in Physics and Chemistry*, North-Holland Publishing, Amsterdam, 1981.

4 Scattering Mechanisms

In a semiconductor the charge carriers are scattered by various mechanisms [4.1–4.6]. They are scattered by phonons, which are the quasiparticles of the lattice vibrations (Sec. 4.1), alloy disorder in composites (Sec. 4.2), ionized dopants (Sec. 4.3), impact ionization (Sec. 4.4), or microscopically roughness of interfaces between different materials (Sec. 4.5). The scattering processes are characterized by their respective transition rates which are calculated with Fermi's Golden Rule [4.3]

$$S_{n',n}(\mathbf{k}'|\mathbf{k}) = \frac{2\pi}{\hbar}|M_{n',n}(\mathbf{k}'|\mathbf{k})|^2\delta\left(\varepsilon_{n'}(\mathbf{k}') - \varepsilon_n(\mathbf{k}) - \varepsilon^{\mathrm{trans}}_{n',n}(\mathbf{k}'|\mathbf{k})\right) , \tag{4.1}$$

where M is the matrix element of the interaction and $\varepsilon^{\mathrm{trans}}$ the corresponding energy transfer. In general the transition rate has a very complex form and different levels of approximation are used (e.g. [4.5,4.7–4.10]). Since ab initio calculations of the transition rate are still difficult, the quality of the various models is estimated by comparison with models of fewer simplifications and by comparison with fundamental experiments. To obtain CPU efficient models applicable in device engineering in this work sometimes even stringent approximations are used, as long as the models reasonably reproduce the results of the fundamental experiments and of the more elaborate models.

4.1 Phonon Scattering

In Refs. [4.7, 4.11] an electron-phonon scattering model for an analytical band structure is given which reproduces experimental drift velocity and diffusion data for relaxed silicon. This phonon model was used in conjunction with a full-band structure under the assumption of constant matrix elements in Ref. [4.12]. This approach has the advantage of retaining the low-field and warm-electron properties of the analytical model, because at low energies the analytical model reproduces the main features of the full-band model. On the other hand, the hot-electron properties are improved by the better description of the band structure at higher energies. In the meantime the ansatz of constant matrix ele-

Table 4.1: *Material parameters for silicon and germanium [4.7, 4.20]*

	Si	Ge
ρ	$2.33\ \mathrm{g/cm^3}$	$5.32\ \mathrm{g/cm^3}$
u_l	$9.05 \cdot 10^5 \mathrm{cm/s}$	$5.40 \cdot 10^5 \mathrm{cm/s}$
ϵ_{si}	$11.7 \cdot \epsilon_0$	$16.0 \cdot \epsilon_0$
Ξ_e	8.90 eV	8.79 eV
Ξ_h	5.12 eV	7.40 eV

ments was proven to be correct by comparison with more detailed phonon models [4.5, 4.9, 4.13–4.16]. Due to the constant matrix elements very CPU efficient scattering algorithms can be used (cf. Sec. 5.7) and the resultant transition rates are velocity randomizing allowing to construct efficient estimators for the drift mobility and other transport parameters (cf. Sec. 2.3) [4.17–4.19].

In this work the ansatz of Ref. [4.12] is used in conjunction with the transition rates of Refs. [4.7, 4.11] for silicon and germanium. The transition rates for SiGe are obtained by linear interpolation of the silicon and germanium cases and the parameters are the ones of Ref. [4.20], where the parameters for electrons in silicon have been slightly modified to achieve better results for II.

The transition rate for the interaction with optical and intervalley acoustic phonons reads

$$\begin{aligned} S^{\mathrm{op}}_{\eta,n',n}(\mathbf{k}'|\mathbf{k}) &= (1-x_{\mathrm{Ge}})\frac{\pi(D_t K_\eta^{\mathrm{Si}})^2}{\Omega\rho^{\mathrm{Si}}\omega_\eta^{\mathrm{Si}}}\left(n(\hbar\omega_\eta^{\mathrm{Si}})+\frac{1}{2}\pm\frac{1}{2}\right) \\ &\quad \delta\left(\varepsilon_{n'}(\mathbf{k}')-\varepsilon_n(\mathbf{k})\pm\hbar\omega_\eta^{\mathrm{Si}}\right) r\left(\eta,\nu(\mathbf{k}'),\nu(\mathbf{k})\right) \\ &+ x_{\mathrm{Ge}}\frac{\pi(D_t K_\eta^{\mathrm{Ge}})^2}{\Omega\rho^{\mathrm{Ge}}\omega_\eta^{\mathrm{Ge}}}\left(n(\hbar\omega_\eta^{\mathrm{Ge}})+\frac{1}{2}\pm\frac{1}{2}\right) \\ &\quad \delta\left(\varepsilon_{n'}(\mathbf{k}')-\varepsilon_n(\mathbf{k})\pm\hbar\omega_\eta^{\mathrm{Ge}}\right) r\left(\eta,\nu(\mathbf{k}'),\nu(\mathbf{k})\right) , \end{aligned} \tag{4.2}$$

where ρ is the mass density (Tab. 4.1), $D_t K_\eta$ the coupling constant, ω_η the phonon frequency, and r the selection rule of the phonon mode η (Tab. 4.2). The occupation number for the phonons is given by the Bose-Einstein distribution

$$n(\varepsilon) = \left(\exp\left(\frac{\varepsilon}{k_{\mathrm{B}}T_0}\right)-1\right)^{-1} , \tag{4.3}$$

where T_0 is the lattice temperature assuming that the phonon system is in equilibrium. The transition rate for scattering with intravalley acoustic phonons is

Table 4.2: *Modes, coupling constants, energies, and selection rules of inelastic phonon scattering [4.7, 4.11, 4.16, 4.20]*

η	Mode	D_tK	$\hbar\omega$	D_tK	$\hbar\omega$	r
		[10^8eV/cm]	[meV]	[10^8eV/cm]	[meV]	
		Si		Ge		
			Electrons			
1	TA	0.470	12.1	0.479	5.60	g
2	LA	0.740	18.5	0.772	8.60	g
3	LO	10.23	62.0	9.280	37.0	g
4	TA	0.280	19.0	0.283	9.90	f
5	LA	1.860	47.4	1.940	28.0	f
6	TO	1.860	58.6	1.690	32.5	f
			Holes			
1		9.910	63.3	3.500	37.0	a

given in the equipartition, elastic approximation by [4.4]

$$\begin{aligned} S^{\mathrm{ac}}_{n',n}(\mathbf{k}'|\mathbf{k}) &= (1-x_{\mathrm{Ge}})\frac{2\pi k_{\mathrm{B}}T_0(\Xi^{\mathrm{Si}})^2}{\Omega\hbar\rho^{\mathrm{Si}}(u_{\mathrm{l}}^{\mathrm{Si}})^2}\delta\left(\varepsilon_{n'}(\mathbf{k}')-\varepsilon_n(\mathbf{k})\right)\delta_{\nu_{n'}(\mathbf{k}'),\nu_n(\mathbf{k})} \\ &+ x_{\mathrm{Ge}}\frac{2\pi k_{\mathrm{B}}T_0(\Xi^{\mathrm{Ge}})^2}{\Omega\hbar\rho^{\mathrm{Ge}}(u_{\mathrm{l}}^{\mathrm{Ge}})^2}\delta\left(\varepsilon_{n'}(\mathbf{k}')-\varepsilon_n(\mathbf{k})\right)\delta_{\nu_{n'}(\mathbf{k}'),\nu_n(\mathbf{k})}\,. \end{aligned} \tag{4.4}$$

Ξ is the deformation potential and u_{l} the longitudinal sound velocity (Tab. 4.1). The valley index ν for all conduction bands is defined as

$$\nu_n(\mathbf{k}) = \begin{cases} 1 : & k_x^* \geq |k_y^*| \;\wedge\; k_x^* \geq |k_z^*| \\ 2 : & -k_x^* \geq |k_y^*| \;\wedge\; -k_x^* \geq |k_z^*| \\ 3 : & k_y^* \geq |k_x^*| \;\wedge\; k_x^* \geq |k_z^*| \\ 4 : & -k_y^* \geq |k_x^*| \;\wedge\; -k_x^* \geq |k_z^*| \\ 5 : & k_z^* \geq |k_x^*| \;\wedge\; k_x^* \geq |k_y^*| \\ 6 : & -k_z^* \geq |k_x^*| \;\wedge\; -k_x^* \geq |k_y^*| \end{cases}, \tag{4.5}$$

where the transformation (5.21) has been used. In the case of the valence bands it reads

$$\nu_n(\mathbf{k}) = 1\,. \tag{4.6}$$

Three different selection rules are used. In the case of a g-phonon the electron is scattered out of its initial valley into the opposite valley on the same principal axis of the KS. In the case of an f-phonon the final valley is one of the four equivalent valleys on the principal axes perpendicular to the initial one. Although the partitioning of the BZ into valleys is strictly correct only for the bottom of

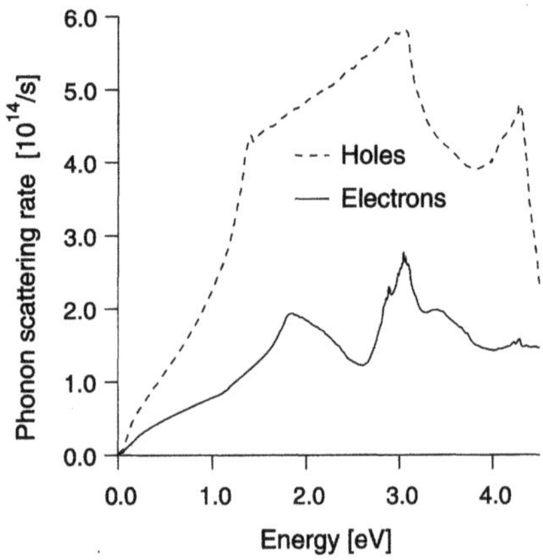

Figure 4.1: Phonon scattering rate of electrons and holes in RSi at room temperature

the lowest conduction band, the selection rules are used in the whole BZ for all conduction bands. The valence bands are not partitioned into valleys and the a-phonons are independent of the valley index. In Fig. 4.1 the resultant phonon scattering rate of electron and holes for relaxed silicon is shown which is more or less proportional to the density of states (DOS) (Fig. 5.9).

In Fig. 4.2 the low-field drift mobility limited by phonon scattering as a function of the lattice temperature is shown together with experimental data for relaxed silicon [4.21–4.23]. As expected, good agreement with the experimental data is found for electrons, because the phonon parameters in Tabs. 4.1 and 4.2 differ only slightly from the ones in Ref. [4.11]. In the case of holes good agreement is found for high lattice temperatures, while in the case of low temperatures a discrepancy occurs between simulation and experiment. This deviation is explained in Ref. [4.20] by nonohmic transport phenomena, whereas C. Jacoboni suggested that the poor quality of the silicon at the time being might be the reason for the lower experimental mobility [4.24]. The drift velocity in the $\langle 111 \rangle$ direction is shown in Fig. 4.3 and good agreement with the experimental data is found for both carrier types [4.25–4.27].

4.2 Alloy Scattering

Random disorder in the SiGe alloy gives rise to an additional scattering process, which is considered here in the formulation developed by Harrison and Hauser [4.28]. Including intravalley and intervalley scattering the transition rate of alloy

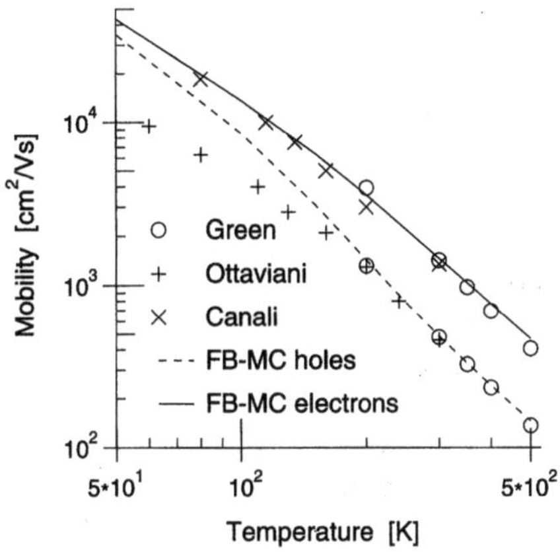

Figure 4.2: Simulated phonon limited drift mobility for electrons and holes in RSi and experimental data (Canali et al. [4.21], Ottaviani et al. [4.22], Green [4.23])

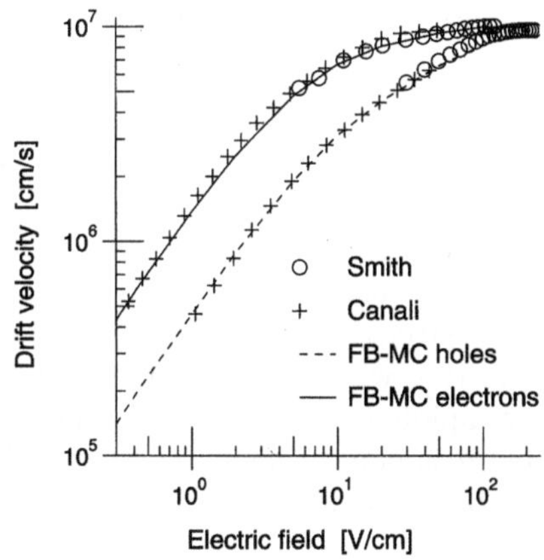

Figure 4.3: Simulated phonon limited drift velocity of electrons and holes in RSi for an electric field in ⟨111⟩ direction and experimental data (Canali et al. [4.25], Smith et al. [4.26, 4.27])

scattering reads

$$S_{n',n}^{\mathrm{alloy}}(\mathbf{k}'|\mathbf{k}) = \frac{\pi a_\perp a_\parallel^2 U^2}{4\Omega\hbar}(1 - x_{\mathrm{Ge}})x_{\mathrm{Ge}}\delta\left(\varepsilon_{n'}(\mathbf{k}') - \varepsilon_n(\mathbf{k})\right) , \tag{4.7}$$

where $a_\perp$ is the lattice constant in growth direction and $a_\parallel$ parallel to the SiGe layer, which are calculated self-consistently with the band structure for the given germanium content [4.29–4.31]. The alloy scattering potential U is 0.75eV for electrons and 0.70eV for holes [4.20].

4.3 Impurity Scattering

Scattering by ionized impurities is based on the model by Brooks and Herring [4.4, 4.32]. Neglecting umklapp processes and interband scattering the transition rate reads

$$S^{\mathrm{BH}}_{n',n}(\mathbf{k}'|\mathbf{k}) = \frac{2\pi q^4 (N_{\mathrm{acc}} + N_{\mathrm{don}})}{\Omega \hbar \epsilon^2(x_{\mathrm{Ge}})((\beta^{\mathrm{sc}}(x_{\mathrm{Ge}}))^2 + (\mathbf{k}' - \mathbf{k})^2)^2} \delta\left(\varepsilon_{n'}(\mathbf{k}') - \varepsilon_n(\mathbf{k})\right) \delta_{n',n} \delta_{\nu_{n'}(\mathbf{k}'),\nu_n(\mathbf{k})} , \tag{4.8}$$

where N_{acc} and N_{don} are the acceptor and donor concentration, respectively, and the dielectric constant of SiGe is obtained by linear interpolation

$$\epsilon(x_{\mathrm{Ge}}) = (1 - x_{\mathrm{Ge}})\epsilon_{\mathrm{Si}} + x_{\mathrm{Ge}}\epsilon_{\mathrm{Ge}} . \tag{4.9}$$

The dielectric constants of silicon and germanium are given in Tab. 4.1. This is the only parameter of the impurity scattering model, which depends on the germanium content. The inverse screening length β^{sc} is calculated in the Debye formulation [4.4]

$$(\beta^{\mathrm{sc}}(x_{\mathrm{Ge}}))^2 = \frac{q^2 n}{\epsilon(x_{\mathrm{Ge}}) k_{\mathrm{B}} T_0} + \frac{q^2 p}{\epsilon(x_{\mathrm{Ge}}) k_{\mathrm{B}} T_0} . \tag{4.10}$$

n and p are the electron and hole density, respectively.

Scattering by impurities is very CPU intensive, because the matrix element of the transition rate (4.8) strongly depends on the momentum transfer $\mathbf{k}' - \mathbf{k}$ favoring small-angle scattering with a very high scattering rate. On the other hand, the scattering rate rapidly decreases at high energies and compared to the rate of phonon scattering it is only of importance at low energies for which the first conduction band can be approximated by a nonparabolic spherical band [4.4]

$$\frac{\hbar^2 \mathbf{k}^2}{2 m_n^{\mathrm{DOS}}} = \varepsilon_n(\mathbf{k})(1 + \alpha_n^{\mathrm{NP}} \varepsilon_n(\mathbf{k})) , \tag{4.11}$$

where m_n^{DOS} and α_n^{NP} are the DOS mass and the nonparabolicity parameter of the approximation for the nth band, respectively (Tab. 4.3) [4.33]. Impurity scattering in the higher conduction bands is neglected. Despite being rather crude the above approximation is also applied to the first two valence bands.

In order to improve the CPU efficiency impurity scattering is further approximated by an isotropic-elastic process which has the same microscopic relaxation time as Eq. (4.8) (cf. Sec. 2.3) [4.19, 4.34–4.36]

$$S^{\mathrm{ABH}}_{n',n}(\mathbf{k}'|\mathbf{k}) \approx \frac{1}{\tau^{\mathrm{ABH}}{}_n(\varepsilon_n(\mathbf{k}))} \frac{\delta\left(\varepsilon_{n'}(\mathbf{k}') - \varepsilon_n(\mathbf{k})\right) \delta_{n',n} \delta_{\nu_{n'}(\mathbf{k}'),\nu_n(\mathbf{k})}}{\frac{\Omega}{(2\pi)^3} \int_{\mathrm{BZ}} \delta\left(\varepsilon_n(\mathbf{k}') - \varepsilon_n(\mathbf{k})\right) \delta_{\nu_n(\mathbf{k}'),\nu_n(\mathbf{k})} d^3 k'} , \tag{4.12}$$

Table 4.3: *Parameters of the analytical band approximation*

n	m^{DOS}	α^{NP}
	m_0	[eV]
1c	0.32	0.5
1v	0.689	0.669
2v	0.269	0.86

where τ^{ABH} is the microscopic relaxation time calculated with Eqs. (2.51), (4.8), and (4.11). Since the microscopic relaxation times are the same with and without this approximation, the low-field mobility remains unchanged by this approximation. Furthermore, comparison of simulations with and without the above approximation yielded no significant difference in the nonlinear transport regime [4.35, 4.36].

Since the Brooks-Herring model does not correctly describe the mobility at high doping concentrations for various reasons [4.37–4.39], here an empirical correction of the model is used

$$S_{n',n}^{\mathrm{CBH}}(\mathbf{k}'|\mathbf{k}) = \zeta(N_{\mathrm{acc}}, N_{\mathrm{don}}) S_{n',n}^{\mathrm{BH}}(\mathbf{k}'|\mathbf{k}) \ , \tag{4.13}$$

where $\zeta(N_{\mathrm{acc}}, N_{\mathrm{don}})$ is chosen in such a way that the simulated mobility exactly reproduces the following Caughey-Thomas expressions for the minority and majority mobility of both carrier types in relaxed silicon [4.20, 4.40]

$$\mu = \mu_{\mathrm{min}}^{\mathrm{CT}} + \frac{\mu_{\mathrm{max}}^{\mathrm{CT}} - \mu_{\mathrm{min}}^{\mathrm{CT}}}{1 + \left(\frac{N_{\mathrm{acc}} + N_{\mathrm{don}}}{N_{\mathrm{ref}}^{\mathrm{CT}}}\right)^{\alpha^{\mathrm{CT}}}} \ . \tag{4.14}$$

The parameters of this model given in Tab. 4.4 were determined by matching experimental results [4.41–4.48]. The resultant ζ is used in SiGe without further modifications. In Figs. 4.4 and 4.5 the majority and minority drift mobility at room temperature is shown for electrons and holes in relaxed silicon together with the corresponding correction factor ζ.

Experimental data [4.49] and simulation results for the in-plane mobility of majority electrons in strained SiGe grown on relaxed silicon by molecular beam epitaxy (MBE) and low pressure chemical vapor deposition (LPCVD) are given in Tab. 4.5. In Tab. 4.6 the hole drift mobility and experimental results [4.50] are shown for lowly doped relaxed SiGe. In Tab. 4.7 the in-plane hole drift mobility and experimental results [4.51] are given for highly doped strained SiGe grown on relaxed silicon. In all three cases good agreement of simulations and experiments is obtained verifying our approach to alloy and impurity scattering.

Table 4.4: *Mobility parameters of the Caughey-Thomas expression for relaxed silicon (Majority electrons/holes: E/H, minority electrons/holes: e/h) [4.20, 4.41–4.47]*

Typ	$\mu_{\mathrm{max}}^{\mathrm{CT}}$	$\mu_{\mathrm{min}}^{\mathrm{CT}}$	$N_{\mathrm{ref}}^{\mathrm{CT}}$	α^{CT}
	[cm^2/Vs]	[cm^2/Vs]	[cm^{-3}]	
E	1430	74.5	$8.6 \cdot 10^{16}$	0.77
e	1430	200	$5.3 \cdot 10^{16}$	0.68
H	480	47.0	$1.7 \cdot 10^{17}$	0.79
h	480	122.3	$1.4 \cdot 10^{17}$	0.70

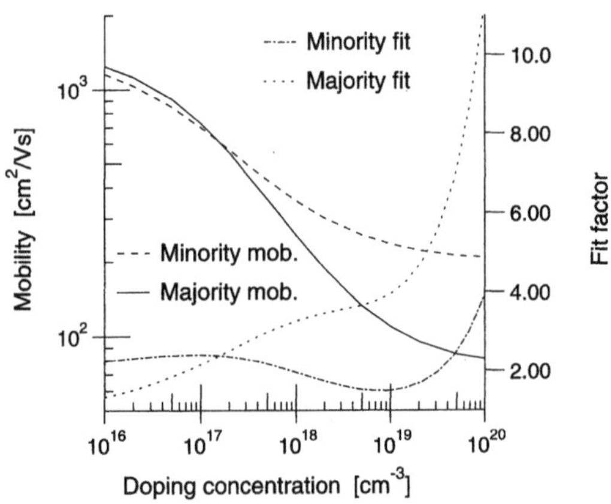

Figure 4.4: Electron majority and minority drift mobility at room temperature and the accordant heuristic fit factor ζ for RSi

4.4 Impact Ionization by Electrons

In the devices investigated in this work II occurs only in regions of relaxed silicon, and the II scattering model is only developed for relaxed silicon.

The process of II involves two electrons [4.53]. A primary electron in a conduction band interacts via Coulomb forces with another electron in a valence band and the latter electron is transfered to a conduction band resulting in a secondary hole in the valence band. During the process the total energy and quasi-momentum are conserved. The calculation of the scattering rate of this process is very complex, because two particles are involved. In addition, the exact evaluation of the matrix elements of the Coulomb interaction based on first principles is very difficult and different levels of approximation were applied to

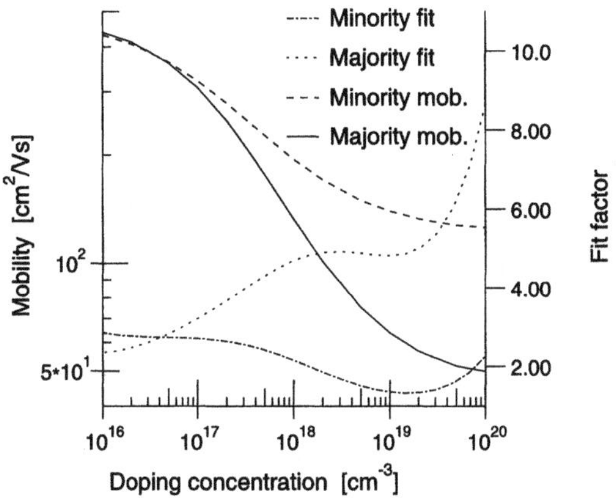

Figure 4.5: Hole majority and minority drift mobility at room temperature and the accordant heuristic fit factor ζ for RSi

Table 4.5: *In-plane mobility of majority electrons in strained SiGe grown on relaxed silicon at* 295K *[4.49]*

x_{Ge}	N_{don}	μ_{exp}	μ_{cal}	Fab.
%	[cm^{-3}]	[cm^2/Vs]	[cm^2/Vs]	
0	$6.60 \cdot 10^{16}$	830	858	MBE
0	$2.55 \cdot 10^{17}$	491	501	MBE
0	$1.88 \cdot 10^{18}$	193	195	MBE
0	$7.57 \cdot 10^{18}$	120	119	MBE
9.0	$8.14 \cdot 10^{17}$	173	180	MBE
9.0	$8.41 \cdot 10^{18}$	73	87	MBE
18.0	$5.12 \cdot 10^{18}$	111	91	MBE
18.0	$1.36 \cdot 10^{19}$	70	75	MBE
0	$2.10 \cdot 10^{17}$	645	547	LPCVD
5.0	$6.30 \cdot 10^{18}$	97	101	LPCVD
9.0	$7.90 \cdot 10^{18}$	80	88	LPCVD
20.0	$7.60 \cdot 10^{17}$	165	157	LPCVD

evaluate the scattering rate [4.12,4.14,4.53–4.59]. A different approach was taken in Ref. [4.60]. By matching experimental results and MC simulations a heuristic model function for the scattering rate was determined. Later it turned out that this scattering rate agrees well with theoretical calculations after adjusting

Table 4.6: *Drift mobility of majority holes in relaxed SiGe at* 300K *[4.52]*

x_{Ge}	N_{acc}	N_{don}	μ_{exp}	μ_{cal}
%	$[10^{15}\mathrm{cm}^{-3}]$	$[10^{15}\mathrm{cm}^{-3}]$	$[\mathrm{cm}^2/\mathrm{Vs}]$	$[\mathrm{cm}^2/\mathrm{Vs}]$
0	1.10	0.06	462	475
6.3	0.94	0.21	376	359
0	25.0	0.87	406	400
6.1	13.0	0.48	327	327
12.7	14.0	1.20	303	286

Table 4.7: *In-plane mobility of majority holes in strained SiGe grown on relaxed silicon at* 300K *[4.51]*

x_{Ge}	N_{acc}	μ_{exp}	μ_{cal}	Fab.
%	$[10^{19}\mathrm{cm}^{-3}]$	$[\mathrm{cm}^2/\mathrm{Vs}]$	$[\mathrm{cm}^2/\mathrm{Vs}]$	
0	1.90	65	57	MBE
8.0	0.46	70	74	MBE
8.0	2.75	45	50	MBE
17.0	0.69	65	71	MBE
17.0	3.00	53	51	MBE
0	4.52	50	52	LPCVD
17.4	6.05	44	46	LPCVD

unknown parameters of the latter models [4.15].

The heuristic approach of Ref. [4.60] is used in this work. The exclusively energy dependent scattering rate of Ref. [4.56] is used to model II. To match experimental results this scattering rate is multiplied with a global scaling factor being the only adjustable parameter of the model. The resultant scattering rate reads for electrons in relaxed silicon [4.16]

$$S^{\mathrm{II}}(\varepsilon) = \begin{cases} 1.49 \cdot 10^{11}\frac{1}{s}\left(\frac{\varepsilon}{\mathrm{eV}} - 1.128\right)^3 & : \quad 1.128\mathrm{eV} < \varepsilon < 1.750\mathrm{eV} \\ 1.13 \cdot 10^{12}\frac{1}{s}\left(\frac{\varepsilon}{\mathrm{eV}} - 1.572\right)^2 & : \quad 1.750\mathrm{eV} < \varepsilon \end{cases} . \tag{4.15}$$

The state after scattering is chosen with a uniform probability density on an equienergy surface for the final energy $\varepsilon_n^{\mathrm{II}}(\mathbf{k})$ resulting in a simple transition rate

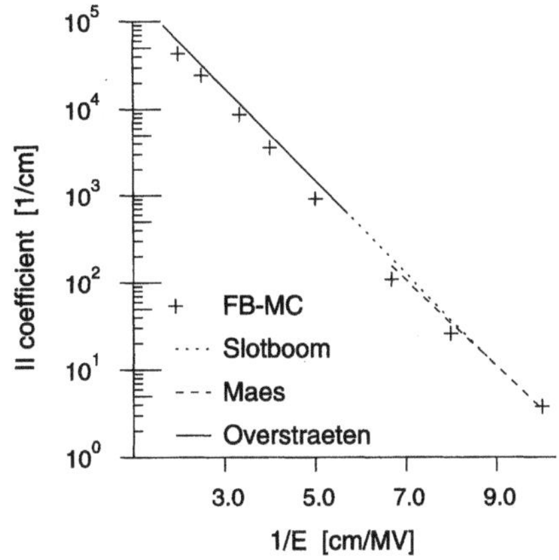

Figure 4.6: Simulated II coefficient for electrons and experimental results for RSi at room temperature (Slotboom et al. [4.61], Maes et al. [4.62], and van Overstraeten et al. [4.63])

for II

$$S^{\mathrm{II}}_{n',n}(\mathbf{k}',\mathbf{k}) = \frac{S^{\mathrm{II}}(\varepsilon_n(\mathbf{k}))}{\sum_{n'} D_{n'}(\varepsilon^{\mathrm{II}}_n(\mathbf{k}))}\delta\left(\varepsilon_{n'}(\mathbf{k}') - \varepsilon^{\mathrm{II}}_n(\mathbf{k})\right) , \tag{4.16}$$

where the sum of the DOS (5.31) in the denominator runs over all conduction bands. The assumption that all three particles (primary electron, secondary electron and hole) have the same energy after scattering gives the final energy [4.14]

$$\varepsilon^{\mathrm{II}}_n(\mathbf{k}) = \frac{\varepsilon_n(\mathbf{k}) - \varepsilon^{\mathrm{gap}}}{3} . \tag{4.17}$$

$\varepsilon^{\mathrm{gap}}$ is the energy of the band gap. The final states of the secondary particles are randomly selected in the same fashion as the final state of the primary electron. This CPU efficient approximation, which does not conserve quasi-momentum, can be used, because the impact of the exact final state after II scattering on transport in devices is rather small in most cases.

The II coefficient gives the number of secondary electrons generated by a primary electron per unit length. It has been measured by different groups under homogeneous bulk conditions as a function of the electric field. The experimental data are shown in Fig. 4.6 together with simulation results and good agreement between simulation and experiment is obtain over four orders of magnitude. In contrast to the II coefficient the quantum yield is measured in a bulk system with zero electric field into which the electrons are injected with a given initial energy. The quantum yield is defined as the number of secondary electrons generated by a primary electron. In Fig. 4.7 results from experiment and simulation are shown and again good agreement is found. It is worth noting that these results (Figs. 4.6

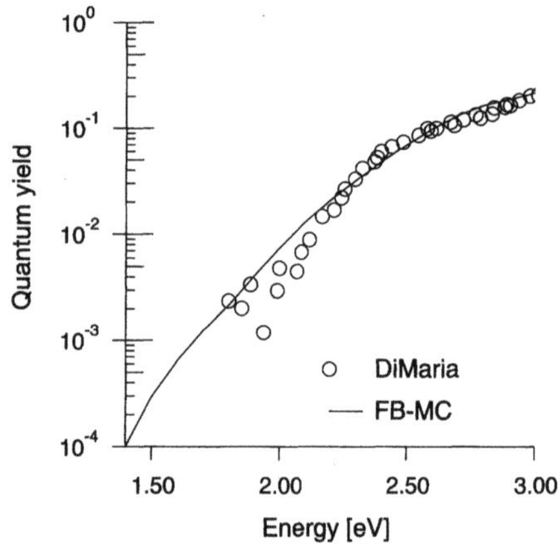

Figure 4.7: Simulated quantum yield for electrons and experimental results by DiMaria et al. [4.64] at room temperature for RSi

and 4.7) were obtained by fitting only the global scaling parameter of the II rate without modifying any other parameter of the transport model [4.16].

Modeling of II by a primary hole is not attempted in this work due to a lack of essential experimental data and problems with modeling of the hole-phonon scattering at high energies.

4.5 Surface Roughness Scattering

In addition to the bulk scattering mechanisms mentioned before particles are scattered when they impinge upon the microscopically rough Si/SiO_2-interface [4.65]. The full quantum mechanical treatment of this effect including the simulation of the quasi 2D particle gas forming at the material interface is beyond the scope of this work [4.66–4.68]. Here, a simple approach consisting of reflective and diffusive scattering is used [4.69–4.71]. Whenever a particle hits the Si/SiO_2-interface it is either reflected with the probability $1 - P_{\mathrm{diff}}$ or scattered diffusively with the probability P_{diff}. The reflection must not necessarily be specular as explained in Sec. 5.6 [4.72]. In the case of diffusive scattering the particle is elastically scattered back into the silicon with a uniform probability distribution on an equienergy surface. The probability for a diffusive scattering event is 14% in the case of electrons and 13% in the case of holes.

In Fig. 4.8 the average channel mobility versus the effective field in an inversion layer is shown for electrons and holes in relaxed silicon. The effective field for electrons is defined as

$$E_{\mathrm{eff}} = \frac{q}{\epsilon_{\mathrm{si}}}(\frac{1}{2}N_{\mathrm{inv}} + N_{\mathrm{depl}}) \ , \tag{4.18}$$

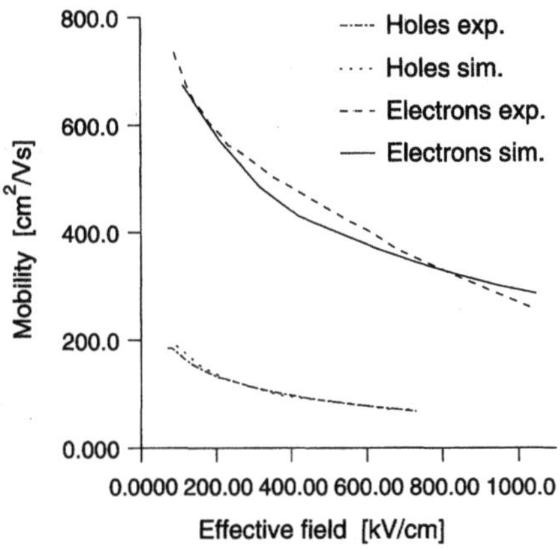

Figure 4.8: Simulated inversion layer mobility for electrons and holes in RSi at room temperature and experimental results by Takagi et al. for an acceptor concentration of $2{\cdot}10^{16}/\mathrm{cm}^3$ (electrons) and a donor concentration of $1.6{\cdot}10^{16}/\mathrm{cm}^3$ (holes) [4.73]

where N_{inv} and N_{depl} are the inversion and depletion layer sheet density, respectively [4.74]. The effective field for holes reads [4.75]

$$E_{\mathrm{eff}} = \frac{q}{\epsilon_{\mathrm{si}}}\left(\frac{1}{3}N_{\mathrm{inv}} + N_{\mathrm{depl}}\right) . \tag{4.19}$$

In addition, in Fig. 4.8 experimental results are shown [4.73]. In the case of holes excellent agreement between simulation and measurement is obtained. In the case of electrons for which the variation of the mobility is larger than in the case of holes the deviation is somewhat larger. Due to a lack of experimental and theoretical data the surface roughness scattering model is used with the same parameters also in the case of SiGe.

References

[4.1] W. Shockley, *Electrons and Holes in Semiconductors*, van Nostrand, Princeton, New Jersey, 1950.

[4.2] J.M. Ziman, *Electrons and Phonons*, Clarendon Press, Oxford, 1960.

[4.3] O. Madelung, *Introduction to Solid State Theory*, Springer, Berlin, 1978.

[4.4] C. Jacoboni and P. Lugli, *The Monte Carlo Method for Semiconductor Device Simulation*, Springer, Wien, 1989.

[4.5] K. Hess, Ed., *Monte Carlo Device Simulation: Full Band and Beyond*, Kluwer, Boston, 1991.

[4.6] C. Moglestue, *Monte Carlo Simulation of Semiconductor Devices*, Chapman & Hall, London, 1993.

[4.7] C. Jacoboni and L. Reggiani, "The Monte Carlo method for the solution of charge transport in semiconductors with application to covalent materials", *Rev. Mod. Phys.*, vol. 55, pp. 645–705, 1983.

[4.8] M. V. Fischetti and S. E. Laux, "Monte Carlo analysis of electron transport in small semiconductor devices including band–structure and space–charge effects", *Phys. Rev. B*, vol. 38, pp. 9721–9745, 1988.

[4.9] A. Abramo, L. Baudry, R. Brunetti, R. Castagne, M. Charef, F. Dessenne, P. Dollfus, R. Dutton, W. L. Engl, R. Fauquembergue, C. Fiegna, M. V. Fischetti, S. Galdin, N. Goldsman, M. Hackel, C. Hamaguchi, K. Hess, K. Hennacy, P. Hesto, J. M. Higman, T. Iizuka, C. Jungemann, Y.Kamakura, H. Kosina, T. Kunukiyo, S. E. Laux, H. Lin, C. Maziar, H. Mizuno, H. J. Peifer, S. Ramaswamy, N. Sano, P. G. Scrobohaci, S. Selberherr, M. Takenaka, T. Tang, J. L. Thobel, R. Thoma, K. Tomizawa, M. Tomizawa, T. Vogelsang, S. Wang, X. Wang, C.Yao, P. D. Yoder, and A. Yoshii, "A comparison of numerical solutions of the Boltzmann transport equation for high-energy electron transport silicon", *IEEE Trans. Electron Devices*, vol. 41, no. 9, pp. 1646–1654, 1994.

[4.10] M. V. Fischetti, S. E. Laux, and E. Crabbe, "Understanding hot-electron transport in silicon devices: Is there a shortcut?", *J. Appl. Phys.*, vol. 78, pp. 1058–1087, 1995.

[4.11] R. Brunetti, C. Jacoboni, F. Nava, L. Reggiani, G. Bosman, and R. J. J. Zijlstra, "Diffusion coefficient of electrons in silicon", *J. Appl. Phys.*, vol. 52, pp. 6713–6722, 1981.

[4.12] J. Y. Tang and K. Hess, "Impact ionization of electrons in silicon (steady state)", *J. Appl. Phys.*, vol. 54, pp. 5139–5144, 1983.

[4.13] P. D. Yoder, V. D. Natoli, and R. M. Martin, "Ab initio analysis of the electron-phonon interaction in silicon", *J. Appl. Phys.*, vol. 73, pp. 4378–4383, 1993.

[4.14] T. Kunikiyo, M. Takenaka, Y. Kamakura, M. Yamaji, H. Mizuno, M. Morifuji, K. Taniguchi, and C. Hamaguchi, "A Monte Carlo simulation of anisotropic electron transport in silicon including full band structure and anisotropic impact–ionization model", *J. Appl. Phys.*, vol. 75, pp. 297–312, 1994.

[4.15] M. V. Fischetti and S. E. Laux, "Monte Carlo simulation of electron transport in Si: The first 20 years", in *Proc. ESSDERC*, Bologna, 1996, vol. 26, pp. 813–820.

[4.16] C. Jungemann, S. Keith, F. M. Bufler, and B. Meinerzhagen, "Effects of band structure and phonon models on hot electron transport in silicon", *Electrical Engineering*, vol. 79, pp. 99–101, 1996.

[4.17] P. J. Price, "Monte Carlo calculation of electron transport in solids", *Semiconductors and Semimetals*, vol. 14, pp. 249–309, 1979.

[4.18] C. Jungemann, S. Yamaguchi, and H. Goto, "Efficient full band Monte Carlo hot carrier simulation for silicon devices", in *Proc. ESSDERC*, Bologna, 1996, vol. 26, pp. 821–824.

[4.19] C. Jungemann, S. Keith, M. Bartels, and B. Meinerzhagen, "Efficient full-band Monte Carlo simulation of silicon devices", *IEICE Trans. on Electronics*, vol. E82-C, no. 6, pp. 870–879, 1999.

[4.20] F. M. Bufler, "Full–band Monte Carlo simulation of electrons and holes in strained Si and SiGe", Dissertation, Universität Bremen, Bremen, 1997, (H. Utz Verlag Wissenschaft, München: 1998).

[4.21] C. Canali, C. Jacoboni, F. Nava, G. Ottaviani, and A. Alberigi-Quaranta, "Electron drift velocity in silicon", *Phys. Rev. B*, vol. 12, pp. 2265–2284, 1975.

[4.22] G. Ottaviani, L. Reggiani, C. Canali, F. Nava, and A. Alberigi-Quaranta, "Hole drift velocity in silicon", *Phys. Rev. B*, vol. 12, pp. 3318–3329, 1975.

[4.23] M. A. Green, "Intrinsic concentration, effective densities of states, and effective mass in silicon", *J. Appl. Phys.*, vol. 67, pp. 2944–2954, 1990.

[4.24] C. Jacoboni, ", 1998, private communication.

[4.25] C. Canali, G. Ottaviani, and A. Alberigi-Quaranta, "Drift velocity of electrons and holes and associated anisotropic effects in silicon", *J. Phys. Chem. Solids*, vol. 32, pp. 1707–1720, 1971.

[4.26] P. M. Smith, M. Inoue, and J. Frey, "Electron velocity in Si and GaAs at very high electric fields", *Appl. Phys. Lett.*, vol. 37, pp. 797–798, 1980.

[4.27] P. M. Smith and J. Frey, "High-field transport of holes in silicon", *Appl. Phys. Lett.*, vol. 39, pp. 332–333, 1981.

[4.28] J. W. Harrison and J. R. Hauser, "Alloy scattering in ternary III-V compounds", *Phys. Rev. B*, vol. 13, pp. 5347–5350, 1976.

[4.29] M. M. Rieger, "Elektronische Bandparameter in $Si_{1-x}Ge_x$-Legierungen", Diplomarbeit, TU München, 1997.

[4.30] M. M. Rieger and P. Vogl, "Electronic-band parameters in strained $Si_{1-x}Ge_x$ alloys on $Si_{1-y}Ge_y$ substrates", *Phys. Rev. B*, vol. 48, pp. 14276–14287, 1993.

[4.31] M. M. Rieger and P. Vogl, "Electronic-band parameters in strained $Si_{1-x}Ge_x$ alloys on $Si_{1-y}Ge_y$ substrates", *Phys. Rev. B*, vol. 50, pp. 8138, 1994, Erratum.

[4.32] H. Brooks, "Scattering by ionized impurities in semiconductors", *Phys. Rev.*, vol. 83, pp. 879, 1951.

[4.33] F. M. Bufler, P. Graf, and B. Meinerzhagen, ", Abschlußbericht zum BMBF–Förderungsvorhaben 01 M 2416 A: Monte-Carlo-Bauelementsimulator für Si/SiGe-Heterobipolartransistoren.

[4.34] S. Jallepalli, M. Rashed, W.-K. Shih, C. M. Maziar, and A. F. Tasch, Jr., "A full-band Monte Carlo model for hole transport in silicon", *J. Appl. Phys.*, vol. 81, pp. 2250–2255, 1997.

[4.35] P. Graf, F. M. Bufler, B. Meinerzhagen, and C. Jungemann, "A comprehensive SiGe Monte Carlo model for transient 2D simulations of HBTs", in *IEDM Tech. Dig.*, 1997, pp. 881–884.

[4.36] H. Kosina, "A method to reduce small-angle scattering in Monte Carlo device analysis", *IEEE Trans. Electron Devices*, vol. 46, no. 6, pp. 1196–1200, 1999.

[4.37] D. Chattopadhyay and H. J. Queisser, "Electron scattering by ionized impurities in semiconductors", *Rev. Mod. Phys.*, vol. 53, pp. 745–768, 1981.

[4.38] M. V. Fischetti, "Effect of the electron-plasmon interaction on the electron mobility in silicon", *Phys. Rev. B*, vol. 44, pp. 5527–5534, 1991.

[4.39] H. Kosina and G. Kaiblinger-Grujin, "Ionized-impurity scattering of majority electrons in silicon", *Solid–State Electron.*, vol. 42, pp. 331–338, 1998.

[4.40] D. M. Caughey and R. E. Thomas, "Carrier mobilities in silicon empirically related to doping and field", *Proc. IEEE*, vol. 55, pp. 2192–2193, 1967.

[4.41] J. Dziewior and D. Silber, "Minority-carrier diffusion coefficients in highly doped silicon", *Appl. Phys. Lett.*, vol. 35, pp. 170–172, 1979.

[4.42] W. R. Thurber and B. S. Carpenter, "Boron determination in silicon by the nuclear track technique", *J. Electrochem. Soc.*, vol. 125, pp. 654–657, 1978.

[4.43] W. R. Thurber, R. L. Mattis, and Y. M. Liu, "Resistivity-dopant density relationship for boron-doped silicon", *J. Electrochem. Soc.*, vol. 127, pp. 2291–2294, 1980.

[4.44] W. R. Thurber, R. L. Mattis, and Y. M Liu, "Resistivity-dopant density relationship for phosphorus-doped silicon", *J. Electrochem. Soc.*, vol. 127, pp. 1807–1812, 1980.

[4.45] S. E. Swirhun, Y.-H. Kwark, and R. M. Swanson, "Measurement of electron lifetime and electron mobility, band-gap narrowing in heavily doped p-type silicon", in *IEDM Tech. Dig.*, 1986, pp. 24–27.

[4.46] S. E. Swirhun, J. A. del Alamo, and R. M. Swanson, "Measurement of hole mobility in heavily doped n-type silicon", *IEEE Electron Device Lett.*, vol. 7, no. 3, pp. 168–171, 1986.

[4.47] A. B. Sproul, M. A. Green, and A. W. Stephens, "Accurate determination of minority carrier- and lattice scattering-mobility in silicon from photoconductance decay", *J. Appl. Phys.*, vol. 72, pp. 4161–4171, 1992.

[4.48] C. Jungemann, B. Heinemann, K. Tittelbach-Helmrich, and B. Meinerzhagen, "An accurate, experimentally verified electron minority carrier mobility model for Si and SiGe", in *IEDM Tech. Dig.*, San Francisco (USA), 2000, pp. 101–104.

[4.49] F. M. Bufler, P. Graf, B. Meinerzhagen, B. Adeline, M. M. Rieger, H. Kibbel, and G. Fischer, "Low– and high–field electron–transport parameters for unstrained and strained $Si_{1-x}Ge_x$", *IEEE Electron Device Lett.*, vol. 18, no. 6, pp. 264–266, 1997.

[4.50] K. Tittelbach-Helmrich and P. Gaworzewski, ", unpublished, 1997.

[4.51] F. M Bufler, P. Graf, B. Meinerzhagen, G. Fischer, and H. Kibbel, "Hole transport investigation in unstrained and strained SiGe", *J. Vac. Sci. Technol. B*, vol. 16, pp. 1667–1669, 1998.

[4.52] P. Gaworzewski, K. Tittelbach-Helmrich, U. Penner, and N. V. Abrosimov, "Electrical properties of lightly doped *p*–type silicon–germanium single crystals", *J. Appl. Phys.*, vol. 83, pp. 5258–5263, 1998.

[4.53] E. O. Kane, "Electron scattering by pair production in silicon", *Phys. Rev.*, vol. 159, pp. 624–631, 1967.

[4.54] D. J. Robbins, "Aspects of the theory of impact ionization (1)", *Phys. Status Solidi B*, vol. 97, pp. 9–50, 1980.

[4.55] D. J. Robbins, "Aspects of the theory of impact ionization (2)", *Phys. Status Solidi B*, vol. 97, pp. 387–406, 1980.

[4.56] R. Thoma, H. J. Peifer, W. L. Engl, W. Quade, R. Brunetti, and C. Jacoboni, "An improved impact–ionization model for high–energy electron transport in Si with Monte Carlo simulation", *J. Appl. Phys.*, vol. 69, pp. 2300–2311, 1991.

[4.57] J. Bude and K. Hess, "Impact ionization in semiconductors: Effects of high electric fields and high scattering rates", *Phys. Rev. B*, vol. 45, pp. 10958–10964, 1992.

[4.58] N. Sano and A. Yoshii, "Impact ionization theory consistent with a realistic band structure of silicon", *Phys. Rev. B*, vol. 45, pp. 4171–4179, 1992.

[4.59] M. V. Fischetti, N. Sano, S. E. Laux, and K. Natori, "Full-band-structure theory of high-field transport and impact ionization of electrons and holes in Ge, Si, and GaAs", *IEEE J. Tech. Comp. Aided Design*, , no. 3, 1997.

[4.60] E. Cartier, M. V. Fischetti, E. A. Eklund, and F. R. McFeely, "Impact ionization in silicon", *Appl. Phys. Lett.*, vol. 62, pp. 3339–3341, 1993.

[4.61] J. W. Slotboom, G. Streutker, G. J. T. Davids, and P. B. Hartog, "Surface impact ionization in Silicon devices", in *IEDM*, 1987, pp. 494–497.

[4.62] W. Maes, K. de Meyer, and R. Van Overstraeten, "Impact ionization in silicon: A review and update", *Solid–State Electron.*, vol. 33, pp. 705–718, 1990.

[4.63] R. van Overstraeten and H. de Man, "Measurement of the ionization rates in diffused silicon p-n junctions", *Solid–State Electron.*, vol. 13, pp. 583–608, 1970.

[4.64] D. J. DiMaria, T. N. Theis, J. R. Kirtley, F. L. Pesavento, and D. W. Dong, "Electron heating in silicon dioxide and off-stoichiometric silicon dioxide films", *J. Appl. Phys.*, vol. 57, pp. 1214–1238, 1985.

[4.65] Y. C. Cheng, "Electron mobility in an MOS inversion layer", in *Proceedings of the 3rd Conference on Solid State Devices*, Tokyo, 1971, pp. 173–180.

[4.66] T. Ando, A. B. Fowler, and F. Stern, "Electronic properties of two-dimensional systems", *Rev. Mod. Phys.*, vol. 54, pp. 437–672, 1982.

[4.67] M. V. Fischetti and S. E. Laux, "Monte Carlo study of electron transport in silicon inversion layers", *Phys. Rev. B*, vol. 48, pp. 2244–2274, 1993.

[4.68] C. Jungemann, A. Emunds, and W. L. Engl, "Simulation of linear and nonlinear electron transport in homogeneous silicon inversion layers", *Solid–State Electron.*, vol. 36, pp. 1529–1540, 1993.

[4.69] K. Fuchs and H. H. Wills, "The conductivity of thin metallic films according to the electron theory of metals", *Proc. Camb. Phil. Soc.*, vol. 34, pp. 100–108, 1938.

[4.70] H.-J. Peifer, "Monte–Carlo Simulation des Hochenergietransports von Elektronen in submikron MOS–Strukturen", Doctor thesis, RWTH Aachen, Aachen, 1992, Augustinus Buchhandlung.

[4.71] E. Sangiorgi and M. R. Pinto, "A semi-empirical model of surface scattering for Monte Carlo simulation of silicon n-MOSFET's", *IEEE Trans. Electron Devices*, vol. 39, no. 2, pp. 356–361, 1992.

[4.72] C. Jungemann, S. Keith, and B. Meinerzhagen, "Full-band Monte Carlo simulation of a 0.12μm-Si-PMOSFET with and without a strained SiGe-channel", in *IEDM Tech. Dig.*, San Francisco (USA), 1998, pp. 897–900.

[4.73] S. Takagi, A. Toriumi, M. Iwase, and H. Tango, "On the universality of inversion layer mobility in Si MOSFET's: Part I–Effects of substrate impurity concentration", *IEEE Trans. Electron Devices*, vol. 41, pp. 2357–2362, 1994.

[4.74] A. G. Sabnis and J. T. Clemens, "Characterization of the electron mobility in the inverted $\langle 100 \rangle$ Si surface", in *IEDM Tech. Dig.*, 1979, pp. 18–22.

[4.75] N. D. Arora and G. SH. Gildenblat, "A semi-empirical model of the MOSFET inversion layer mobility for low-temperature operation", *IEEE Trans. Electron Devices*, vol. 34, no. 1, pp. 89–93, 1987.

5 Full-Band Structure

The complicated dependence of the energy on the wave vector makes it impossible to capture all details of the band structure by analytical approximations and the full details of the band structure (full-band) are included in the MC model based on a numerical representation of the band structure [5.1]. The basic properties and symmetries of the band structure of RSi are discussed in the first section of this chapter. The more general case of strained SiGe follows in the next section. The grid and the interpolation method for the energy in the KS are developed in the third section. Based on this grid efficient methods for the calculation of the density of states are discussed in the fourth section and a formulation of the mass tensor consistent with an unstructured tetrahedral grid is given in the fifth section. Methods for the motion of particles in the KS are presented in the sixth section and CPU efficient methods for the selection of the final state are given in the seventh section.

5.1 Basic Properties of the Band Structure of Relaxed Silicon

Relaxed silicon (RSi) has a diamond lattice structure and is invariant under translations in the RS of the form

$$\mathbf{r}' = \mathbf{r} + i\mathbf{a}_1^{\mathrm{RSi}} + j\mathbf{a}_2^{\mathrm{RSi}} + k\mathbf{a}_3^{\mathrm{RSi}} \; , \tag{5.1}$$

where i, j and k are integers [5.2,5.3]. The basis vectors of the direct lattice are

$$\mathbf{a}_1^{\mathrm{RSi}} = \frac{a_0^{\mathrm{RSi}}}{2}\begin{pmatrix}0\\1\\1\end{pmatrix} \; , \quad \mathbf{a}_2^{\mathrm{RSi}} = \frac{a_0^{\mathrm{RSi}}}{2}\begin{pmatrix}1\\0\\1\end{pmatrix} \; , \quad \mathbf{a}_3^{\mathrm{RSi}} = \frac{a_0^{\mathrm{RSi}}}{2}\begin{pmatrix}1\\1\\0\end{pmatrix} \; . \tag{5.2}$$

a_0^{RSi} denotes the lattice constant of RSi ($a_0^{\mathrm{RSi}} = 0.543\mathrm{nm}$). The respective general reciprocal lattice vector in KS reads

$$\mathbf{G}_{i,j,k} = i\mathbf{b}_1^{\mathrm{RSi}} + j\mathbf{b}_2^{\mathrm{RSi}} + k\mathbf{b}_3^{\mathrm{RSi}} \; . \tag{5.3}$$

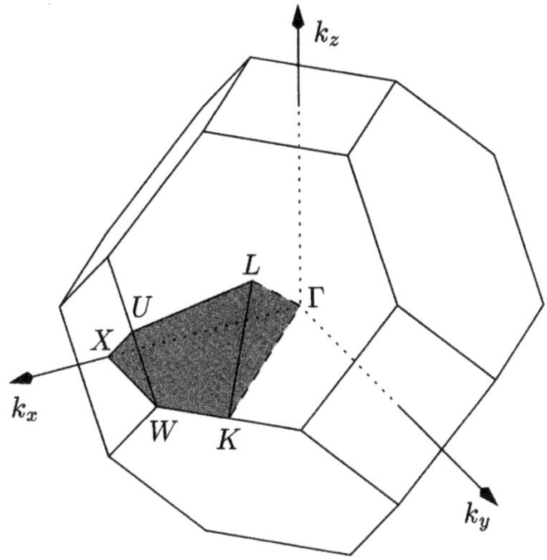

Figure 5.1: The first BZ of Si and the first irreducible wedge

The basis vectors of the reciprocal lattice are

$$\mathbf{b}_1^{\mathrm{RSi}} = \frac{2\pi}{a_0^{\mathrm{RSi}}} \begin{pmatrix} -1 \\ 1 \\ 1 \end{pmatrix} , \quad \mathbf{b}_2^{\mathrm{RSi}} = \frac{2\pi}{a_0^{\mathrm{RSi}}} \begin{pmatrix} 1 \\ -1 \\ 1 \end{pmatrix} , \quad \mathbf{b}_3^{\mathrm{RSi}} = \frac{2\pi}{a_0^{\mathrm{RSi}}} \begin{pmatrix} 1 \\ 1 \\ -1 \end{pmatrix} . \tag{5.4}$$

with ${\mathbf{a}_i^{\mathrm{RSi}}}^{\mathrm{T}} \mathbf{b}_j^{\mathrm{RSi}} = 2\pi\delta_{i,j}$.

The unit cell of the reciprocal lattice, the Brillouin zone (BZ) contains all points nearest to the enclosed reciprocal lattice vector. Its surface is given by the planes which are perpendicular to the reciprocal lattice vectors pointing from the center of the cell to the 14 nearest reciprocal lattice vectors and which bisect these vectors (Fig. 5.1). The 14 reciprocal lattice vectors nearest to the origin of the reciprocal lattice are the eight vectors in $\langle 111 \rangle$-direction with a length of $\sqrt{3} \cdot 2\pi/a_0^{\mathrm{RSi}}$ and the six vectors in $\langle 100 \rangle$-direction with a length of $2 \cdot 2\pi/a_0^{\mathrm{RSi}}$. The 14 faces of the first BZ are

$$|k_x| + |k_y| + |k_z| = \frac{3}{2} \frac{2\pi}{a_0^{\mathrm{RSi}}} , \quad |k_x| = \frac{2\pi}{a_0^{\mathrm{RSi}}} , \quad |k_y| = \frac{2\pi}{a_0^{\mathrm{RSi}}} , \quad |k_z| = \frac{2\pi}{a_0^{\mathrm{RSi}}} . \tag{5.5}$$

Due to the periodicity in RS the Bloch functions and the energy bands are periodic in the KS with respect to the reciprocal lattice and it is sufficient to consider only the first BZ (reduced zone scheme [5.3]). All wave vectors outside of the first BZ can be mapped into the first BZ by subtracting a reciprocal lattice vector

$$\mathbf{k}^{\mathrm{1st\ BZ}} = \mathbf{k} - \mathbf{G}_{i,j,k} , \tag{5.6}$$

where $\mathbf{G}_{i,j,k}$ points to the center of the BZ which contains the wave vector $\mathbf{k}$.

Due to additional symmetries of the crystal (inversion, rotation) the energy bands are invariant under the following eight spatial reflections

$$\varepsilon_n(k_x, k_y, k_z) = \varepsilon_n(|k_x|, |k_y|, |k_z|) \tag{5.7}$$

and six permutations

$$\begin{aligned}\varepsilon_n(k_x, k_y, k_z) &= \varepsilon_n(k_z, k_x, k_y) = \varepsilon_n(k_y, k_z, k_x) = \\ \varepsilon_n(k_x, k_z, k_y) &= \varepsilon_n(k_y, k_x, k_z) = \varepsilon_n(k_z, k_y, k_x) \; .\end{aligned} \tag{5.8}$$

The eight spatial reflections and the six permutations can be combined in any way resulting in 48 different symmetry operations. Thus, the energy band information is redundant in the full BZ and the BZ can be reduced to an irreducible wedge containing only 1/48th of the full BZ. The first irreducible wedge has six corners (Fig. 5.1)

$$\begin{aligned}\Gamma &= \frac{2\pi}{a_0^{\mathrm{RSi}}}\begin{pmatrix}0\\0\\0\end{pmatrix}, \quad L = \frac{2\pi}{a_0^{\mathrm{RSi}}}\begin{pmatrix}\frac{1}{2}\\\frac{1}{2}\\\frac{1}{2}\end{pmatrix}, \quad K = \frac{2\pi}{a_0^{\mathrm{RSi}}}\begin{pmatrix}\frac{3}{4}\\\frac{3}{4}\\0\end{pmatrix}, \\ W &= \frac{2\pi}{a_0^{\mathrm{RSi}}}\begin{pmatrix}1\\\frac{1}{2}\\0\end{pmatrix}, \quad X = \frac{2\pi}{a_0^{\mathrm{RSi}}}\begin{pmatrix}1\\0\\0\end{pmatrix}, \quad U = \frac{2\pi}{a_0^{\mathrm{RSi}}}\begin{pmatrix}1\\\frac{1}{4}\\\frac{1}{4}\end{pmatrix},\end{aligned} \tag{5.9}$$

and five faces

$$\begin{aligned}k_x + k_y + k_z &= \frac{3}{2}\frac{2\pi}{a_0^{\mathrm{RSi}}} , \\ k_x &= \frac{2\pi}{a_0^{\mathrm{RSi}}} , \\ k_x &= k_y , \\ k_y &= k_z , \\ k_z &= 0 .\end{aligned} \tag{5.10}$$

The first two planes are part of the surface of the BZ whereas the last three ones are inside of it.

The linear mapping corresponding to the symmetry within the BZ reads

$$\mathbf{k}' = \hat{T}_j\,\mathbf{k} \quad \text{with} \quad \varepsilon_n(\mathbf{k}') = \varepsilon_n(\mathbf{k}) , \tag{5.11}$$

where the matrix $\hat{T}_j$ can be any matrix of the above 48 operations. With this equation any wave vector $\mathbf{k}$ of the first irreducible wedge can be mapped into the jth irreducible wedge. The inverse transformation is given by

$$\mathbf{k} = \hat{T}_j^{\mathrm{T}}\,\mathbf{k}' \quad \text{with} \quad \hat{T}_j^{\mathrm{T}} = \hat{T}_j^{-1} . \tag{5.12}$$

The $\nabla_{\mathbf{k}}$-operator of the first irreducible wedge is transformed into the jth irreducible wedge by [5.4]

$$\nabla_{\mathbf{k}'} = \hat{T}_j \, \nabla_{\mathbf{k}} \ . \tag{5.13}$$

This results in the following transformation for the group velocity (2.6)

$$\mathbf{v}_n(\mathbf{k}') = \hat{T}_j \, \mathbf{v}_n(\mathbf{k}) \tag{5.14}$$

and the inverse mass tensor (2.7)

$$\hat{m}_n^{-1}(\mathbf{k}') = \hat{T}_j \, \hat{m}_n^{-1}(\mathbf{k}) \, \hat{T}_j^{\mathrm{T}} \ . \tag{5.15}$$

5.2 Basic Properties of the Band Structure of Strained SiGe

The lattice constants of strained $\mathrm{Si}_{1-x}\mathrm{Ge}_x$ pseudomorphically grown on relaxed $\mathrm{Si}_{1-y}\mathrm{Ge}_y$ differ in the growth direction $a_\perp(x_{\mathrm{Ge}}, y_{\mathrm{Ge}})$ and parallel to the SiGe layer $a_\parallel(x_{\mathrm{Ge}}, y_{\mathrm{Ge}})$ and their numerical values are given in [5.5, 5.6]. The distortion of the direct lattice compared to the one of RSi is given by linear mapping with the symmetric matrix

$$\hat{S}(x_{\mathrm{Ge}}, y_{\mathrm{Ge}}) = \begin{pmatrix} \frac{a_\perp(x_{\mathrm{Ge}}, y_{\mathrm{Ge}})}{a_0^{\mathrm{RSi}}} & 0 & 0 \\ 0 & \frac{a_\parallel(x_{\mathrm{Ge}}, y_{\mathrm{Ge}})}{a_0^{\mathrm{RSi}}} & 0 \\ 0 & 0 & \frac{a_\parallel(x_{\mathrm{Ge}}, y_{\mathrm{Ge}})}{a_0^{\mathrm{RSi}}} \end{pmatrix} , \tag{5.16}$$

and the basis vectors of the strained direct lattice read

$$\mathbf{a}_i(x_{\mathrm{Ge}}, y_{\mathrm{Ge}}) = \hat{S}(x_{\mathrm{Ge}}, y_{\mathrm{Ge}}) \mathbf{a}_i^{\mathrm{RSi}} \ . \tag{5.17}$$

The basis vectors of the reciprocal lattice are obtained by the inverse transformation

$$\mathbf{b}_i(x_{\mathrm{Ge}}, y_{\mathrm{Ge}}) = \hat{S}^{-1}(x_{\mathrm{Ge}}, y_{\mathrm{Ge}}) \mathbf{b}_i^{\mathrm{RSi}} \ . \tag{5.18}$$

Compared to the case of RSi the symmetry of the lattice is reduced by a factor of three. The energy is invariant only under two permutations

$$\varepsilon_n(k_x, k_y, k_z) = \varepsilon_n(k_x, k_z, k_y) \tag{5.19}$$

and the eight spatial reflections

$$\varepsilon_n(k_x, k_y, k_z) = \varepsilon_n(|k_x|, |k_y|, |k_z|) \ . \tag{5.20}$$

The irreducible wedge is therefore about three times as large as the one of RSi. It is given by the irreducible wedge of RSi and its two permutations (k_y, k_x, k_z),

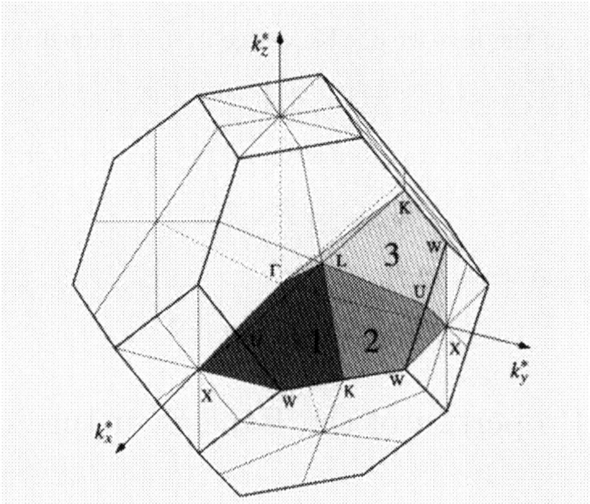

Figure 5.2: The first BZ of strained SiGe, its first irreducible wedge, and the three corresponding wedges of RSi

(k_z, k_x, k_y) and subsequent transformation with $\hat{S}^{-1}(x_{\mathrm{Ge}}, y_{\mathrm{Ge}})$. In order to simplify the calculations in the KS a transformed KS is introduced

$$\mathbf{k}' = \hat{S}(x_{\mathrm{Ge}}, y_{\mathrm{Ge}})\ \mathbf{k}\ , \tag{5.21}$$

which corresponds to the KS of RSi. The irreducible wedge is divided into three irreducible wedges of RSi (Fig. 5.2) and a wedge index w is defined

$$w = \begin{cases} 1 & \text{for the} \quad (k_x^*, k_y^*, k_z^*) \quad \text{permutation} \\ 2 & \text{for the} \quad (k_y^*, k_x^*, k_z^*) \quad \text{permutation} \\ 3 & \text{for the} \quad (k_z^*, k_x^*, k_y^*) \quad \text{permutation} \end{cases} . \tag{5.22}$$

All calculations are performed in the first irreducible wedge of RSi. To this end, the data of the two other wedges of RSi are transformed into the first wedge by the symmetry operation $\hat{T}_w^{\mathrm{T}}$

$$\begin{aligned} \hat{T}_1 &= \begin{pmatrix} 1 & 0 & 0 \\ 0 & 1 & 0 \\ 0 & 0 & 1 \end{pmatrix} , \\ \hat{T}_2 &= \begin{pmatrix} 0 & 1 & 0 \\ 1 & 0 & 0 \\ 0 & 0 & 1 \end{pmatrix} , \\ \hat{T}_3 &= \begin{pmatrix} 0 & 0 & 1 \\ 0 & 1 & 0 \\ 1 & 0 & 0 \end{pmatrix} . \end{aligned} \tag{5.23}$$

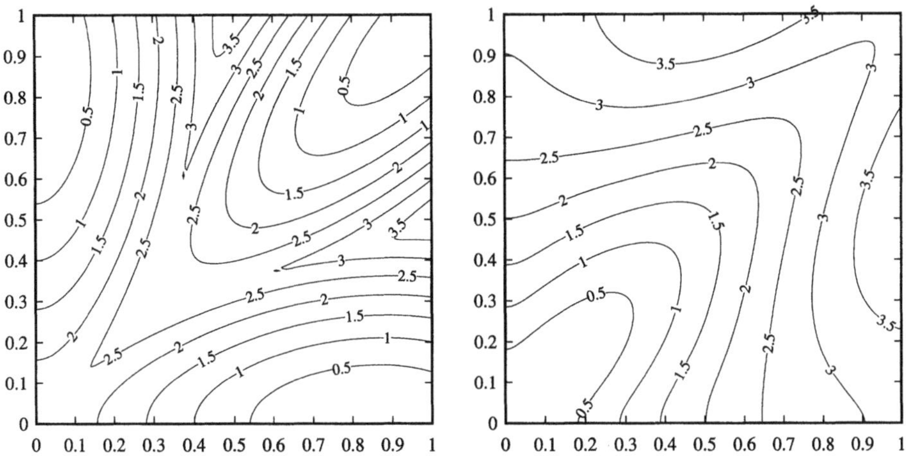

Figure 5.3: Contour plot of the energy for the first conduction band (left graph) and for the first valence band (right graph) in the k_x, k_y-plane for RSi (wave vector in units of $2\pi/a_0^{\mathrm{RSi}}$, energy in eV)

The relation between a wave vector **k** in the irreducible wedge of SiGe and a wave vector $\mathbf{k}^*$ in the first irreducible wedge of RSi is given by

$$\mathbf{k} = \hat{S}^{-1}(x_{\mathrm{Ge}}, y_{\mathrm{Ge}})\ \hat{T}_w\ \mathbf{k}^* \ . \tag{5.24}$$

5.3 k-Space Grid

The band energy of the first four conduction bands and the first three valence bands is calculated with the nonlocal empirical pseudopotential method [5.5–5.7]. In Fig. 5.3 contour plots of the energy for the first conduction and valence bands of RSi are shown. The complicated nature of the band structure does not allow for a simple analytical representation of the band energy. Therefore, the irreducible wedge is discretized and the band energy is interpolated over the resultant grid. Different schemes have been developed for the interpolation of the band structure [5.1, 5.8–5.21]. Here, piecewise linear interpolation in conjunction with a tetrahedral grid is used, because it makes the calculations in the KS numerically efficient, stable, and simple [5.8, 5.12].

The KS grid is generated only for the first irreducible wedge of RSi (Fig. 5.1), of which the shape is exactly resolved by the nonuniform tetrahedral grid and which is completely filled with tetrahedra. This grid is expanded to the case of the full irreducible wedge (Fig. 5.2) with Eq. (5.24). For the below calculations the parts of the energy bands within the second and third wedges are mapped onto the grid of the first wedge. The three different parts of the energy bands are distinguished within the first wedge by the wedge index w.

For each energy band an unstructured nonuniform tetrahedral grid is generated with a method described further below. The grid of the nth energy band has N_n^{nod} grid nodes and N_n^{tet} tetrahedra. The band energy of the kth grid node $\varepsilon_{n,w,k}^{\mathrm{nod}}$ is calculated with the nonlocal empirical pseudopotential method for the corresponding wave vector

$$\mathbf{k}_{n,w,k}^{\mathrm{nod}} = \hat{S}^{-1}(x_{\mathrm{Ge}}, y_{\mathrm{Ge}})\hat{T}_w \mathbf{k}_{n,k}^{*,\mathrm{nod}} . \tag{5.25}$$

The ith tetrahedron of the nth band is given by the four indices of the corresponding grid nodes $t(n,w,i,1)$, $t(n,w,i,2)$, $t(n,w,i,3)$ and $t(n,w,i,4)$ (Fig. 5.4). The four indices are sorted in such a way that the corresponding band energies are in ascending order $(\varepsilon_{n,w,t(n,w,i,1)}^{\mathrm{nod}} < \varepsilon_{n,w,t(n,w,i,2)}^{\mathrm{nod}} < \varepsilon_{n,w,t(n,w,i,3)}^{\mathrm{nod}} < \varepsilon_{n,w,t(n,w,i,4)}^{\mathrm{nod}})$[1,2]. Within a tetrahedron the energy is linearly interpolated

$$\varepsilon_n(\mathbf{k}) = \varepsilon_{n,w,t(n,w,i,1)}^{\mathrm{nod}} + (\mathbf{v}_{n,w,i}^{\mathrm{tet}})^{\mathrm{T}} \hbar(\mathbf{k} - \mathbf{k}_{n,w,t(n,w,i,1)}^{\mathrm{nod}}) . \tag{5.26}$$

The velocity $\mathbf{v}_{n,w,i}^{\mathrm{tet}}$ is calculated in such a way that the interpolated energy reproduces the exact band energy at the four corners of the tetrahedron

$$\begin{aligned}
\varepsilon_{n,w,t(n,w,i,2)}^{\mathrm{nod}} &= \varepsilon_{n,w,t(n,w,i,1)}^{\mathrm{nod}} \\
&\quad + (\mathbf{v}_{n,w,i}^{\mathrm{tet}})^{\mathrm{T}} \hbar(\mathbf{k}_{n,w,t(n,w,i,2)}^{\mathrm{nod}} - \mathbf{k}_{n,w,t(n,w,i,1)}^{\mathrm{nod}}) , \\
\varepsilon_{n,w,t(n,w,i,3)}^{\mathrm{nod}} &= \varepsilon_{n,w,t(n,w,i,1)}^{\mathrm{nod}} \\
&\quad + (\mathbf{v}_{n,w,i}^{\mathrm{tet}})^{\mathrm{T}} \hbar(\mathbf{k}_{n,w,t(n,w,i,3)}^{\mathrm{nod}} - \mathbf{k}_{n,w,t(n,w,i,1)}^{\mathrm{nod}}) , \\
\varepsilon_{n,w,t(n,w,i,4)}^{\mathrm{nod}} &= \varepsilon_{n,w,t(n,w,i,1)}^{\mathrm{nod}} \\
&\quad + (\mathbf{v}_{n,w,i}^{\mathrm{tet}})^{\mathrm{T}} \hbar(\mathbf{k}_{n,w,t(n,w,i,4)}^{\mathrm{nod}} - \mathbf{k}_{n,w,t(n,w,i,1)}^{\mathrm{nod}}) .
\end{aligned} \tag{5.27}$$

Each of the four faces of a tetrahedron is shared with an adjacent tetrahedron as long as the face does not belong to the surface of the irreducible wedge. Since the energies of the three grid nodes constituting the interface between two contiguous tetrahedra are the same for both tetrahedra, the energy interpolation is continuous within the irreducible wedge.

The error of the linear interpolation is minimized by adapting a nonuniform unstructured tetrahedral grid to each energy band of the band structure [5.12, 5.15, 5.17, 5.19]. Each grid is created starting with the six nodes of the irreducible wedge of RSi (5.9) plus the minimum of the first conduction band

[1]Degenerate energies are numerically removed by adding randomly small quantities of energy in the order of a few tenths of a μeV.

[2]Although the ith tetrahedron is the same for all three wedge indices, the different energies require an individual ordering of the tetrahedron's nodes for each wedge index.

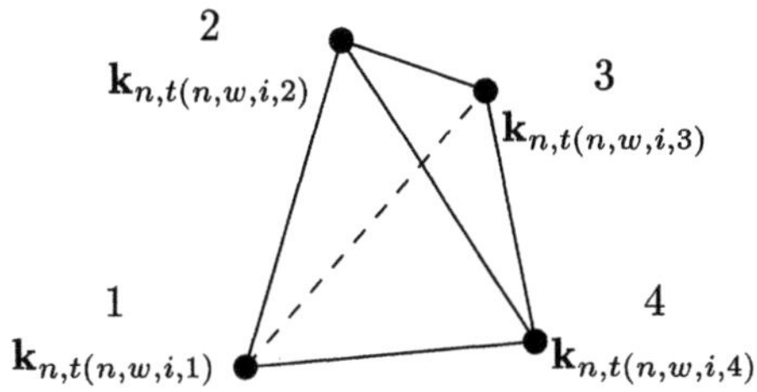

Figure 5.4: A tetrahedron of the **k**-space grid

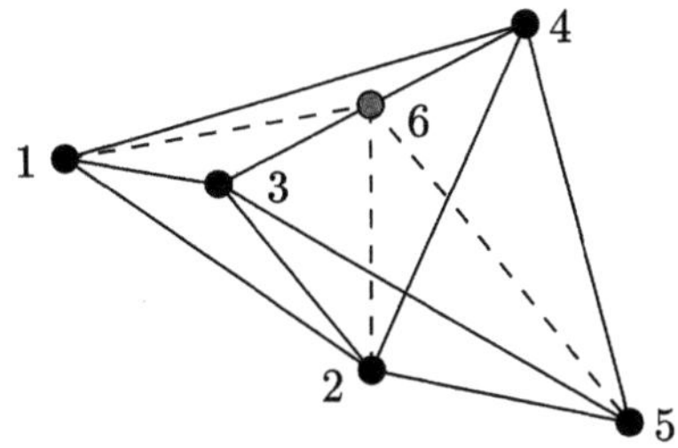

Figure 5.5: Splitting of an edge and creation of new tetrahedra

($M = 2\pi/a_0^{\mathrm{RSi}}(0.85,0,0)^{\mathrm{T}}$). The initial four tetrahedra are given in the transformed KS by

$$\{\Gamma, W, L, K\},\ \{\Gamma, W, L, M\}, \{X, W, L, M\},\ \{X, W, L, U\}\ . \tag{5.28}$$

Subsequently, the grid is refined by splitting edges of the tetrahedra. A new node is inserted in the middle of an edge, and all tetrahedra containing this particular edge are split. In Fig. 5.5 this is illustrated. The edge between grid point 3 and 4 is split by inserting the new grid point 6. This divides the two tetrahedra {1,2,3,4} and {2,3,4,5} into four new tetrahedra: {1,2,3,6}, {1,2,6,4}, {2,3,6,5} and {2,6,4,5}.

The edges to be split are selected according to the following rules. The error in the linear energy interpolation along an edge must be less than 2 % of the energy at the center of the edge or less than 50 μeV, whatever is larger. This criterion is checked for five points, which are uniformly distributed on the edge, and for all three wedges. Furthermore, the maximum length of an edge of a tetrahedron must be less than 0.1 times $2\pi/a_0^{\mathrm{RSi}}$ in the transformed KS. Among the edges violating these conditions the longest one is marked for splitting. Before the marked edge is split, it is checked, whether the volume quality of one of the tetrahedra containing this edge is less than 5 %. If this is the case, the longest edge of the tetrahedron with a volume quality of less than 5 % is split instead of the marked one. This ensures convergence of the algorithm and reduces the loss in volume quality. The volume quality $Q_{n,i}^{*,\mathrm{tet}}$ is defined in the transformed KS by

$$Q_{n,i}^{*,\mathrm{tet}} = 6\sqrt{2}\frac{V_{n,i}^{*,\mathrm{tet}}}{(h_{\mathrm{max},n,i}^{*,\mathrm{tet}})^3}\ , \tag{5.29}$$

where $h_{\mathrm{max},n,i}^{*,\mathrm{tet}}$ is the maximum edge length of the tetrahedron [5.15]. Its volume $V_{n,i}^{*,\mathrm{tet}}$ is given by a scalar triple product

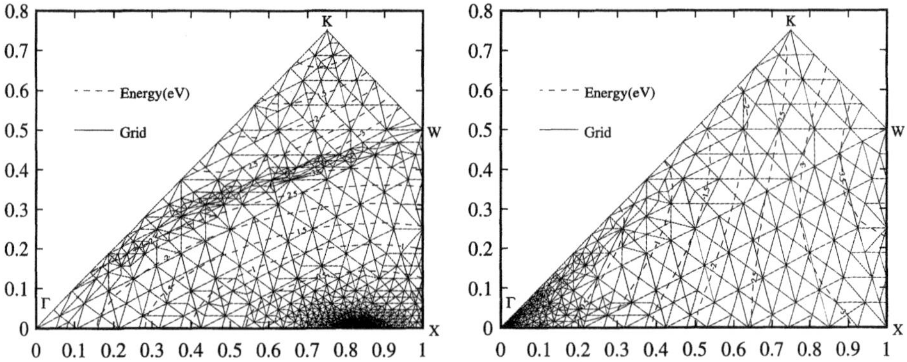

Figure 5.6: Nonuniform adaptive tetrahedral mesh of the first conduction band (left) and of the first valence band (right) in the k_x, k_y-plane for RSi (wave vector in units of $2\pi/a_0^{\mathrm{RSi}}$)

$$
\begin{aligned}
V_{n,i}^{*,\mathrm{tet}} = & \frac{1}{6}|(\mathbf{k}_{n,t(n,1,i,2)}^{*,\mathrm{nod}} - \mathbf{k}_{n,t(n,1,i,1)}^{*,\mathrm{nod}})^{\mathrm{T}}((\mathbf{k}_{n,t(n,1,i,3)}^{*,\mathrm{nod}} - \mathbf{k}_{n,t(n,1,i,1)}^{*,\mathrm{nod}}) \\
& \times(\mathbf{k}_{n,t(n,1,i,4)}^{*,\mathrm{nod}} - \mathbf{k}_{n,t(n,1,i,1)}^{*,\mathrm{nod}}))| \, , \qquad (5.30)
\end{aligned}
$$

where the volume does not depend on the wedge index and $w = 1$ has been used. The refinement is stopped when all edges satisfy the above criteria. The same grid is used for all configurations of x_{Ge} and y_{Ge}. In order to ensure that the grid yields an accurate interpolation for all technically important configurations, the grid is optimized simultaneously for ($x_{\mathrm{Ge}} = 0\%, y_{\mathrm{Ge}} = 0\%$), ($x_{\mathrm{Ge}} = 40\%, y_{\mathrm{Ge}} = 0\%$), and ($x_{\mathrm{Ge}} = 0\%, y_{\mathrm{Ge}} = 40\%$) [5.21].

In Fig. 5.6 the grids of the first conduction and valence band in the k_x, k_y-plane are shown. Due to the criterion for the relative error in energy the grids are extremely dense in the vicinity of the band minima. This is necessary for the calculation of low-field transport parameters at low temperatures ($\geq 50\,\mathrm{K}$) which are mainly determined by the band structure at low energies. The shortest edge of all tetrahedra is 0.0003 times $2\pi/a_0^{\mathrm{RSi}}$ while the longest one is 0.1 times $2\pi/a_0^{\mathrm{RSi}}$ demonstrating the flexibility of the grid generation. Another advantage of this kind of grid generation is that the refinement of the grid is locally contained as shown in Fig. 5.6. In Table 5.1 the number of nodes and tetrahedra per grid is given together with the average volume quality.

As mentioned above the energy interpolation is continuous within the first irreducible wedge (5.9). Due to the linear transformations (5.6) and (5.11) this is also the case in all other irreducible wedges. Between two irreducible wedges sharing an interface that lies within the BZ or contains the point X the energy is continuous, because the grid in the interface is invariant with respect to the symmetry operation mapping one of the two adjacent wedges into the other. This

Table 5.1: *Number of grid nodes N^{nod} and tetrahedra N^{tet} per conduction and valence band and the average volume quality $\overline{Q}^{tet}$*

Band	N^{nod}	N^{tet}	$\overline{Q}^{\mathrm{tet}}$
4c	2763	12439	0.2401
3c	2761	12499	0.2394
2c	3051	13494	0.2361
1c	6572	29735	0.2625
1v	3082	13111	0.2630
2v	2209	9354	0.2634
3v	1284	5226	0.2850

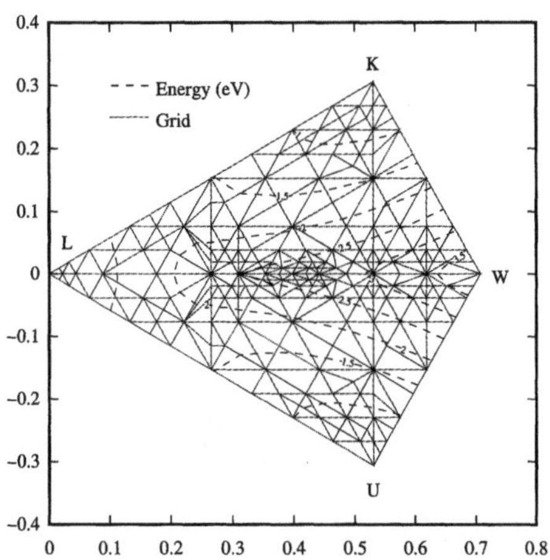

Figure 5.7: Symmetrical mesh of the $LUWK$-surface for the first conduction band of RSi (wave vector in units of $2\pi/a_0^{\mathrm{RSi}}$)

is not the case for the $LUWK$-surface. Although the wedges on both sides of the interface share the same L- and W-points, the U- and K-points are interchanged. Nevertheless the real band energy is continuous in this interface, because it is symmetric with respect to the LW-line. To reproduce this symmetry with the interpolated band energy the grid must be symmetric on the $LUWK$-surface with respect to the LW-line. The initial grid (5.28) has this symmetry property and whenever an edge in the $LUWK$-surface on one side of the LW-line is split, the corresponding edge on the other side is also split. The symmetrical grid of the $LUWK$-surface is shown in Fig. 5.7. The resultant grid of the irreducible wedge ensures a continuous energy interpolation over the whole KS [5.17, 5.19].

During the grid generation the band energy has to be evaluated very often and it is not possible to calculate the band energy with the very CPU intensive nonlocal empirical pseudopotential method. Instead, the band energy is calculated with the nonlocal empirical pseudopotential method for an equidistant grid with a spacing of 0.00833 times $2\pi/a_0^{\mathrm{RSi}}$ in the transformed KS and interpolated with the method described in Ref. [5.10]. Around each node of the equidistant grid the band energy is expanded into a Taylor series up to the second-order. The necessary derivatives (group velocity and inverse mass tensor) are calculated with finite differences. Within a cube of the equidistant grid the band energy is evaluated by interpolating the Taylor series expansions of the eight corners of the cube with trilinear weights. After the tetrahedral grid has been generated the band energy is calculated anew for each grid node with the nonlocal empirical pseudopotential method to remove errors by the above interpolation method.

5.4 Calculation of the Density of States

The DOS for one energy band and one spin direction reads [5.3]

$$\begin{aligned} D_n(\varepsilon) &= \frac{1}{(2\pi)^3} \int\limits_{BZ} \delta(\varepsilon - \varepsilon_n(\mathbf{k})) d^3k \\ &= \frac{16}{(2\pi)^3 \hbar} \int\limits_{A_n(\varepsilon)} \frac{d^2k}{|\mathbf{v}_n(\mathbf{k})|} , \end{aligned} \tag{5.31}$$

where $A_n(\varepsilon)$ is the equienergy surface for the given energy and band in the irreducible wedge. Due to the symmetry of the band structure it is sufficient to run the integral only over the irreducible wedge and the factor 16 accounts for all similar wedges in the BZ. Since the energy is linearly interpolated (5.26), the velocity is constant and the equienergy surface is plane within a tetrahedron. The contribution of the ith tetrahedron to the DOS is

$$D_{n,w,i}^{\mathrm{tet}}(\varepsilon) = \frac{1}{(2\pi)^3 \hbar} \frac{A_{n,w,i}^{\mathrm{tet}}(\varepsilon)}{|\mathbf{v}_{n,w,i}^{\mathrm{tet}}|} , \tag{5.32}$$

where $A_{n,w,i}^{\mathrm{tet}}(\varepsilon)$ is the area of the intersection of the equienergy surface and the ith tetrahedron in the nth energy band and wth part of the irreducible wedge. The area is zero for energies outside of the energy interval of the tetrahedron.

Depending on the energy the intersection is either a triangle or a quadrangle (Fig. 5.8) [5.8]. The points of intersection of the plane and the edges of the ith tetrahedron are

$$\begin{aligned} \mathbf{c}_{n,w,i,j,k}^{\mathrm{tet}}(\varepsilon) &= \mathbf{k}_{n,w,t(n,w,i,j)}^{\mathrm{nod}} \\ &+ \frac{\varepsilon - \varepsilon_{n,w,t(n,w,i,j)}^{\mathrm{nod}}}{(\mathbf{v}_{n,w,i}^{\mathrm{tet}})^{\mathrm{T}} \hbar \left(\mathbf{k}_{n,w,t(n,w,i,k)}^{\mathrm{nod}} - \mathbf{k}_{n,w,t(n,w,i,j)}^{\mathrm{nod}} \right)} \end{aligned}$$

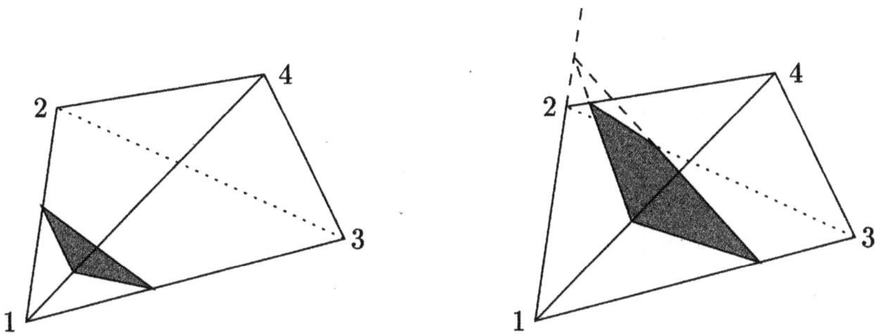

Figure 5.8: The different intersections of a plane with a tetrahedron

$$\left(\mathbf{k}^{\mathrm{nod}}_{n,w,t(n,w,i,k)} - \mathbf{k}^{\mathrm{nod}}_{n,w,t(n,w,i,j)}\right) , \tag{5.33}$$

with $j, k = 1, 2, 3, 4$, $j \neq k$, and $\mathbf{c}^{\mathrm{tet}}_{n,w,i,j,k}(\varepsilon) = \mathbf{c}^{\mathrm{tet}}_{n,w,i,k,j}(\varepsilon)$.

For energies between $\varepsilon^{\mathrm{nod}}_{n,w,t(n,w,i,1)}$ and $\varepsilon^{\mathrm{nod}}_{n,w,t(n,w,i,2)}$ the intersection is given by the triangle $\{\mathbf{c}^{\mathrm{tet}}_{n,w,i,1,2}(\varepsilon), \mathbf{c}^{\mathrm{tet}}_{n,w,i,1,3}(\varepsilon), \mathbf{c}^{\mathrm{tet}}_{n,w,i,1,4}(\varepsilon)\}$ (left graph in Fig. 5.8) of which the area is

$$\begin{aligned} A^{\mathrm{tet}}_{n,w,i}(\varepsilon) &= \frac{1}{2}\left|\left(\mathbf{c}^{\mathrm{tet}}_{n,w,i,1,3}(\varepsilon) - \mathbf{c}^{\mathrm{tet}}_{n,w,i,1,2}(\varepsilon)\right) \times \left(\mathbf{c}^{\mathrm{tet}}_{n,w,i,1,4}(\varepsilon) - \mathbf{c}^{\mathrm{tet}}_{n,w,i,1,2}(\varepsilon)\right)\right| \\ &= \alpha^{\mathrm{tet}}_{n,w,i}\left(\varepsilon - \varepsilon^{\mathrm{nod}}_{n,w,t(n,w,i,1)}\right)^2 . \end{aligned} \tag{5.34}$$

Between $\varepsilon^{\mathrm{nod}}_{n,w,t(n,w,i,2)}$ and $\varepsilon^{\mathrm{nod}}_{n,w,t(n,w,i,3)}$ the intersection is a quadrangle (right graph in Fig. 5.8). Its area is obtained by subtracting the area of the triangle $\{\mathbf{c}^{\mathrm{tet}}_{n,w,i,2,1}(\varepsilon), \mathbf{c}^{\mathrm{tet}}_{n,w,i,2,3}(\varepsilon), \mathbf{c}^{\mathrm{tet}}_{n,w,i,2,4}(\varepsilon)\}$ from the area of the triangle $\{\mathbf{c}^{\mathrm{tet}}_{n,w,i,1,2}(\varepsilon), \mathbf{c}^{\mathrm{tet}}_{n,w,i,1,3}(\varepsilon), \mathbf{c}^{\mathrm{tet}}_{n,w,i,1,4}(\varepsilon)\}$

$$\begin{aligned} A^{\mathrm{tet}}_{n,w,i}(\varepsilon) &= \alpha^{\mathrm{tet}}_{n,w,i}\left(\varepsilon - \varepsilon^{\mathrm{nod}}_{n,w,t(n,w,i,1)}\right)^2 \\ &\quad -\frac{1}{2}\left|\left(\mathbf{c}^{\mathrm{tet}}_{n,w,i,2,3}(\varepsilon) - \mathbf{c}^{\mathrm{tet}}_{n,w,i,2,1}(\varepsilon)\right) \times \left(\mathbf{c}^{\mathrm{tet}}_{n,w,i,2,4}(\varepsilon) - \mathbf{c}^{\mathrm{tet}}_{n,w,i,2,1}(\varepsilon)\right)\right| \\ &= \alpha^{\mathrm{tet}}_{n,w,i}\left(\varepsilon - \varepsilon^{\mathrm{nod}}_{n,w,t(n,w,i,1)}\right)^2 - \beta^{\mathrm{tet}}_{n,w,i}\left(\varepsilon - \varepsilon^{\mathrm{nod}}_{n,w,t(n,w,i,2)}\right)^2 . \end{aligned} \tag{5.35}$$

The case of energies between $\varepsilon^{\mathrm{nod}}_{n,w,t(n,w,i,3)}$ and $\varepsilon^{\mathrm{nod}}_{n,w,t(n,w,i,4)}$ is similar to the first case. The intersection is given by the triangle $\{\mathbf{c}^{\mathrm{tet}}_{n,w,i,4,1}(\varepsilon), \mathbf{c}^{\mathrm{tet}}_{n,w,i,4,2}(\varepsilon), \mathbf{c}^{\mathrm{tet}}_{n,w,i,4,3}(\varepsilon)\}$ (left graph in Fig. 5.8) of which the area is

$$\begin{aligned} A^{\mathrm{tet}}_{n,w,i}(\varepsilon) &= \frac{1}{2}\left|\left(\mathbf{c}^{\mathrm{tet}}_{n,w,i,4,3}(\varepsilon) - \mathbf{c}^{\mathrm{tet}}_{n,w,i,4,1}(\varepsilon)\right) \times \left(\mathbf{c}^{\mathrm{tet}}_{n,w,i,4,2}(\varepsilon) - \mathbf{c}^{\mathrm{tet}}_{n,w,i,4,1}(\varepsilon)\right)\right| \\ &= \gamma^{\mathrm{tet}}_{n,w,i}\left(\varepsilon - \varepsilon^{\mathrm{nod}}_{n,w,t(n,w,i,4)}\right)^2 . \end{aligned} \tag{5.36}$$

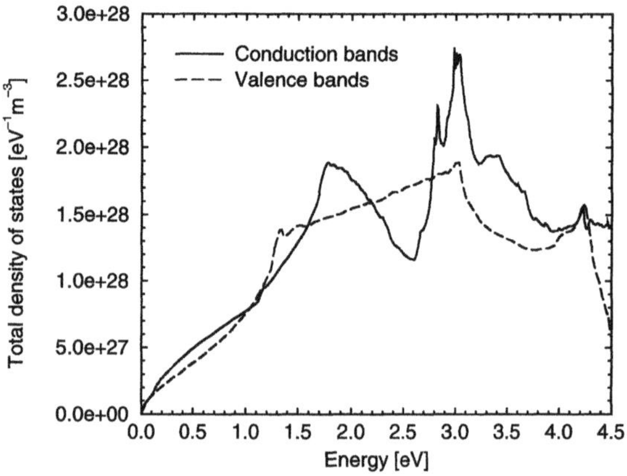

Figure 5.9: Total density of states for one spin direction in RSi

For the calculation of the area of the intersection only the coefficients $\alpha_{n,w,i}^{\text{tet}}$, $\beta_{n,w,i}^{\text{tet}}$ and $\gamma_{n,w,i}^{\text{tet}}$ are required which have to be calculated only once. The DOS now reads

$$D_n(\varepsilon) = \sum_{w=1}^{3} D_{n,w}(\varepsilon) \tag{5.37}$$

with

$$D_{n,w}(\varepsilon) = 16 \sum_{i=1}^{N_n^{\text{tet}}} D_{n,w,i}^{\text{tet}}(\varepsilon) \ . \tag{5.38}$$

Thus, the evaluation of the DOS is simple and CPU efficient. Furthermore, it is exact and consistent with the interpolation of the band energy in contrast to other methods [5.10,5.14,5.22]. The DOS summed over all conduction and valence bands, respectively, is shown in Fig. 5.9 for RSi.

The effective DOS in the conduction bands can be evaluated by [5.2]

$$N_{\text{c}} = \sum_{n \in \text{c.b.}} \int_0^\infty D_n(\varepsilon) \exp\left(-\frac{\varepsilon}{k_{\text{B}} T_0}\right) d\varepsilon \ , \tag{5.39}$$

where the sum extends over all conduction bands. In the case of the valence bands it reads

$$N_{\text{v}} = \sum_{n \in \text{v.b.}} \int_0^\infty D_n(\varepsilon) \exp\left(-\frac{\varepsilon}{k_{\text{B}} T_0}\right) d\varepsilon \ , \tag{5.40}$$

and the sum extends over all valence bands. The intrinsic carrier density is given by

$$n_{\text{i}} = \sqrt{N_{\text{c}} N_{\text{v}}} \exp\left(-\frac{E_{\text{c}} - E_{\text{v}}}{2 k_{\text{B}} T_0}\right) \ . \tag{5.41}$$

5.5 Mass Tensor Evaluation

The inverse mass tensor (2.7) is the derivative of the group velocity (2.6) in KS [5.3]. Due to the linear interpolation (5.26) the velocity is constant within a tetrahedron and discontinuous between two contiguous tetrahedra. The inverse mass tensor is therefore Dirac-function-like. To circumvent the problem of sampling such a kind of quantity, the inverse mass tensor $\hat{m}_n^{-1}(\mathbf{k})$ is averaged over the volume

$$\begin{aligned} V_{n,w,i}^{\text{tet}} &= \frac{1}{6}|(\mathbf{k}_{n,w,t(n,w,i,2)}^{\text{nod}} - \mathbf{k}_{n,w,t(n,w,i,1)}^{\text{nod}})^{\text{T}}((\mathbf{k}_{n,w,t(n,w,i,3)}^{\text{nod}} - \mathbf{k}_{n,w,t(n,w,i,1)}^{\text{nod}}) \\ &\times (\mathbf{k}_{n,w,t(n,w,i,4)}^{\text{nod}} - \mathbf{k}_{n,w,t(n,w,i,1)}^{\text{nod}}))| \end{aligned} \quad (5.42)$$

of the ith tetrahedron resulting in a constant inverse mass tensor $(\hat{m}_{n,w,i}^{\text{tet}})^{-1}$ for each tetrahedron

$$(\hat{m}_{n,w,i}^{\text{tet}})^{-1} = \frac{1}{V_{n,w,i}^{\text{tet}}} \int\limits_{V_{n,w,i}^{\text{tet}}} \hat{m}_n^{-1}(\mathbf{k})\, d^3k\ . \quad (5.43)$$

Based on the theorem of Gauss [5.23] it can be shown that the integral of the inverse mass tensor over the volume of the tetrahedron equals the integral of the group velocity over the surface $S_{n,w,i}^{\text{tet}}$ of the tetrahedron:

$$\int\limits_{V_{n,w,i}^{\text{tet}}} \hat{m}_n^{-1}(\mathbf{k})\, d^3k = \frac{1}{\hbar} \oint\limits_{S_{n,w,i}^{\text{tet}}} \mathbf{v}_n(\mathbf{k})\, d^2\mathbf{k}^{\text{T}}\ . \quad (5.44)$$

Each tetrahedron has four adjacent tetrahedra with which it shares an interface. Since the velocity is discontinuous in these interfaces, the surface integral is evaluated with the arithmetical mean of the velocities on both sides of the interface:

$$\oint\limits_{S_{n,w,i}^{\text{tet}}} \mathbf{v}_n(\mathbf{k})\, d^2\mathbf{k}^{\text{T}} = \sum_{k=1}^{4} \frac{\mathbf{v}_{n,w,j(n,w,i,k)}^{\text{tet}} + \mathbf{v}_{n,w,i}^{\text{tet}}}{2} (\mathbf{S}_{n,w,i,k}^{\text{tet}})^{\text{T}}\ , \quad (5.45)$$

where $j(n,w,i,k)$ is the index of the adjacent tetrahedron that does not contain the grid node $t(n,w,i,k)$. The outwardly facing surface area vectors $\mathbf{S}_{n,w,i,k}^{\text{tet}}$ read

$$\mathbf{S}_{n,w,i,1}^{\text{tet}} = -\frac{1}{2} \frac{(\mathbf{k}_{n,w,t(n,w,i,1)}^{\text{nod}} - \mathbf{k}_{n,w,t(n,w,i,2)}^{\text{nod}})^{\text{T}} \mathbf{n}_{n,w,i,1}^{\text{tet}}}{|(\mathbf{k}_{n,w,t(n,w,i,1)}^{\text{nod}} - \mathbf{k}_{n,w,t(n,w,i,2)}^{\text{nod}})^{\text{T}} \mathbf{n}_{n,w,i,1}^{\text{tet}}|} \mathbf{n}_{n,w,i,1}^{\text{tet}}\ ,$$

$$\mathbf{S}_{n,w,i,2}^{\text{tet}} = -\frac{1}{2} \frac{(\mathbf{k}_{n,w,t(n,w,i,2)}^{\text{nod}} - \mathbf{k}_{n,w,t(n,w,i,3)}^{\text{nod}})^{\text{T}} \mathbf{n}_{n,w,i,2}^{\text{tet}}}{|(\mathbf{k}_{n,w,t(n,w,i,2)}^{\text{nod}} - \mathbf{k}_{n,w,t(n,w,i,3)}^{\text{nod}})^{\text{T}} \mathbf{n}_{n,w,i,2}^{\text{tet}}|} \mathbf{n}_{n,w,i,2}^{\text{tet}}\ ,$$

$$\mathbf{S}^{\text{tet}}_{n,w,i,3} = -\frac{1}{2}\frac{\left(\mathbf{k}^{\text{nod}}_{n,w,t(n,w,i,3)} - \mathbf{k}^{\text{nod}}_{n,w,t(n,w,i,4)}\right)^{\mathrm{T}}\mathbf{n}^{\text{tet}}_{n,w,i,3}}{\left|\left(\mathbf{k}^{\text{nod}}_{n,w,t(n,w,i,3)} - \mathbf{k}^{\text{nod}}_{n,w,t(n,w,i,4)}\right)^{\mathrm{T}}\mathbf{n}^{\text{tet}}_{n,w,i,3}\right|}\mathbf{n}^{\text{tet}}_{n,w,i,3} ,$$

$$\mathbf{S}^{\text{tet}}_{n,w,i,4} = -\frac{1}{2}\frac{\left(\mathbf{k}^{\text{nod}}_{n,w,t(n,w,i,4)} - \mathbf{k}^{\text{nod}}_{n,w,t(n,w,i,1)}\right)^{\mathrm{T}}\mathbf{n}^{\text{tet}}_{n,w,i,4}}{\left|\left(\mathbf{k}^{\text{nod}}_{n,w,t(n,w,i,4)} - \mathbf{k}^{\text{nod}}_{n,w,t(n,w,i,1)}\right)^{\mathrm{T}}\mathbf{n}^{\text{tet}}_{n,w,i,4}\right|}\mathbf{n}^{\text{tet}}_{n,w,i,1} , \quad (5.46)$$

where the normal vectors are

$$\begin{aligned}
\mathbf{n}^{\text{tet}}_{n,w,i,1} &= \left(\mathbf{k}^{\text{nod}}_{n,w,t(n,w,i,2)} - \mathbf{k}^{\text{nod}}_{n,w,t(n,w,i,3)}\right) \\
&\quad \times\left(\mathbf{k}^{\text{nod}}_{n,w,t(n,w,i,3)} - \mathbf{k}^{\text{nod}}_{n,w,t(n,w,i,4)}\right) , \\
\mathbf{n}^{\text{tet}}_{n,w,i,2} &= \left(\mathbf{k}^{\text{nod}}_{n,w,t(n,w,i,3)} - \mathbf{k}^{\text{nod}}_{n,w,t(n,w,i,4)}\right) \\
&\quad \times\left(\mathbf{k}^{\text{nod}}_{n,w,t(n,w,i,4)} - \mathbf{k}^{\text{nod}}_{n,w,t(n,w,i,1)}\right) , \\
\mathbf{n}^{\text{tet}}_{n,w,i,3} &= \left(\mathbf{k}^{\text{nod}}_{n,w,t(n,w,i,4)} - \mathbf{k}^{\text{nod}}_{n,w,t(n,w,i,1)}\right) \\
&\quad \times\left(\mathbf{k}^{\text{nod}}_{n,w,t(n,w,i,1)} - \mathbf{k}^{\text{nod}}_{n,w,t(n,w,i,2)}\right) , \\
\mathbf{n}^{\text{tet}}_{n,w,i,4} &= \left(\mathbf{k}^{\text{nod}}_{n,w,t(n,w,i,1)} - \mathbf{k}^{\text{nod}}_{n,w,t(n,w,i,2)}\right) \\
&\quad \times\left(\mathbf{k}^{\text{nod}}_{n,w,t(n,w,i,2)} - \mathbf{k}^{\text{nod}}_{n,w,t(n,w,i,3)}\right) . \qquad (5.47)
\end{aligned}$$

This results in a symmetric inverse mass tensor consistent with the nonuniformly discretized band structure for each tetrahedron

$$(\hat{m}^{\text{tet}}_{n,w,i})^{-1} = \frac{1}{\hbar V^{\text{tet}}_{n,w,i}}\sum_{k=1}^{4}\frac{\mathbf{v}^{\text{tet}}_{n,w,j(n,w,i,k)} + \mathbf{v}^{\text{tet}}_{n,w,i}}{2}\mathbf{S}^{\mathrm{T}}_{n,w,i,k} . \qquad (5.48)$$

In Fig. 5.10 the trace of the inverse mass tensor times 1/3 averaged over an equienergy surface and the first four conduction bands and the first three valence bands of RSi is shown.

5.6 Particle Motion in Phase-Space

During the MC simulation a particle is accelerated by an external force (2.5) according to Newton's second law of motion (2.3). Assuming that the force is constant within a PS simplex, Eq. (2.3) can be integrated within a tetrahedron

$$\hbar\mathbf{k}_l(t) = \hbar\mathbf{k}_l(t_0) + \mathbf{F}(t - t_0) , \qquad (5.49)$$

where $\mathbf{k}_l$ is the lth particle's wave vector. In order to calculate the time until the next transition from one tetrahedron into another one Eq. (5.49) is first transformed into the KS of RSi with Eq. (5.21)

$$\hbar\mathbf{k}'_l(t) = \hbar\mathbf{k}'_l(t_0) + \mathbf{F}'(t - t_0) \qquad (5.50)$$

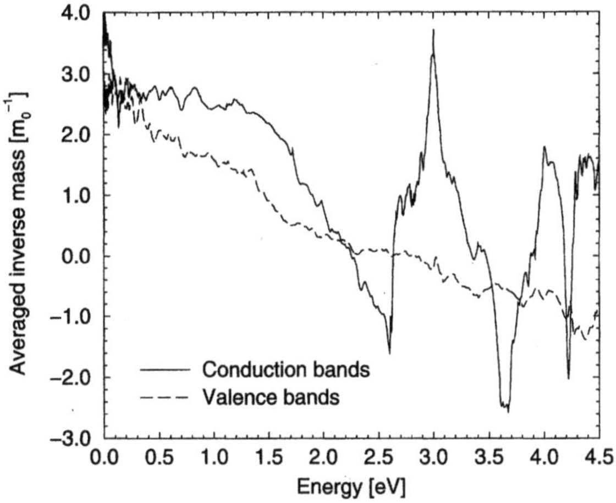

Figure 5.10: Inverse mass averaged over an equienergy surface, all three directions and the first four conduction bands or the first three valence bands of RSi

with

$$\mathbf{F}' = \hat{S}(x_{\mathrm{Ge}}, y_{\mathrm{Ge}})\mathbf{F} \ , \tag{5.51}$$

and then transformed from the jth irreducible wedge of RSi into the first irreducible wedge with Eq. (5.12), because the grid data is only given for this wedge

$$\begin{aligned} \hbar\mathbf{k}_l^*(t) &= \hat{T}_j^{\mathrm{T}}\hbar\mathbf{k}_l'(t) = \hat{T}_j^{\mathrm{T}}\hbar\mathbf{k}_l'(t_0) + \hat{T}_j^{\mathrm{T}}\mathbf{F}'(t-t_0) \\ &= \hbar\mathbf{k}_l^*(t_0) + \mathbf{F}^*(t-t_0) \ , \end{aligned} \tag{5.52}$$

where it is assumed that the particle is in the jth irreducible wedge. Since the wave vector (5.52) is linear in time and the four faces of the tetrahedron are planes, the calculation of the time, when the particle path and one of the tetrahedron's faces intersect, is straight forward. Based on the Hessian forms of the four faces these times are

$$t_1 - t_0 = -\frac{(\mathbf{S}_{n,i,1}^{*,\mathrm{tet}})^{\mathrm{T}}\hbar\left(\mathbf{k}_l^*(t_0) - \mathbf{k}_{n,i,2}^{*,\mathrm{nod}}\right)}{(\mathbf{S}_{n,i,1}^{*,\mathrm{tet}})^{\mathrm{T}}\mathbf{F}^*} \ , \tag{5.53}$$

$$t_2 - t_0 = -\frac{(\mathbf{S}_{n,i,2}^{*,\mathrm{tet}})^{\mathrm{T}}\hbar\left(\mathbf{k}_l^*(t_0) - \mathbf{k}_{n,i,3}^{*,\mathrm{nod}}\right)}{(\mathbf{S}_{n,i,2}^{*,\mathrm{tet}})^{\mathrm{T}}\mathbf{F}^*} \ , \tag{5.54}$$

$$t_3 - t_0 = -\frac{(\mathbf{S}_{n,i,3}^{*,\mathrm{tet}})^{\mathrm{T}}\hbar\left(\mathbf{k}_l^*(t_0) - \mathbf{k}_{n,i,4}^{*,\mathrm{nod}}\right)}{(\mathbf{S}_{n,i,3}^{*,\mathrm{tet}})^{\mathrm{T}}\mathbf{F}^*} \ , \tag{5.55}$$

$$t_4 - t_0 = -\frac{(\mathbf{S}_{n,i,4}^{*,\mathrm{tet}})^{\mathrm{T}}\hbar\left(\mathbf{k}_l^*(t_0) - \mathbf{k}_{n,i,1}^{*,\mathrm{nod}}\right)}{(\mathbf{S}_{n,i,4}^{*,\mathrm{tet}})^{\mathrm{T}}\mathbf{F}^*} \tag{5.56}$$

with

$$\begin{aligned}
\mathbf{S}^{*,\mathrm{tet}}_{n,i,1} &= -\frac{1}{2}\frac{(\mathbf{k}^{*,\mathrm{nod}}_{n,t(n,1,i,1)}-\mathbf{k}^{*,\mathrm{nod}}_{n,t(n,1,i,2)})^{\mathrm{T}}\mathbf{n}^{*,\mathrm{tet}}_{n,i,1}}{|(\mathbf{k}^{*,\mathrm{nod}}_{n,t(n,1,i,1)}-\mathbf{k}^{*,\mathrm{nod}}_{n,t(n,1,i,2)})^{\mathrm{T}}\mathbf{n}^{*,\mathrm{tet}}_{n,i,1}|}\mathbf{n}^{*,\mathrm{tet}}_{n,i,1}\ , \\
\mathbf{S}^{*,\mathrm{tet}}_{n,i,2} &= -\frac{1}{2}\frac{(\mathbf{k}^{*,\mathrm{nod}}_{n,t(n,1,i,2)}-\mathbf{k}^{*,\mathrm{nod}}_{n,t(n,1,i,3)})^{\mathrm{T}}\mathbf{n}^{*,\mathrm{tet}}_{n,i,2}}{|(\mathbf{k}^{*,\mathrm{nod}}_{n,t(n,1,i,2)}-\mathbf{k}^{*,\mathrm{nod}}_{n,t(n,1,i,3)})^{\mathrm{T}}\mathbf{n}^{*,\mathrm{tet}}_{n,i,2}|}\mathbf{n}^{*,\mathrm{tet}}_{n,i,2}\ , \\
\mathbf{S}^{*,\mathrm{tet}}_{n,i,3} &= -\frac{1}{2}\frac{(\mathbf{k}^{*,\mathrm{nod}}_{n,t(n,1,i,3)}-\mathbf{k}^{*,\mathrm{nod}}_{n,t(n,1,i,4)})^{\mathrm{T}}\mathbf{n}^{*,\mathrm{tet}}_{n,i,3}}{|(\mathbf{k}^{*,\mathrm{nod}}_{n,t(n,1,i,3)}-\mathbf{k}^{*,\mathrm{nod}}_{n,t(n,1,i,4)})^{\mathrm{T}}\mathbf{n}^{*,\mathrm{tet}}_{n,i,3}|}\mathbf{n}^{*,\mathrm{tet}}_{n,i,3}\ , \\
\mathbf{S}^{*,\mathrm{tet}}_{n,i,4} &= -\frac{1}{2}\frac{(\mathbf{k}^{*,\mathrm{nod}}_{n,t(n,1,i,4)}-\mathbf{k}^{*,\mathrm{nod}}_{n,t(n,1,i,1)})^{\mathrm{T}}\mathbf{n}^{*,\mathrm{tet}}_{n,i,4}}{|(\mathbf{k}^{*,\mathrm{nod}}_{n,t(n,1,i,4)}-\mathbf{k}^{*,\mathrm{nod}}_{n,t(n,1,i,1)})^{\mathrm{T}}\mathbf{n}^{*,\mathrm{tet}}_{n,i,4}|}\mathbf{n}^{*,\mathrm{tet}}_{n,i,1}\ ,
\end{aligned} \tag{5.57}$$

and

$$\begin{aligned}
\mathbf{n}^{*,\mathrm{tet}}_{n,i,1} &= (\mathbf{k}^{*,\mathrm{nod}}_{n,t(n,1,i,2)}-\mathbf{k}^{*,\mathrm{nod}}_{n,t(n,1,i,3)})\times(\mathbf{k}^{*,\mathrm{nod}}_{n,t(n,1,i,3)}-\mathbf{k}^{*,\mathrm{nod}}_{n,t(n,1,i,4)})\ , \\
\mathbf{n}^{*,\mathrm{tet}}_{n,i,2} &= (\mathbf{k}^{*,\mathrm{nod}}_{n,t(n,1,i,3)}-\mathbf{k}^{*,\mathrm{nod}}_{n,t(n,1,i,4)})\times(\mathbf{k}^{*,\mathrm{nod}}_{n,t(n,1,i,4)}-\mathbf{k}^{*,\mathrm{nod}}_{n,t(n,1,i,1)})\ , \\
\mathbf{n}^{*,\mathrm{tet}}_{n,i,3} &= (\mathbf{k}^{*,\mathrm{nod}}_{n,t(n,1,i,4)}-\mathbf{k}^{*,\mathrm{nod}}_{n,t(n,1,i,1)})\times(\mathbf{k}^{*,\mathrm{nod}}_{n,t(n,1,i,1)}-\mathbf{k}^{*,\mathrm{nod}}_{n,t(n,1,i,2)})\ , \\
\mathbf{n}^{*,\mathrm{tet}}_{n,i,4} &= (\mathbf{k}^{*,\mathrm{nod}}_{n,t(n,1,i,1)}-\mathbf{k}^{*,\mathrm{nod}}_{n,t(n,1,i,2)})\times(\mathbf{k}^{*,\mathrm{nod}}_{n,t(n,1,i,2)}-\mathbf{k}^{*,\mathrm{nod}}_{n,t(n,1,i,3)})\ .
\end{aligned} \tag{5.58}$$

The smallest positive solution of Eqs. (5.53)-(5.56) gives the time, when the particle moves into the next tetrahedron through the respective interface. To avoid a search in the unstructured grid for the adjacent tetrahedron a list is used which contains the four contiguous tetrahedra for each tetrahedron. In the case that the particle moves through an interface being part of the surface of the BZ, the particle's wave vector is folded back into the first BZ with Eq. (5.6).

The motion of a particle in the RS is described by Eq. (2.4). Because of the constant group velocity within a tetrahedron of the KS a particle moves along a path consisting of pieces of straight lines in analogy to Eq. (5.49)

$$\mathbf{r}_l(t) = \mathbf{r}_l(t_0) + \hat{T}_j \mathbf{v}^{\mathrm{tet}}_{n,w,i}(t-t_0)\ , \tag{5.59}$$

where $\mathbf{r}_l$ is the lth particle's position in the RS and j the index of the wedge containing $\mathbf{k}'$. Above it was assumed that the force is constant. This is only the case within a primitive of the RS grid (cf. Sec. 6.1). Whenever a particles arrives at an interface in the RS the particle is stopped and the forces are updated on entering the new RS primitive. The primitives of the RS are rectangles and the time of intersection of the particle path and the rectangle's edge is the smallest positive solution of the following four equations

$$t_5 - t_0 = \frac{x_m - x_l(t_0)}{\mathbf{e}_x^{\mathrm{T}} \hat{T}_j \mathbf{v}_{n,w,i}^{\mathrm{tet}}} , \tag{5.60}$$

$$t_6 - t_0 = \frac{x_{m+1} - x_l(t_0)}{\mathbf{e}_x^{\mathrm{T}} \hat{T}_j \mathbf{v}_{n,w,i}^{\mathrm{tet}}} , \tag{5.61}$$

$$t_7 - t_0 = \frac{y_k - y_l(t_0)}{\mathbf{e}_y^{\mathrm{T}} \hat{T}_j \mathbf{v}_{n,w,i}^{\mathrm{tet}}} , \tag{5.62}$$

$$t_8 - t_0 = \frac{y_{k+1} - y_l(t_0)}{\mathbf{e}_y^{\mathrm{T}} \hat{T}_j \mathbf{v}_{n,w,i}^{\mathrm{tet}}} , \tag{5.63}$$

where the particle is in the rectangle $\Omega_{m,k}$ of the RS grid and x_m, y_k are the corresponding grid lines.

The combined particle motion within a PS simplex (Eqs. (5.49), (5.59)) is linear and the time for the next change of the PS primitive is given by the smallest positive solution of Eqs. (5.53)-(5.56) and (5.60)-(5.63). The resultant piecewise linear particle path in PS is the exact solution of the equations of motion. Thus, the problems related to higher order interpolation schemes in conjunction with numerical integration (e.g. [5.10]) are avoided and the method is highly CPU efficient [5.12, 5.13].

At certain boundaries in the RS the particles are reflected. During the reflection the quasi momentum parallel to the interface and the energy of the particle should be conserved

$$\mathbf{k}^{\mathrm{a}}(\lambda) = \mathbf{k}^{\mathrm{b}} + \lambda \mathbf{n} \quad \wedge \quad \varepsilon_n(\mathbf{k}^{\mathrm{a}}(\lambda)) = \varepsilon_n(\mathbf{k}^{\mathrm{b}}) , \tag{5.64}$$

where $\mathbf{k}^{\mathrm{b}}$ is the wave vector before the reflection and $\mathbf{k}^{\mathrm{a}}$ afterwards. The vector $\mathbf{n}$ is perpendicular to the interface and of unit length. The velocity should be altered in such a way that the particle does not cross the boundary

$$\left(\mathbf{v}_n^{\mathrm{T}}(\mathbf{k}^{\mathrm{a}}(\lambda))\mathbf{n}\right) \left(\mathbf{v}_n^{\mathrm{T}}(\mathbf{k}^{\mathrm{b}})\mathbf{n}\right) < 0 . \tag{5.65}$$

One way to satisfy these constraints is specular reflection. The component of the particle's wave vector perpendicular to the interface is inverted (Fig. 5.11, example a)

$$\mathbf{k}^{\mathrm{a}}\left(-2\mathbf{n}^{\mathrm{T}}\mathbf{k}^{\mathrm{b}}\right) = \mathbf{k}^{\mathrm{b}} - 2\left(\mathbf{n}^{\mathrm{T}}\mathbf{k}^{\mathrm{b}}\right)\mathbf{n} . \tag{5.66}$$

Due to the symmetry of the band structure the energy of the particle is conserved and the component of the particle's velocity perpendicular to the interface is inverted.

In the case of a complex band structure Eqs. (5.64) and (5.65) have more than one root (Fig. 5.11, example b). In analogy to classical mechanics the particle is

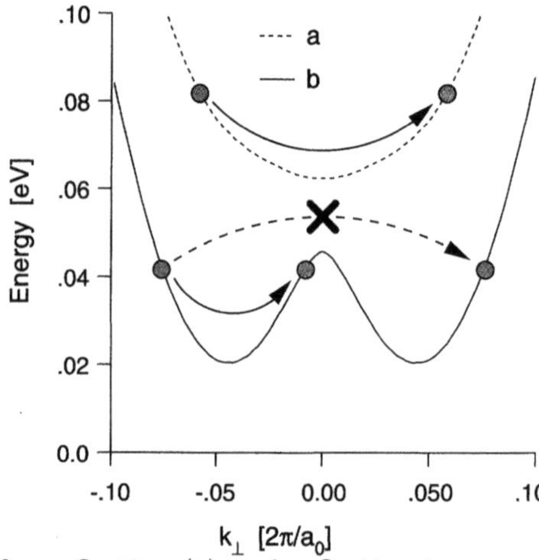

Figure 5.11: Specular reflection (a) and reflection in a complex band (b). $k_\perp$ is the component of the wave vector perpendicular to the interface

reflected into the first state satisfying the constraints in the direction of decreasing energy in the KS. Substituting λ by

$$\lambda = -\mu \frac{\mathbf{n}^{\mathrm{T}} \mathbf{v}_n(\mathbf{k}^{\mathrm{b}})}{|\mathbf{n}^{\mathrm{T}} \mathbf{v}_n(\mathbf{k}^{\mathrm{b}})|} \tag{5.67}$$

the smallest positive root μ of Eq. (5.64) gives the final state. In this case also the velocity parallel to the interface might change [5.24]. In the case of holes this effect has a strong impact on the mobility in a confining potential.

A somewhat similar situation occurs, when a particle moves from one RS primitive into the next one. Due to the assumption of a constant band structure relative to the band edge within an RS primitive (cf. Sec. 6.1), the band energy might change abruptly at the interface between the two primitives. The state of the particle in the new band structure is calculated with energy and parallel wave vector conservation as above. To this end, the particle is placed in the new band structure at the position $\mathbf{k}^{',b}$, which is the wave vector in the transformed KS before the transition. This new particle state conserves the parallel momentum, but it does not conserve energy. Therefore, the particle is moved in the KS perpendicular to the interface into the direction, into which the difference in energy is reduced, in analogy to the reflection process described above. The first state, which satisfies energy conservation, is selected as the new state. If no state is found, which conserves the energy, the particle is reflected by the energy barrier as described above [5.21]. This procedure is CPU efficient, because the KS grid is the same for the old and new band structure. Thus, the very CPU intensive search for the tetrahedron, which contains $\mathbf{k}^{',b}$ is avoided. In addition, the difference in energy between two adjacent band structures is small and the new state is in most cases found in the vicinity of the original wave vector.

5.7 Selection of a Final State in k-Space

In the case of a scattering event the scattering process, the final band, and one of the 48 irreducible wedges of RSi are determined with the methods described in Sec. 3.2. Due to the special formulation of the different scattering mechanisms the final wave vector $\mathbf{k}$ in the wth irreducible wedge is selected with a probability density $p_{n,w}(\mathbf{k},\varepsilon)$ that is uniform on an equienergy surface regardless of the scattering type

$$p_{n,w}(\mathbf{k},\varepsilon) = \frac{\delta\left(\varepsilon_{n,w}(\mathbf{k}) - \varepsilon\right)}{\int_{\text{wedge}} \delta\left(\varepsilon_{n,w}(\mathbf{k}') - \varepsilon\right) d^3k'}\,, \tag{5.68}$$

where ε denotes the already determined final energy of the particle, n the final band, and w the final wedge. The probability $P^{\text{tet}}_{n,w,i}(\varepsilon)$ of the ith tetrahedron to be selected reads with Eq. (5.32)

$$\begin{aligned} P^{\text{tet}}_{n,w,i}(\varepsilon) &= \int\limits_{V^{\text{tet}}_{n,w,i}} p_n(\mathbf{k},\varepsilon) d^3k = \frac{\int_{V^{\text{tet}}_{n,w,i}} \delta\left(\varepsilon_n(\mathbf{k}) - \varepsilon\right) d^3k}{\int_{\text{wedge}} \delta\left(\varepsilon_n(\mathbf{k}') - \varepsilon\right) d^3k'} \\ &= \frac{D^{\text{tet}}_{n,w,i}(\varepsilon)}{\sum_{i=1}^{N_n^{\text{tet}}} D^{\text{tet}}_{n,w,i}(\varepsilon)}\,. \end{aligned} \tag{5.69}$$

The final tetrahedron can be selected with the direct method or with the rejection method (cf. Sec. 3.1). In the case of the direct method the probability $P^{\text{tet}}_{n,w,i}(\varepsilon)$ has to be evaluated for each tetrahedron of the grid and the number of tetrahedra can be as large as 29735 (cf. Table 5.1). On the other hand, the probability $P^{\text{tet}}_{n,w,i}(\varepsilon)$ is zero for most of these tetrahedra, because they do not contain the final energy. This makes it possible to restrict the search for the final tetrahedron to those tetrahedra overlapping with the final energy [5.10]. To this end the final energy is divided into consecutive intervals and for each interval a list is built which contains all tetrahedra that overlap with the interval. In Fig. 5.12 the number of tetrahedra per energy interval $N^{\text{tet}}_{n,w}(\varepsilon)$ is shown. Since the width of the tetrahedra $(\varepsilon_{n,w,t(n,w,i,4)} - \varepsilon_{n,w,t(n,w,i,1)})$ increases with energy due to the relative error criterion of the algorithm for the grid generation, the width of the energy intervals is set to roughly one half of the average width of the tetrahedra overlapping with the energy interval (Fig. 5.13). Compared to energy intervals with a smaller width this saves memory space without much increase in the number of tetrahedra overlapping with the intervals. Although these lists reduce the maximum number of final tetrahedra, their number is still too large to use the direct method for the selection of a final tetrahedron in an efficient way.

The evaluation of the probability $P^{\text{tet}}_{n,w,i}(\varepsilon)$ for all possible final tetrahedra is avoided by using the rejection method [5.17]. Moreover, the rejection technique requires only a quantity which is proportional to the probability (5.69) and here the DOS for one tetrahedron $D^{\text{tet}}_{n,w,i}(\varepsilon)$ is used. This has the advantage of circumventing the evaluation of the denominator in Eq. (5.69). For the rejection

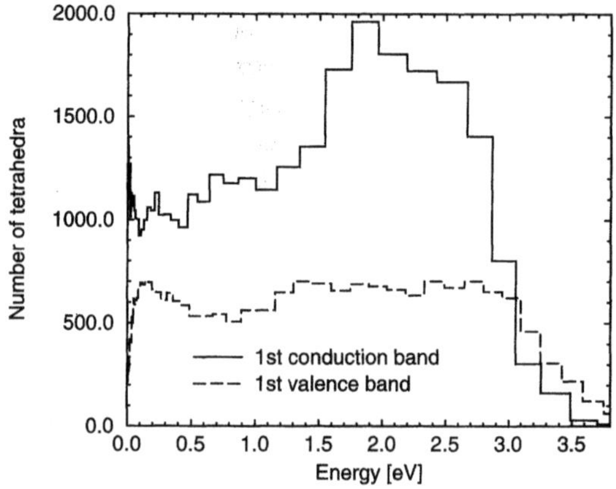

Figure 5.12: Number of tetrahedra per energy interval for RSi

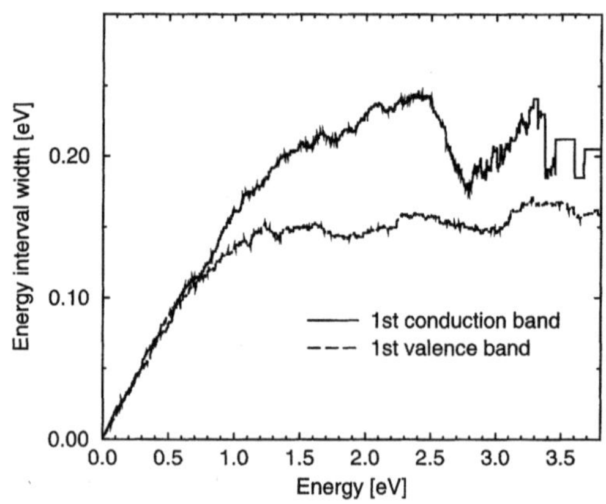

Figure 5.13: Width of the energy intervals for RSi

technique an upper bound of the DOS $D^{\mathrm{tet}}_{n,w,i}(\varepsilon)$ is required. In order to avoid the CPU intensive search for the maximum DOS over all tetrahedra during the MC simulation a table of upper bounds $\overline{D}_{n,w}(\varepsilon)$ is used. This table is built only once during the initialization of the program for consecutive intervals of the final energy $[\varepsilon_j, \varepsilon_{j+1})$ of 1meV width. $\overline{D}_{n,w}(\varepsilon)$ is the upper bound of the DOS for the interval containing the final energy ε

$$\overline{D}_{n,w}(\varepsilon) = \max_{l,\varepsilon_j \leq \varepsilon' < \varepsilon_{j+1}} [D^{\mathrm{tet}}_{n,w,l}(\varepsilon')] \quad \text{with} \quad j : \varepsilon_j \leq \varepsilon < \varepsilon_{j+1} \; . \tag{5.70}$$

In Fig. 5.14 a flowchart of this procedure is shown.

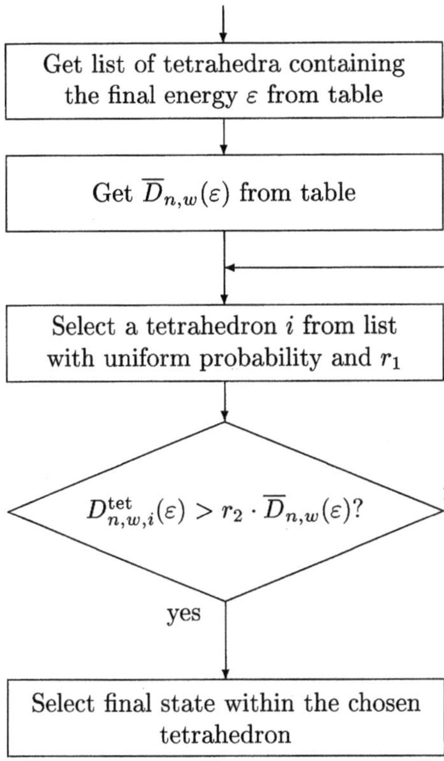

Figure 5.14: Rejection technique for the selection of the final state in the irreducible wedge. r_1 and r_2 are uniformly distributed random numbers between zero and one

The efficiency of the rejection method depends on the number of rejection steps. The expected value of the number of rejection steps (3.9) $N^{\text{reject}}_{n,w}(\varepsilon)$ is given by

$$\mathrm{E}\left\{N^{\text{reject}}_{n,w}(\varepsilon)\right\} = \frac{\overline{D}_{n,w}(\varepsilon) N^{\text{tet}}_{n,w}(\varepsilon)}{D_{n,w}(\varepsilon)} \ . \tag{5.71}$$

Since the average number of rejection steps (Fig. 5.15) is much smaller than the number of tetrahedra containing the final energy $N^{\text{tet}}_{n,w}(\varepsilon)$ (Fig. 5.12) and since the CPU times of one rejection step and of the evaluation of the probability $P^{\text{tet}}_{n,w,i}(\varepsilon)$ are similar, the rejection technique is much more efficient than the direct method.

Finally, the wave vector within the tetrahedron has to be determined. Depending on the final energy the intersection of the final energy with the tetrahedron is either a triangle or a quadrangle (Fig. 5.8). In the latter case the quadrangle is split into two triangles: $\{\mathbf{c}^{\text{tet}}_{n,w,i,2,3}(\varepsilon), \mathbf{c}^{\text{tet}}_{n,w,i,1,3}(\varepsilon), \mathbf{c}^{\text{tet}}_{n,w,i,1,4}(\varepsilon)\}$ and $\{\mathbf{c}^{\text{tet}}_{n,w,i,1,4}(\varepsilon),$

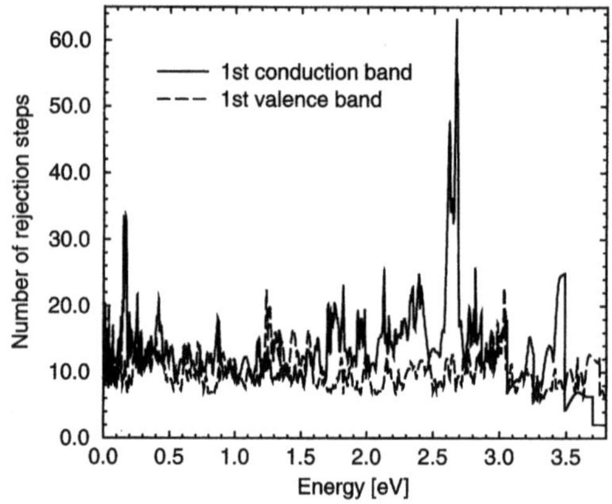

Figure 5.15: Average number of rejection steps for the first conduction and valence bands of RSi

$\mathbf{c}^{\text{tet}}_{n,w,i,2,4}(\varepsilon)$, $\mathbf{c}^{\text{tet}}_{n,w,i,2,3}(\varepsilon)\}$. Since the group velocity is constant within a tetrahedron, the probability of one of the two triangles to contain the final state is proportional to its area and one triangle is selected with the direct method. After the final triangle has been decided, a final state is chosen within the triangle with a uniform probability density

$$\mathbf{k} = \mathbf{a} + \lambda_1(\mathbf{b} - \mathbf{a}) + \lambda_2(\mathbf{c} - \mathbf{a}) \ , \tag{5.72}$$

where the vectors **a**, **b**, and **c** point to the three corners of the selected triangle (e.g. $\{\mathbf{a}, \mathbf{b}, \mathbf{c}\} = \{\mathbf{c}^{\text{tet}}_{n,w,i,2,3}(\varepsilon), \mathbf{c}^{\text{tet}}_{n,w,i,1,3}(\varepsilon), \mathbf{c}^{\text{tet}}_{n,w,i,1,4}(\varepsilon)\}$). The random numbers λ_1 and λ_2 are given by

$$\begin{aligned} \lambda_1 &= 1 - \sqrt{1 - r_1} \ , \\ \lambda_2 &= r_2(1 - \lambda_1) \end{aligned} \tag{5.73}$$

with two random numbers r_1 and r_2 distributed uniformly between zero and one.

References

[5.1] K. Hess, Ed., *Monte Carlo Device Simulation: Full Band and Beyond*, Kluwer, Boston, 1991.

[5.2] S. M. Sze, *Physics of Semiconductors Devices*, Wiley, New York, 1981.

[5.3] O. Madelung, *Introduction to Solid State Theory*, Springer, Berlin, 1978.

[5.4] E. Klingbeil, *Tensorrechnung für Ingenieure*, vol. 197 of *Hochschultaschenbücher*, Bibliographisches Institut, Mannheim, 1966.

[5.5] M. M. Rieger and P. Vogl, "Electronic-band parameters in strained $Si_{1-x}Ge_x$ alloys on $Si_{1-y}Ge_y$ substrates", *Phys. Rev. B*, vol. 48, pp. 14276–14287, 1993.

[5.6] M. M. Rieger and P. Vogl, "Electronic-band parameters in strained $Si_{1-x}Ge_x$ alloys on $Si_{1-y}Ge_y$ substrates", *Phys. Rev. B*, vol. 50, pp. 8138, 1994, Erratum.
[5.7] M. L. Cohen and J. R. Chelikowsky, *Electronic Structure and Optical Properties of Semiconductors*, Springer, New York, 2nd edition, 1989.
[5.8] G. Lehmann and M. Taut, "On the numerical calculation of the density of states and related properties", *Phys. Status Solidi B*, vol. 54, pp. 469–477, 1972.
[5.9] J. Y. Tang, H. Shichijo, K. Hess, and G. J. Iafrate, "Band-structure dependent impact ionization in silicon and gallium arsenide", *Journal de Physique*, vol. 42, pp. 63–69, 1981.
[5.10] M. V. Fischetti and S. E. Laux, "Monte Carlo analysis of electron transport in small semiconductor devices including band–structure and space–charge effects", *Phys. Rev. B*, vol. 38, pp. 9721–9745, 1988.
[5.11] G. Wiesenekker and E. J. Baerends, "Quadratic integration over the three-dimensional Brillouin zone", *J. Phys.: Condens. Matter*, vol. 3, pp. 6721–6742, 1991.
[5.12] R. K. Smith and J. Bude, "Highly efficient full band Monte Carlo simulations", in *Proceedings of the International Workshop on Computational Electronics*, Leeds, Aug. 1993, pp. 224–230.
[5.13] J. Bude and R. K. Smith, "Phase-space simplex Monte Carlo for semiconductor transport", *Semicond. Sci. Technol.*, vol. 9, pp. 840–843, 1994.
[5.14] T. Kunikiyo, M. Takenaka, Y. Kamakura, M. Yamaji, H. Mizuno, M. Morifuji, K. Taniguchi, and C. Hamaguchi, "A Monte Carlo simulation of anisotropic electron transport in silicon including full band structure and anisotropic impact–ionization model", *J. Appl. Phys.*, vol. 75, pp. 297–312, 1994.
[5.15] E. X. Wang, M. D. Giles, S. Yu, F. A. Leon, A. Hiroki, and S. Odanaka, "Recursive M–tree method for 3–D adaptive tetrahedral mesh refinement and its application to Brillouin zone discretization", in *Proc. SISPAD*, Tokyo, Sept. 1996, pp. 67–68.
[5.16] M. Yamaji, K. Taniguchi, and C. Hamaguchi, "Multi-band Monte Carlo method using anisotropic-analytical multi-band model", in *Proc. SISPAD*, Tokyo, Sept. 1996, pp. 63–64.
[5.17] C. Jungemann, M. Bartels, S. Keith, and B. Meinerzhagen, "Efficient methods for Hall factor and transport coefficient evaluation for electrons and holes in Si and SiGe based on a full-band structure", in *Proc. IWCE*, Osaka (Japan), 1998, pp. 104–107.
[5.18] B. Fischer and K. R. Hofmann, "Discretization of the Brillouin zone by an Octree/Delaunay method with application to full-band Monte Carlo transport simulation", in *Proc. SISPAD*, Leuven (Belgium), 1998, pp. 181–184.
[5.19] C. Jungemann, S. Keith, M. Bartels, and B. Meinerzhagen, "Efficient full-band Monte Carlo simulation of silicon devices", *IEICE Trans. on Electronics*, vol. E82-C, no. 6, pp. 870–879, 1999.
[5.20] B. Fischer, "A full-band Monte Carlo charge transport model for nanoscale silicon devices including strain", Doctor thesis, University Hannover, Hannover, 1999.
[5.21] C. Jungemann, S. Keith, and B. Meinerzhagen, "Full-band Monte Carlo device simulation of a Si/SiGe-HBT with a realistic Ge profile", *IEICE Trans. on Electronics*, vol. E83-C, no. 8, pp. 1228–1234, 2000.
[5.22] G. Gilat and L. J. Raubenheimer, "Accurate numerical method for calculation frequency-distribution functions in solids", *Phys. Rev.*, vol. 144, pp. 390–395,

1966.
[5.23] I. N. Bronstein and K. A. Semendjajew, *Taschenbuch der Mathematik,* B. G. Teubner, Stuttgart, 1991.
[5.24] C. Jungemann, S. Keith, and B. Meinerzhagen, "Full-band Monte Carlo simulation of a 0.12μm-Si-PMOSFET with and without a strained SiGe-channel", in *IEDM Tech. Dig.*, San Francisco (USA), 1998, pp. 897–900.

6 Device Simulation

In the case of a device simulation the BTE has to be solved self-consistently with the Poisson equation for the electric field [6.1]. To this end the RS is discretized with a tensor-product grid as described in the first section of this chapter. The material parameters, like the germanium concentration, doping, etc, are defined on this grid together with the boundary conditions. The germanium-dependent band edges are given in the next section. The discrete Poisson equation is presented next and in the fourth section the self-consistent solution of the BTE and Poisson is discussed. The extension to a nonlinear Poisson equation based on the zero-current approximation for one carrier type is given in the following section. Nonself-consistent MC simulations, where the electric field is calculated with a momentum-based method, are introduced in the sixth section. A method for the enhancement of rare events (e.g. impact ionization) is discussed in the next section. In the following two sections methods for the evaluation of terminal currents and inclusion of contact resistances are presented. Finally, the normalization of physical quantities is discussed in the last section of this chapter.

6.1 Device Discretization

The device structure is discretized with a nonuniform tensor-product grid. In order to reduce the computational burden only a two-dimensional grid is used in the RS (Fig. 6.1). The grid consists of N_x lines in x-direction and N_y lines in y-direction which do not terminate within the discretized region. The number of grid nodes is $N_x \cdot N_y$. The cells of the RS grid are rectangles $\Omega_{i,j}$ with $x_i \leq x < x_{i+1}$ and $y_j \leq y < y_{j+1}$. Within a primitive of the grid the doping, dielectric constant, and other simulation parameters are assumed to be constant, whereas the lattice temperature is assumed to be constant in the whole device. The cells can be of the following four types: SiGe (**S**), Oxide (**O**), Vacuum (**V**), and Void (**N**). Regions of the void type are used to impose boundary conditions of the Neumann-type for the displacement density and particle flux on the interfaces they share with other regions. Particles cannot enter regions of this type and all material parameters are zero including the dielectric constant.

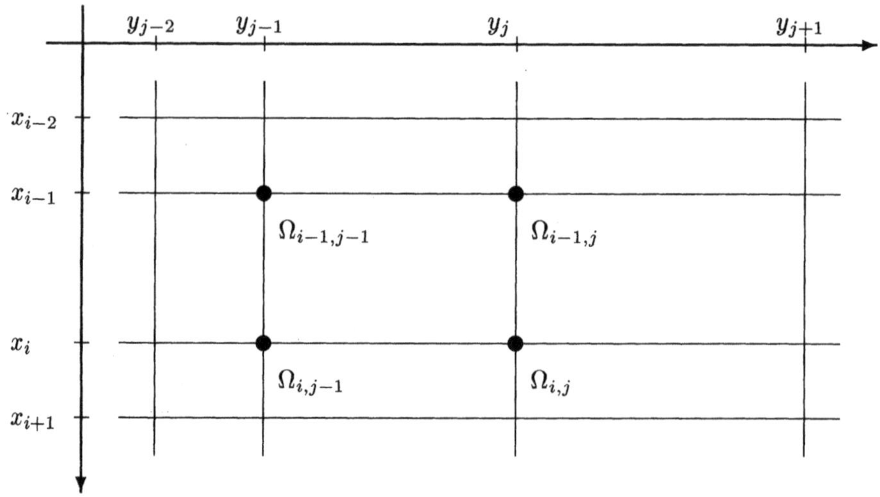

Figure 6.1: The two-dimensional nonuniform tensor product grid of the real space

The SiGe material within a cell is characterized by the germanium content x_{Ge}, which depends only on the x-position, where it is assumed that the device is grown in pseudomorphic layers by epitaxy in x-direction on relaxed silicon ($y_{\mathrm{Ge}} = 0$) as it is usually the case in SiGe HBTs [6.2–6.4] or PMOS devices with a buried strained SiGe channel [6.5–6.7]. The other important case considered here is the one of a strained silicon layer pseudomorphically grown on a relaxed SiGe buffer [6.8, 6.9]. The band structure relative to the band edges is assumed to be constant within the cell $\Omega_{i,j}$ with $\varepsilon_n(\mathbf{r}, \mathbf{k}) = \varepsilon_{n,i}(\mathbf{k})$. In order to limit the number of band structures, each of which requires about 50MBytes of main memory, the germanium profile is approximated for the band structure selection by a stair-case function with N_{B} different values, which are equidistantly distributed between the minimum and maximum germanium content. The band edges on the other hand are linearly interpolated over the cell (see also Sec. 6.2) [6.10]. The two different approximations of the germanium content are shown in Fig. 6.2 for $N_{\mathrm{B}} = 6$.

Particles are only simulated in SiGe regions with the methods described in the previous chapters. Whenever a particle reaches a boundary an appropriate model has to be selected to simulate a contact, a device boundary, or the rough $\mathrm{Si/SiO_2}$-interface [6.11]. These models are based on the following boundary conditions for the particle motion which can be applied to any interface between two cells: catch (**c**), enter (**e**), generate (**g**), periodic (**p**), reflect (**r**), and scatter (**s**), which are depicted in Fig. 6.3. The boundary condition **e** allows a particle to enter the contiguous cell. This is the usual interface condition within a SiGe region of a device (Fig. 6.4). In the case that the band structure changes at the interface, the particle might be reflected if its energy is not large enough for entering the

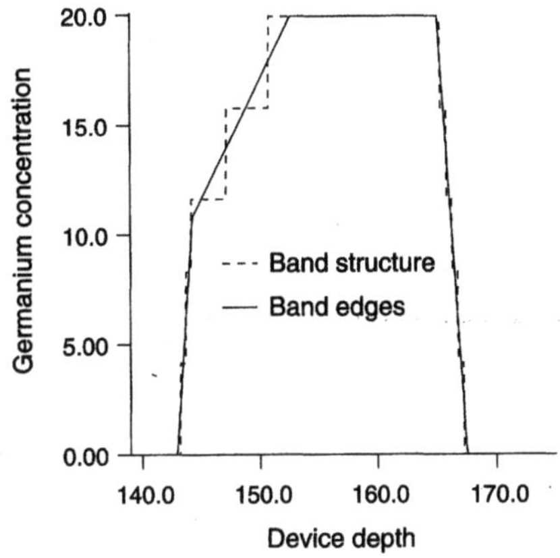

Figure 6.2: An example for the two different approximations of the germanium profile for the band structure and band edges

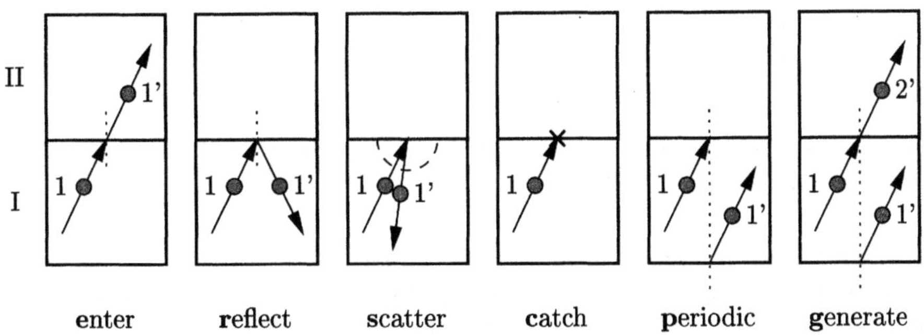

Figure 6.3: The different boundary conditions for a particle at a grid cell boundary

next cell (cf. Sec. 5.6). In the case of an **r** boundary the particle is reflected at the cell boundary (cf. Sec. 5.6) resulting in a Neumann-type boundary condition for the particle flux at a surface of a SiGe region. The **s** boundary condition at which the particle is either reflected or diffusively scattered (cf. Sec. 4.5) is used to simulate the rough Si/SiO_2-interface. The **c** boundary condition removes a particle from the simulation and is used to simulate a contact together with the following two boundary conditions. At a **p** interface the particle is moved to the opposite boundary of the current cell. At a **g** type a particle is reproduced and a copy of the initial particle is placed on the other side of the interface into the contiguous cell. Then the **p** boundary condition is applied to the initial particle.

In Fig. 6.4 boundary conditions for the particle simulation of an exemplary device are shown. In the SiGe bulk particles can freely move from one cell to another. If they hit the Si/SiO_2-interface they are scattered or reflected. They

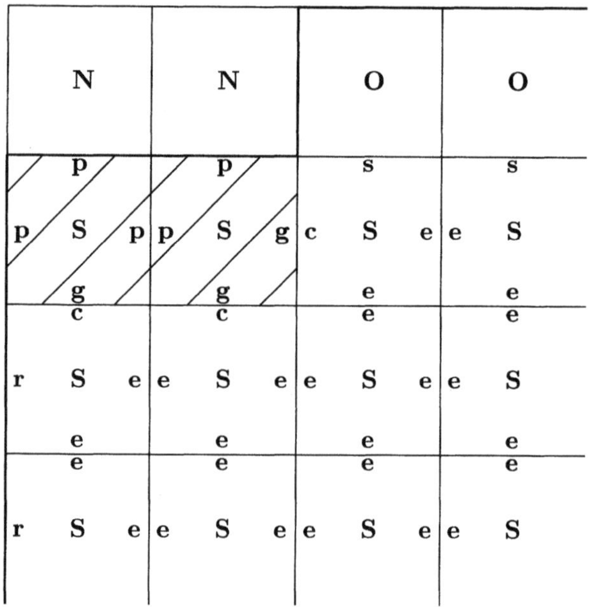

Figure 6.4: Boundary conditions for particles in the real space. The hatched area is an ohmic contact

are also reflected, if they impinge upon the surface of the device. The hatched area in Fig. 6.4 is an ohmic contact [6.11]. Particles in cells belonging to the contact can not leave the contact, because they are enclosed by periodic boundary conditions or generation conditions including a periodic boundary condition. Therefore, the particle density within the contact is kept constant resulting in a boundary condition for the particle density similar to the Dirichlet-type. The doping concentration must be high and homogenous in and around the contact such that the electric field in the vicinity of the contact is small to ensure that the contact is close to equilibrium. The electric field within a contact cell can be chosen to be zero in order to enforce equilibrium conditions for the particle gas in the contact. Whenever a particle in the contact hits a generation boundary condition a new particle is created outside of the contact simulating the outgoing particle flux of an ohmic contact. When a particle hits the outside of the contact it is removed from the simulation and only in equilibrium are incoming and outgoing particle flux the same. Otherwise they differ resulting in a nonzero particle flux contribution to the terminal current. The maximum current of the contact is obtained, when the incoming flux is zero. It is given by

$$I_{\max} = \int_A qN_{\text{dop}} \left\langle \mathbf{v}_\perp^{\mathrm{T}} \Theta \left(\mathbf{v}_\perp^{\mathrm{T}} d\mathbf{r}^2 \right) \right\rangle d\mathbf{r}^2 \ , \tag{6.1}$$

where A is the surface of the contact, $\mathbf{v}_\perp$ is the component of the velocity perpendicular to the contact surface, and the Θ-function is one for positive arguments and zero otherwise. Under equilibrium conditions the expected value of the positive perpendicular velocity is for relaxed silicon at room temperature about $5 \cdot 10^6$cm/s. In order to ensure equilibrium conditions near the contact, the terminal current should not exceed a few percent of the maximum current.

6.2 Band Edges

In a device the band edges are position-dependent due to doping and Ge content (Fig. 6.2). The band shifts of the conduction and valence bands relative to relaxed Si ($y_{\mathrm{Ge}} = 0$) due to the Ge content x_{Ge} are given by [6.12,6.13]

$$\Delta E_{\mathrm{c}}^{\mathrm{Ge}} = -0.196\mathrm{eV}\, x_{\mathrm{Ge}} + 0.396\mathrm{eV}\, x_{\mathrm{Ge}}^2 , \tag{6.2}$$

$$\Delta E_{\mathrm{v}}^{\mathrm{Ge}} = 0.7\mathrm{eV}\, x_{\mathrm{Ge}} . \tag{6.3}$$

In the case of strained Si ($x_{\mathrm{Ge}} = 0$) on relaxed $\mathrm{Si}_{1-y_{\mathrm{Ge}}}\mathrm{Ge}_{y_{\mathrm{Ge}}}$ they read [6.12]

$$\Delta E_{\mathrm{c}}^{\mathrm{Ge}} = -0.5\mathrm{eV}\, y_{\mathrm{Ge}} - 0.155\mathrm{eV}\, y_{\mathrm{Ge}}^2 , \tag{6.4}$$

$$\Delta E_{\mathrm{v}}^{\mathrm{Ge}} = -0.255\mathrm{eV}\, y_{\mathrm{Ge}} + 0.155\mathrm{eV}\, y_{\mathrm{Ge}}^2 . \tag{6.5}$$

The apparent band gap narrowing due to doping is assumed to be

$$\Delta E^{\mathrm{Bgn}} = E_0 \left(\ln \frac{N_{\mathrm{acc}} + N_{\mathrm{don}}}{N_{\mathrm{Ref}}} + \sqrt{\left(\ln \frac{N_{\mathrm{acc}} + N_{\mathrm{don}}}{N_{\mathrm{Ref}}} \right)^2 + 0.5} \right) \tag{6.6}$$

with $E_0 = 6.92$meV and $N_{\mathrm{Ref}} = 1.3 \cdot 10^{17}/\mathrm{cm}^3$ [6.14, 6.15]. In accordance with Ref. [6.16] the above doping dependent apparent bandgap narrowing developed for Si is also used without any modifications in the case of SiGe and symmetrically distributed. The total band offsets read

$$E_{\mathrm{c}} = \Delta E_{\mathrm{c}}^{\mathrm{Ge}} - \frac{1}{2}\Delta E^{\mathrm{Bgn}} , \tag{6.7}$$

$$E_{\mathrm{v}} = -\varepsilon_0^{\mathrm{gap}} + \Delta E_{\mathrm{v}}^{\mathrm{Ge}} + \frac{1}{2}\Delta E^{\mathrm{Bgn}} . \tag{6.8}$$

Since the accuracy of the band gap evaluated with the nonlocal empirical pseudopotential for undoped and relaxed Si is not accurate enough, here, the experimentally determined expression of Ref. [6.17] is used

$$\frac{\varepsilon_0^{\mathrm{gap}}}{\mathrm{eV}} = \begin{cases} 1.17 + 1.059 \cdot 10^{-5} \frac{T_0}{\mathrm{K}} - 6.05 \cdot 10^{-7} \frac{T_0^2}{\mathrm{K}^2} & : \quad T_0 < 190\mathrm{K} \\ 1.1785 - 9.025 \cdot 10^{-5} \frac{T_0}{\mathrm{K}} - 3.05 \cdot 10^{-7} \frac{T_0^2}{\mathrm{K}^2} & : \quad 190\mathrm{K} \le T_0 < 275\mathrm{K} \\ 1.206 - 2.730 \cdot 10^{-4} \frac{T_0}{\mathrm{K}} & : \quad 275\mathrm{K} \le T_0 \end{cases} . \tag{6.9}$$

Table 6.1: *Dielectric constants of the different region types*

Type	ϵ
	$[\epsilon_0]$
Silicon	11.7
Germanium	16.0
Oxide	3.9
Vacuum	1.0
Void	0.0

The position-dependent band edges lead to forces acting on the particles (cf. Eq. (2.5)). The gradient of the band edges is assumed to be constant within a cell $\Omega_{i,j}$ of the grid

$$\nabla_{\mathbf{r}} E_{c/v;\,i,j} = \frac{1}{2} \begin{pmatrix} \frac{E_{c/v;\,i+1,j} - E_{c/v;\,i,j}}{x_{i+1} - x_i} & + & \frac{E_{c/v;\,i+1,j+1} - E_{c/v;\,i,j+1}}{x_{i+1} - x_i} \\ \frac{E_{c/v;\,i,j+1} - E_{c/v;\,i,j}}{y_{j+1} - y_j} & + & \frac{E_{c/v;\,i+1,j+1} - E_{c/v;\,i+1,j}}{y_{j+1} - y_j} \\ & 0 & \end{pmatrix} , \tag{6.10}$$

where $E_{c/v;\,i,j}$ is the value of the band edge on the grid node (x_i, y_j).

6.3 Poisson Equation

In the eddy current free approximation the quasi-static potential $\Psi(\mathbf{r}, t)$ is given by the PE [6.18, 6.19]

$$\nabla_{\mathbf{r}}^{\mathrm{T}} (\epsilon \nabla_{\mathbf{r}} \Psi) = -\varrho , \tag{6.11}$$

where ϱ is the space charge density. The dielectric constant (4.9) is assumed to be position-independent within the cells of the RS grid (Tab. 6.1). At the interfaces between regions the potential must be continuous

$$\Psi^{\mathrm{l}}(\mathbf{r}, t) = \Psi^{\mathrm{r}}(\mathbf{r}, t) . \tag{6.12}$$

Ψ^{l} is the potential on the left side of the interface and Ψ^{r} the one on the right side. The displacement density must satisfy

$$\epsilon^{\mathrm{l}}(\mathbf{r}) E_{\perp}^{\mathrm{l}}(\mathbf{r}, t) + \epsilon^{\mathrm{r}}(\mathbf{r}) E_{\perp}^{\mathrm{r}}(\mathbf{r}, t) = \overline{q} , \tag{6.13}$$

where $E_{\perp}$ is the electric field component perpendicular to the interface with orientation into the region and $\overline{q}$ is the interface charge density. On the surface of the simulation region Neumann-type boundary conditions are used

$$\epsilon(\mathbf{r}) E_{\perp}(\mathbf{r}, t) = 0 \tag{6.14}$$

and on contacts Dirichlet-type boundary conditions

$$\Psi(\mathbf{r},t) = V^{\mathrm{i}} \;, \tag{6.15}$$

where V^{i} is the inner potential of the contact. In the case of SiGe it is given by

$$V^{\mathrm{i}} = V^{\mathrm{a}} + V_{\mathrm{T}}\mathrm{arsinh}\left(\frac{N_{\mathrm{don}} - N_{\mathrm{acc}}}{2n_{\mathrm{i}}}\right) - \Psi_h \;. \tag{6.16}$$

V^{a} is the applied contact voltage and n_{i} the intrinsic carrier density (5.41). The workfunction difference Ψ_h is defined relative to relaxed silicon (RSi)

$$\Phi_h = \frac{1}{2}\left(-\frac{E_{\mathrm{c}} - E_{\mathrm{c}}^{\mathrm{RSi}} + E_{\mathrm{v}} - E_{\mathrm{v}}^{\mathrm{RSi}}}{q} + V_{\mathrm{T}}\ln\frac{N_{\mathrm{c}}N_{\mathrm{v}}^{\mathrm{RSi}}}{N_{\mathrm{v}}N_{\mathrm{c}}^{\mathrm{RSi}}}\right) \;. \tag{6.17}$$

In the case of a metal contact the inner potential reads

$$V^{\mathrm{i}} = V^{\mathrm{a}} + \Phi_{\mathrm{MS}} \;, \tag{6.18}$$

where Φ_{MS} is the appropriate workfunction for a metal contact relative to relaxed silicon [6.20].

The PE (6.11) is discretized on the RS grid (Fig. 6.1) with finite differences and box integration [6.21]. The resultant linear equation for the potentials $\Psi_{i,j}$ reads for the grid node (i,j)

$$\begin{array}{llllllll}
(c_{i-1,j-1} & + & c_{i,j-1}) & (\Psi_{i,j-1} & - & \Psi_{i,j}) & + & \\
(\overline{c}_{i-1,j-1} & + & \overline{c}_{i-1,j}) & (\Psi_{i-1,j} & - & \Psi_{i,j}) & + & \\
(c_{i-1,j} & + & c_{i,j}) & (\Psi_{i,j+1} & - & \Psi_{i,j}) & + & \\
(\overline{c}_{i,j-1} & + & \overline{c}_{i,j}) & (\Psi_{i+1,j} & - & \Psi_{i,j}) & = & Q_{i,j}^{\mathrm{box}}
\end{array} \tag{6.19}$$

with

$$c_{i,j} = \begin{cases} \frac{1}{2}\epsilon_{i,j}\frac{x_{i+1}-x_i}{y_{j+1}-y_j} & \text{for} \quad 1 \le i < N_x \text{ and } 1 \le j < N_y \\ 0 & \text{else} \end{cases} \;, \tag{6.20}$$

$$\overline{c}_{i,j} = \begin{cases} \frac{1}{2}\epsilon_{i,j}\frac{y_{j+1}-y_j}{x_{i+1}-x_i} & \text{for} \quad 1 \le i < N_x \text{ and } 1 \le j < N_y \\ 0 & \text{else} \end{cases} \;, \tag{6.21}$$

and

$$Q_{i,j}^{\mathrm{box}} = \int_{x_i^l}^{x_i^u}\int_{y_j^l}^{y_j^u} q\left(N_{\mathrm{don}}(x,y) - N_{\mathrm{acc}}(x,y)\right)dx\,dy + \int \overline{q}\,ds + Q_{i,j}^{\mathrm{par,box}} \tag{6.22}$$

with

$$x_i^l = \begin{cases} x_1 & \text{for} \quad i = 1 \\ \frac{x_i + x_{i-1}}{2} & \text{for} \quad 1 < i \leq N_x \end{cases} , \tag{6.23}$$

$$x_i^u = \begin{cases} \frac{x_{i+1} + x_i}{2} & \text{for} \quad 1 \leq i < N_x \\ x_{N_x} & \text{for} \quad i = N_x \end{cases} , \tag{6.24}$$

$$y_j^l = \begin{cases} y_1 & \text{for} \quad j = 1 \\ \frac{y_j + y_{j-1}}{2} & \text{for} \quad 1 < j \leq N_y \end{cases} , \tag{6.25}$$

$$y_j^u = \begin{cases} \frac{y_{j+1} + y_j}{2} & \text{for} \quad 1 \leq j < N_y \\ y_{N_y} & \text{for} \quad j = N_y \end{cases} . \tag{6.26}$$

The contribution of the particle charge to the total space charge in the box is calculated with the cell-in-cloud method (bilinear mapping) [6.1, 6.11, 6.22]

$$\begin{aligned} Q_{i,j}^{\text{par,box}} = & \sum_{(k|\mathbf{r}_k \in \Omega_{i-1,j-1})} \frac{Q_k^{\text{sim}}}{w_{\text{dev}}} \frac{x_k - x_{i-1}}{x_i - x_{i-1}} \frac{y_k - y_{j-1}}{y_j - y_{j-1}} \\ + & \sum_{(k|\mathbf{r}_k \in \Omega_{i,j-1})} \frac{Q_k^{\text{sim}}}{w_{\text{dev}}} \frac{x_{i+1} - x_k}{x_{i+1} - x_i} \frac{y_k - y_{j-1}}{y_j - y_{j-1}} \\ + & \sum_{(k|\mathbf{r}_k \in \Omega_{i-1,j})} \frac{Q_k^{\text{sim}}}{w_{\text{dev}}} \frac{x_k - x_{i-1}}{x_i - x_{i-1}} \frac{y_{j+1} - y_k}{y_{j+1} - y_j} \\ + & \sum_{(k|\mathbf{r}_k \in \Omega_{i,j})} \frac{Q_k^{\text{sim}}}{w_{\text{dev}}} \frac{x_{i+1} - x_k}{x_{i+1} - x_i} \frac{y_{j+1} - y_k}{y_{j+1} - y_j} , \end{aligned} \tag{6.27}$$

where Q_k^{sim} is the charge of the kth particle and w_{dev} the widths of the device in the third dimension.

The discrete PE (6.19) already satisfies the boundary conditions (6.12), (6.13), and (6.14). In the case of a Dirichlet-type boundary condition (6.15) Eq. (6.19) is replaced by

$$\Psi_{i,j} = V_{i,j}^{\text{i}} . \tag{6.28}$$

This results in a closed system of linear equations for the potentials on the grid nodes. Introducing the scalar index

$$l = (j-1) \cdot N_x + i \tag{6.29}$$

yields a matrix equation

$$\hat{P}\mathbf{\Psi} = -\mathbf{Q} . \tag{6.30}$$

The discrete Poisson operator $\hat{P}$ has only five nonzero diagonals: $P_{l,l-N_x}$, $P_{l,l-1}$, $P_{l,l}$, $P_{l,l+1}$, $P_{l,l+N_x}$. The bandwidth of the matrix $\hat{P}$ is $2 \cdot N_x + 1$. Since the matrix is diagonal dominant, a pivot search can be avoided during the LU decomposition preserving the band structure of the matrix and the L and U matrices have a bandwidth[1] of $N_x + 1$ [6.11,6.21]. The LU decomposition is calculated only once at the beginning of the simulation. During the simulation the LU decomposition is used to solve Eq. (6.30). This is faster and more accurate than applying iterative methods to Eq. (6.30) (e.g. SLOR, Fast Fourier, Multigrid, ... [6.1,6.23]).

The electric field for each cell (i,j) is calculated with finite differences

$$\mathbf{E}_{i,j} = -\frac{1}{2}\begin{pmatrix} \frac{\Psi_{i+1,j}-\Psi_{i,j}}{x_{i+1}-x_i} & + & \frac{\Psi_{i+1,j+1}-\Psi_{i,j+1}}{x_{i+1}-x_i} \\ \frac{\Psi_{i,j+1}-\Psi_{i,j}}{y_{j+1}-y_j} & + & \frac{\Psi_{i+1,j+1}-\Psi_{i+1,j}}{y_{j+1}-y_j} \\ & 0 & \end{pmatrix} , \tag{6.31}$$

where the electric field components along the x and y-direction have been averaged over the respective edges of the cell. This yields an electric field which is constant within each primitive of the RS grid.

6.4 Self-Consistent Device Simulation

The MC model and the PE have to be solved self-consistently, because the particles are accelerated by the electric field which in turn depends on the particle distribution [6.1]. In Fig. 6.5 the flowchart of an SC-MC simulation is shown. The MC simulation is performed for a given time step length Δt. After the particle simulation the space charge is evaluated and the electrostatic potential is calculated for the time $t = i\Delta t$. Based on this potential again the MC simulation is performed and the particles are advanced in time to $t = (i+1)\Delta t$. The last two steps are repeated, until a given amount of time T_{sim} has been simulated. Although the MC simulation is continuous in time, the electric field is calculated only every Δt. Since the MC simulation is based on the electric field at the beginning of the time step, the method is a forward Euler scheme which is stable only for short time steps [6.1,6.24–6.26]. BDF techniques have been proposed for the SC device simulation [6.27,6.28], but due to the stochastic nature of the MC method these techniques are difficult to implement and not used here.

The stability of MC device simulations is discussed in detail in Ref. [6.25]. Here only a simple analysis of this effect is given for a silicon cube of infinite size homogeneously doped with a donor concentration of N_0 with periodic boundary conditions. In addition, it is assumed that the microscopic relaxation time τ_r is independent of energy and position and that the conduction band is spherical and parabolic with a scalar effective mass $m = 0.32m_0$. In this case the balance equation for the particle current (2.56) reads

$$\mathbf{j} = -\tau_r \frac{\partial \mathbf{j}}{\partial t} - n\mu\mathbf{E} - D\nabla_{\mathbf{r}} n , \tag{6.32}$$

[1]In the case that N_x is larger than N_y the grid nodes are sorted in a different fashion ($l = (i-1)\cdot N_y + j$). This reduces the bandwidth to $N_y + 1$ and therewith the CPU time.

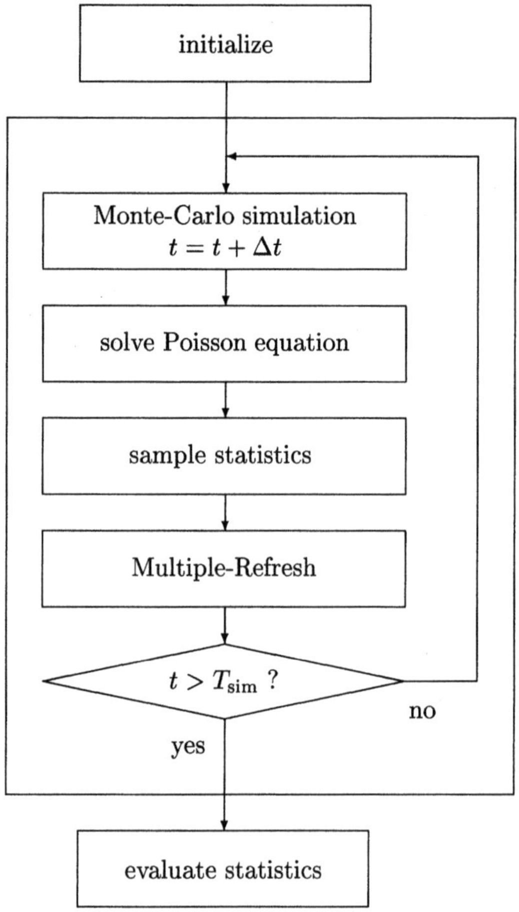

Figure 6.5: Flowchart of the SC-MC device simulation

where the scalar mobility and diffusion constant are given by

$$\mu = \frac{q\tau_{\mathrm{r}}}{m} \quad , \quad D = \frac{\tau_{\mathrm{r}} k_{\mathrm{B}} T_0}{m} \tag{6.33}$$

assuming that the particle gas is close to equilibrium ($m\langle v^2 \rangle_{\mathbf{k}}/2 \approx 3k_{\mathrm{B}}T_0/2$).

Linearization of Eq. (6.32) with respect to the stationary solution

$$\mathbf{j}(\mathbf{r},t) = \mathbf{0} + \delta\mathbf{j}(\mathbf{r},t) \ , \quad n(\mathbf{r},t) = N_0 + \delta n(\mathbf{r},t) \ , \quad \mathbf{E}(\mathbf{r},t) = \mathbf{0} + \delta\mathbf{E}(\mathbf{r},t) \tag{6.34}$$

yields

$$\delta\mathbf{j} = -\tau_{\mathrm{r}} \frac{\partial \delta\mathbf{j}}{\partial t} - N_0 \mu \delta\mathbf{E} - D\nabla_{\mathbf{r}} \delta n \ . \tag{6.35}$$

The corresponding continuity equation is given by

$$\nabla_{\mathbf{r}}^{\mathrm{T}} \delta \mathbf{j} = -\frac{\partial \delta n}{\partial t} , \tag{6.36}$$

and the third Maxwell equation reads for this case

$$\nabla_{\mathbf{r}}^{\mathrm{T}} \delta \mathbf{E} = -\frac{q}{\epsilon_{\mathrm{si}}} \delta n . \tag{6.37}$$

Solving for the particle density yields

$$\left(\frac{\partial^2}{\partial t^2} + \frac{1}{\tau_{\mathrm{r}}} \frac{\partial}{\partial t} + {\omega_{\mathrm{pl}}}^2 \right) \delta n = {\omega_{\mathrm{pl}}}^2 \lambda_{\mathrm{D}}^2 \Delta \delta n . \tag{6.38}$$

The plasma frequency is given by

$${\omega_{\mathrm{pl}}}^2 = \frac{q^2 N_0}{m \epsilon_{\mathrm{si}}} \tag{6.39}$$

and the Debye length by

$$\lambda_{\mathrm{D}}^2 = \frac{k_{\mathrm{B}} T_0 \epsilon_{\mathrm{si}}}{q^2 N_0} . \tag{6.40}$$

Solving Eq. (6.38) for plasma waves of the form (the bar under a symbol denotes a phasor)

$$\delta n(\mathbf{r}, t) = \Re \left\{ \delta \underline{n} \exp \left(\mathrm{i} \omega t - \mathrm{i} \mathbf{k}^{\mathrm{T}} \mathbf{r} \right) \right\} \tag{6.41}$$

yields damped oscillations

$$\omega_{1,2} = \frac{\mathrm{i}}{2\tau_{\mathrm{r}}} \pm \sqrt{{\omega_{\mathrm{pl}}}^2 (1 + \lambda_{\mathrm{D}}^2 k^2) - \frac{1}{4{\tau_{\mathrm{r}}}^2}} . \tag{6.42}$$

In order to analyze the stability of the SC device simulation the Forward Euler scheme (Fig. 6.5) is used to solve the coupled Eqs. (6.35), (6.36) and (6.37) [6.27]. To this end Eq. (6.37) is changed and the charge is calculated at the discrete times $i \Delta t$

$$\nabla_{\mathbf{r}}^{\mathrm{T}} \delta \mathbf{E}(\mathbf{r}, t) = -\frac{q}{\epsilon_{\mathrm{si}}} \delta n(\mathbf{r}, i \Delta t) . \tag{6.43}$$

Solving Eqs. (6.35), (6.36) and (6.43) for plasma waves but retaining the time dependence yields for $i \Delta t \leq t < (i+1) \Delta t$

$$\left(\frac{\partial^2}{\partial t^2} + \frac{1}{\tau_{\mathrm{r}}} \frac{\partial}{\partial t} + {\omega_{\mathrm{pl}}}^2 \lambda_{\mathrm{D}}^2 k^2 \right) \delta n(t) + {\omega_{\mathrm{pl}}}^2 \delta n(i \Delta t) = 0 . \tag{6.44}$$

With the initial conditions

$$\delta n(i \Delta t) = \delta n_i , \quad \frac{\partial \delta n}{\partial t} |_{t = i \Delta t} = \delta \dot{n}_i \tag{6.45}$$

Eq. (6.44) can be solved for $t = (i+1)\Delta t$

$$\begin{pmatrix} \delta n_{i+1} \\ \delta \dot{n}_{i+1} \end{pmatrix} = \begin{pmatrix} S_{1,1} & S_{1,2} \\ S_{2,1} & S_{2,2} \end{pmatrix} \begin{pmatrix} \delta n_i \\ \delta \dot{n}_i \end{pmatrix} . \tag{6.46}$$

The elements of the amplification matrix $\hat{S}$ are

$$\begin{aligned} S_{1,1} &= 1 - \frac{s_\gamma}{s_\beta - s_\alpha}\left(\frac{1-\exp(-s_\alpha \Delta t)}{s_\alpha} - \frac{1-\exp(-s_\beta \Delta t)}{s_\beta}\right) , \\ S_{1,2} &= \frac{\exp(-s_\alpha \Delta t) - \exp(-s_\beta \Delta t)}{s_\beta - s_\alpha} , \\ S_{2,1} &= -\frac{s_\gamma(\exp(-s_\alpha \Delta t) - \exp(-s_\beta \Delta t))}{s_\beta - s_\alpha} , \\ S_{2,2} &= \frac{s_\beta \exp(-s_\beta \Delta t) - s_\alpha \exp(-s_\alpha \Delta t)}{s_\beta - s_\alpha} \end{aligned} \tag{6.47}$$

with

$$\begin{aligned} s_\alpha &= \frac{1}{2\tau_\mathrm{r}} - \sqrt{\frac{1}{4{\tau_\mathrm{r}}^2} - {\omega_\mathrm{pl}}^2 \lambda_\mathrm{D}^2 k^2} , \\ s_\beta &= \frac{1}{2\tau_\mathrm{r}} + \sqrt{\frac{1}{4{\tau_\mathrm{r}}^2} - {\omega_\mathrm{pl}}^2 \lambda_\mathrm{D}^2 k^2} , \\ s_\gamma &= {\omega_\mathrm{pl}}^2 (1 + \lambda_\mathrm{D}^2 k^2) . \end{aligned} \tag{6.48}$$

For short time steps the two eigenvalues $\chi_{1,2}$ of the matrix $\hat{S}$ are given by

$$\chi_{1,2} \approx 1 - \left(\frac{1}{2\tau_\mathrm{r}} \mp \mathrm{i}\sqrt{{\omega_\mathrm{pl}}^2(1+\lambda_\mathrm{D}^2 k^2) - \frac{1}{4{\tau_\mathrm{r}}^2}}\right)\Delta t = 1 + \mathrm{i}\omega_{1,2}\Delta t . \tag{6.49}$$

This result corresponds to the first-order expansion of Eq. (6.41) with respect to time and shows that the Forward Euler scheme is consistent with Eq. (6.38).

The solution of Eq. (6.46) is stable, if the two eigenvalues of the amplification matrix $\hat{S}$ lie within the unit circle [6.1]

$$|\chi_{1,2}| < 1 . \tag{6.50}$$

In the limit of weak scattering ($\Delta t/\tau_\mathrm{r} \ll 1$) this results in the following condition for the time step length [6.25]

$$\Delta t < \frac{2\tau_\mathrm{r}}{1 + {\omega_\mathrm{pl}}^2 {\tau_\mathrm{r}}^2} . \tag{6.51}$$

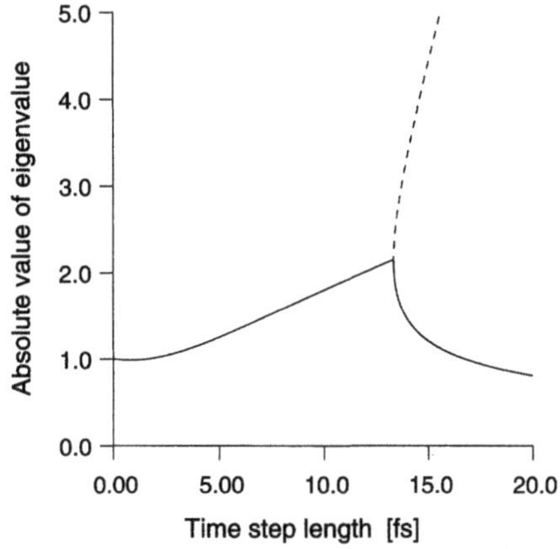

Figure 6.6: Absolute value of the two eigenvalues of the amplification matrix $\hat{S}$ for a donor concentration of $10^{20}/\mathrm{cm}^3$, $m = 0.32m_0$, $\tau_\mathrm{r} = 14.6\mathrm{fs}$, and $k = 0$

Thus, simulations are only stable, if particle scattering is present ($\tau_\mathrm{r} < \infty$).

In Fig. 6.6 the absolute values of the two eigenvalues of the amplification matrix are shown as a function of the time step length for $k = 0$ for which the system has the lowest stability. For time steps longer than 1.6fs the absolute values are larger than one and the system becomes unstable. For time steps shorter than 13.3fs the eigenvalues are complex conjugates and for larger time steps they are real-valued. In Fig. 6.7 the stability limit of Eq. (6.44) is shown for electrons at room temperature. In addition, the stability limit resulting from the sampling theorem [6.29]

$$\Re\{\omega\}\Delta t < \pi \tag{6.52}$$

is shown in Fig. 6.7 which states that a cycle of the plasma oscillations must be resolved by at least two time steps. Comparison of both stability criteria reveals that the limiting factor of the stability is the reduction in the damping of the plasma waves by the Forward Euler scheme. Especially at high donor concentrations for which the inverse relaxation time is small compared to the plasma frequency ($1/{\tau_\mathrm{r}}^2 \ll {\omega_\mathrm{pl}}^2$) the weak scattering reduces the stability considerably compared to the sampling theorem [6.25].

Another problem of the SC device simulation method is self-forces. The charge of all particles is used to calculate the electric field (Eq. (6.27), (6.30), and (6.31)) and this electric field in turn is used to accelerate the particles. Hence, the particles exert forces onto themselves. These self-forces are only partially real (e.g. image forces [6.19]). Some of them are even present in the case of an infinite RS mesh in a homogeneous system [6.1,6.30]. To reduce these spurious self-forces the NEC scheme of Ref. [6.30] is used. But instead of applying this scheme to all particles and calculating a single electric field for all particles, the NEC scheme is applied here only to the particle for which the electric field is calculated. The

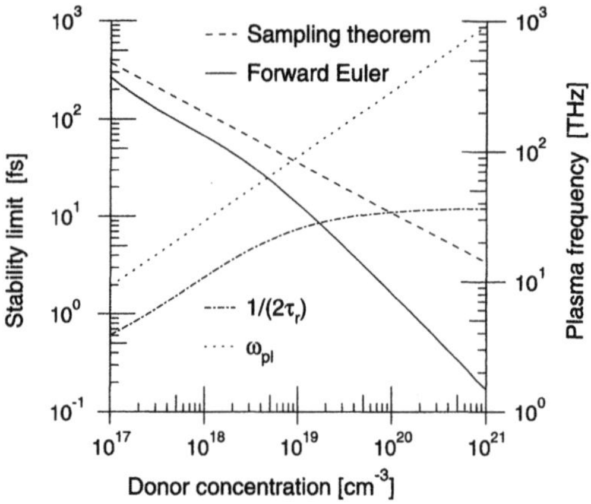

Figure 6.7: Stability limit of the self-consistent device simulation scheme for electrons at room temperature with $k = 0$ based on the Forward Euler scheme and based on the sampling theorem, and the inverse relaxation time and plasma frequency

contribution of the other particles to the electric field is still calculated with the method described in Sec. 6.3.

In the case of the NEC scheme the particle charge is distributed in equal parts to the four grid points of the cell the particle is in. This corresponds to the case of the cell-in-cloud method (6.27), if the particle is in the center of the cell. The electric field for the kth particle being in the cell $(i_k(t), j_k(t))$ is calculated as follows

$$\begin{aligned}\mathbf{E}_{i_k(t),j_k(t),k} &= \mathbf{E}_{i_k(t),j_k(t)}(l\Delta t) - \mathbf{E}^{\text{par}}_{i_k(t),j_k(t),k}(x_k(l\Delta t), y_k(l\Delta t)) \\ &+ \mathbf{E}^{\text{par}}_{i_k(t),j_k(t),k}\left(\frac{x_{i_k(l\Delta t)+1} + x_{i_k(l\Delta t)}}{2}, \frac{y_{j_k(l\Delta t)+1} + y_{j_k(l\Delta t)}}{2}\right) , \quad (6.53)\end{aligned}$$

where $\mathbf{E}_{i,j}(l\Delta t)$ is the electric field of the lth time step calculated as described in Sec. 6.3. $\mathbf{E}^{\text{par}}_{i,j,k}(x, y)$ is the electric field originating from the kth particle. This electric field is calculated considering only the charge of the particle at the position (x, y) and zero bias at the contacts. $(x_k(l\Delta t), y_k(l\Delta t))$ is the position of the kth particle at the beginning of the lth time step. The second term on the right-hand side of Eq. (6.53) removes the contribution of the particle under consideration from the total electric field and the third term reintroduces the electric field of the particle, but this time the particle's field is calculated for the particle's location being the center of the cell.

Solving the PE in each time step for each particle to obtain the respective fields $\mathbf{E}^{\text{par}}_{i,j,k}(x, y)$ would be too CPU intensive. Instead, the particle field is calculated with precomputed Green's functions $\mathbf{G}_{i,j,m,n}$ utilizing the principle of superposition

$$\begin{aligned}
\mathbf{E}^{\mathrm{par}}_{i,j,k}(x_k, y_k) &= Q^{\mathrm{sim}}_k \mathbf{G}_{i,j,i_k,j_k} \frac{x_{i_k+1} - x_k}{x_{i_k+1} - x_{i_k}} \frac{y_{j_k+1} - y_k}{y_{j_k+1} - y_{j_k}} \\
&+ Q^{\mathrm{sim}}_k \mathbf{G}_{i,j,i_k+1,j_k} \frac{x_k - x_{i_k}}{x_{i_k+1} - x_{i_k}} \frac{y_{j_k+1} - y_k}{y_{j_k+1} - y_{j_k}} \\
&+ Q^{\mathrm{sim}}_k \mathbf{G}_{i,j,i_k,j_k+1} \frac{x_{i_k+1} - x_k}{x_{i_k+1} - x_{i_k}} \frac{y_k - y_{j_k}}{y_{j_k+1} - y_{j_k}} \\
&+ Q^{\mathrm{sim}}_k \mathbf{G}_{i,j,i_k+1,j_k+1} \frac{x_k - x_{i_k}}{x_{i_k+1} - x_{i_k}} \frac{y_k - y_{j_k}}{y_{j_k+1} - y_{j_k}} \ . \qquad (6.54)
\end{aligned}$$

The Green's function $\mathbf{G}_{i,j,m,n}$ for the electric field in the grid cell (i, j) and for a unit charge on the mesh node (m, n) is calculated by solving the PE with zero bias at the contacts. To save computer memory the Green's functions are only stored for the first two indices (i, j) in a circle around the grid point (m, n) with the radius $r = v_{\max}\Delta t$, where $v_{\max}$ is the maximum particle velocity, because a particle can not move any further during a time step.

6.5 Nonlinear Poisson Equation

In general both particle types, electrons and holes have to be simulated by the MC method in all semiconductor regions of the device in order to calculate the particle density. But in some cases it can be assumed that one carrier type in a certain region of the device is in equilibrium, where its density is high enough to have an impact on the potential, and its density is well described by a constant quasifermi potential given by the relevant terminal bias (e.g. the electron quasifermi potential in the gate of an NMOSFET given by the gate bias, or the hole quasifermi potential in the base of an npn bipolar transistor given by the base bias) [6.22]. Instead of simulating this particle type in the specified region by the MC method the corresponding particle density is directly evaluated with Eq. (7.132) or (7.133), and the PE (6.11) becomes nonlinear [6.31]. The scheme for the self-consistent solution of the MC model and the PE (Fig. 6.5) does not change, but for each time step the nonlinear PE has to be solved. This can be done with the iterative Newton-Raphson method [6.32]. This scheme requires the LU decomposition of the Jacobian for each iteration, which is very CPU intensive. The computational burden can be reduced by reusing the Jacobian of the previous iteration. Although this scheme does not anymore converge quadratically, it converges in many cases within a few iterations after the initial transients of the stationary MC simulation have died out. Since the LU decomposition is avoided, this algorithm is much faster than the Newton-Raphson method. If the number of iterations for solving the PE within a time step of the SC loop (Fig. 6.5) is larger than ten, the Jacobian is updated. Otherwise the same Jacobian is used for every time step. Only in the beginning of the simulation the full Newton-Raphson method is used to calculate the initial potential.

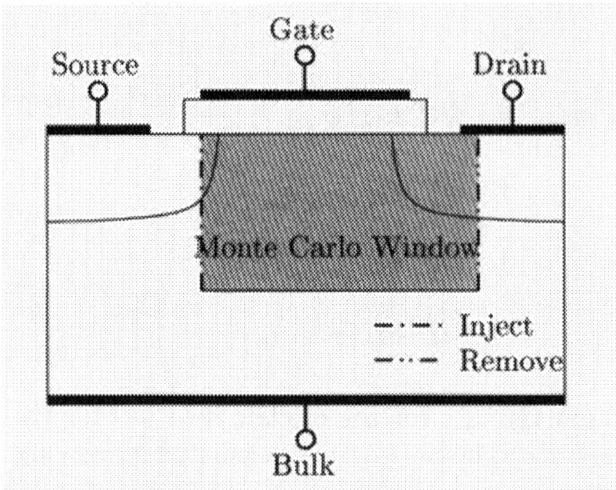

Figure 6.8: Monte-Carlo simulation window of the nonself-consistent method

6.6 Nonself-Consistent Device Simulation

In the case of a nonself-consistent (NSC) device simulation the MC simulation is performed for a given electric field [6.33, 6.34]. This field can be either the result of a previous SC-MC simulation or the result of a different simulation tool like the DD or HD model. The NSC method is mainly used to simulate hot-carrier effects and the simulation domain can be restricted to the area were hot-carrier effects are strong. In Fig. 6.8 a MOSFET is shown. While the whole device is simulated to calculate the electric field, the current density, and the particle density, the NSC-MC simulation is performed only within the marked window [6.33, 6.35]. On the left side of the window the particles are injected and on the right side they are removed. Every time step a particle is injected and the particle charge is given by the product of the current flowing through the injection boundary times the time step length. The injection probability density along the injection boundary is proportional to the component of the current density perpendicular to the injection interface. The particle state in the KS is randomly chosen with a Maxwell-Boltzmann distribution. When a particle hits the right boundary of the simulation window, it is removed. Otherwise the usual boundary conditions apply. The injection boundary is placed in a region where the particles are still close to equilibrium. This is either the source region or a region in the channel far away from the drain junction. The boundary where the particles are removed is placed in the drain region where the number of hot-carriers is negligible. This ensures that all hot-carriers are included in the simulation. Although the NSC method does not correctly reproduce the low energy distribution of the particles, the hot-carrier distribution is well simulated (Sec. 9.2.4). The method is not restricted to the simulation of II and substrate current [6.34, 6.36, 6.37], but can be applied to other hot-carrier phenomena as well [6.38–6.41].

This method has three advantages. First, the time steps of the MC simulation can be longer than in the SC case improving the efficiency of the simulation, because the particle motion is less frequently interrupted. Second, the simulation domain can be reduced excluding large parts of the high particle density regions in the source and drain. This reduces the carrier number or the time which a particle spends within the device. Third, the stability problems discussed in Sec. 6.4 are avoided.

The accuracy of the NSC approach depends on the accuracy of the electric field on which the NSC simulation is based. In Sec. 9.2.4 it is shown by a comparison of substrate current calculations by SC and NSC simulations based on an electric field evaluated by the HD model that the NSC approach yields accurate results for hot-carrier phenomena. In Sec. 9.2.6 the accuracy of the NSC approach is discussed with respect to terminal currents (e.g. drain current) and their noise.

6.7 Statistical Enhancement

The variance of any average over an ensemble (3.48) or time (3.73) of a microscopic quantity is proportional to the variance of the microscopic quantity itself $\sigma^2\{X\}$. Since the relative error (3.59) is proportional to the standard deviation of the average, a reduction in the variance of the microscopic quantity reduces the relative error and thereby increases the simulation efficiency.

By importance sampling the variance of selected microscopic variables can be reduced [6.42–6.45]. The distribution function is scaled with a positive nonzero weighting function

$$p^*(\zeta,t) = \frac{p(\zeta,t)}{w(\zeta)} . \tag{6.55}$$

The scaled distribution function must be normalized as well

$$\int p^*(\zeta,t)d\zeta = \int \frac{p(\zeta,t)}{w(\zeta)}d\zeta = 1 . \tag{6.56}$$

The expected values of the scaled distribution function are defined as

$$\mathrm{E}^*\{X\} = \int X(\zeta)p^*(\zeta,t)d\zeta . \tag{6.57}$$

To recover the original expected value the microscopic quantity has to be modified

$$X^*(\zeta) = w(\zeta)X(\zeta) . \tag{6.58}$$

The expectation of the modified quantity with respect to the scaled distribution function yields the original expected value

$$\mathrm{E}^*\{X^*\} = \mathrm{E}\{X\} . \tag{6.59}$$

Consequently, it is possible to construct unbiased estimators for the expected values (2.25) similar to Eq. (3.46) based on the modified microscopic quantity.

The modified variance of this quantity (6.58) reads

$$\begin{aligned}(\sigma^*\{X^*\})^2 &= \mathrm{E}^*\{(X^*)^2\} - (\mathrm{E}^*\{X^*\})^2 \\ &= \mathrm{E}\{wX^2\} - \mathrm{E}^2\{X\} \\ &\neq \sigma^2\{X\}\,, \end{aligned} \tag{6.60}$$

and it is not equal to the original variance of the microscopic quantity. Thus, by scaling the distribution function it is possible to modify the variance of the estimator without biasing the estimator. By choosing an appropriate weighting function the variance of a microscopic quantity can be reduced. For $w(\zeta) = \mathrm{E}\{X\}/X(\zeta)$ the variance (6.60) vanishes [6.43]. Unfortunately, this special weighting function requires the a priori knowledge of the expected value of which the estimation is the purpose of the simulation. Nevertheless, this example reveals two distinct features of importance sampling. First, it is possible to reduce the variance of an average improving the efficiency of the simulation. Second, the scaling function must be tailored for the quantity of interest. The variance of other quantities will in general increase due to the scaling. Therefore, importance sampling is also called variance redistribution [6.45].

In principle it is possible to reformulate the MC solver for the scaled distribution function [6.46], but these biased MC algorithms easily become instable and often only the expected value of the total particle charge is conserved. But exact conservation of the total particle charge is a prerequisite for an SC device simulation. In the case of a simulation with an unscaled distribution function of a system without charge sources and drains the total charge is constant, because the charge carried by each simulation particle is constant. In the case of a biased MC algorithm the charge is proportional to the statistical weight of the simulation particle. Since the statistical weight does change with the particle state, the total charge might fluctuate and only the expected value of the total charge is conserved. These artificial charge fluctuation might impair the SC solution of the MC method and the PE.

A different approach is to divide the PS into regions with different constant relative weights. Whenever a particle moves from one region into another the particle is either split or removed (gathered) depending on the change in weight at the crossed boundary [6.33, 6.47, 6.48]. Since the distribution function is not known before the actual simulation, it might be necessary to control the total number of particles or the distribution of the particles over the PS making it difficult to apply this method [6.49]. In addition, exact charge conservation is violated in the case that Russian roulette is used to remove particles.

The method used in this work exactly conserves the total particle charge. It falls into the class of population control methods and is also based on a partitioned PS as above, but instead of the particle weight the number of particles is fixed for each region [6.11, 6.44, 6.45, 6.50]. The MC algorithm is not modified and the same for all particles regardless of their statistical weight. This means that the particles can move from one region to another and thereby, the number of particles in a region can change during the MC simulation. To maintain the given particle number in a region the simulation is stopped at fixed time intervals (typically every few femtoseconds). The number of particles is checked and in the case that the number deviates from the given number in the region the particle subensemble of that region is replaced by an ensemble with the desired number of particles. This process is called *Refresh* and it is continually repeated (*Multiple Refresh* (MR)) to maintain the given number of particles in all regions of the PS during the simulation.

The refresh process for one region is independent of all other regions and the following discussion is limited to one region for simplicity. The subensemble of this region before the refresh consists of N^{b} particles, where the ith particle is in the state $\boldsymbol{\zeta}_i^{\mathrm{b}}$ and carries the weight w_i^{b}. The new subensemble after the refresh consists of N^{a} particles with the uniform weight

$$w_j^{\mathrm{a}} = w^{\mathrm{a}} = \frac{\sum_{i=1}^{N^{\mathrm{b}}} w_i^{\mathrm{b}}}{N^{\mathrm{a}}} . \tag{6.61}$$

This ensures conservation of the total statistical weight which is proportional to the total particle charge within the PS region. The new particles are placed in the PS onto the positions of the particles before the refresh $\boldsymbol{\zeta}_i^{\mathrm{b}}$ with the probability

$$P_i^{\mathrm{b}} = \frac{w_i^{\mathrm{b}}}{\sum_{j=1}^{N^{\mathrm{b}}} w_j^{\mathrm{b}}} . \tag{6.62}$$

The states of the new particles are randomly selected with a variant of the direct method called *Comb* [6.45]

$$i^{\mathrm{b}}(j) \ : \ \sum_{k=1}^{i^{\mathrm{b}}(j)-1} P_k^{\mathrm{b}} < \frac{j-1+r}{N^{\mathrm{a}}} \leq \sum_{k=1}^{i^{\mathrm{b}}(j)} P_k^{\mathrm{b}} \qquad \text{for} \quad j = 1, 2, \cdots, N^{\mathrm{a}} , \tag{6.63}$$

where $i^{\mathrm{b}}(j)$ is the index of the old particle state onto which the jth new particle is placed. The random number r uniformly distributed between zero and one is the same for the selection of all N^{a} new particles. In Fig. 6.9 an example for a comb is given. The number of new particles placed on the ith old particle state is $n_i^{\mathrm{a,b}}$. The expected value of the occupation number with respect to the comb is [6.45]

$$\mathrm{E}\{n_i^{\mathrm{a,b}}\} = N^{\mathrm{a}} P_i^{\mathrm{b}} . \tag{6.64}$$

There are only two possible values for the occupation number. It either equals the integer part of its expected value $\lfloor N^{\mathrm{a}} P_i^{\mathrm{b}} \rfloor$ or it equals the integer part plus one $\lceil N^{\mathrm{a}} P_i^{\mathrm{b}} \rceil$

$$\lfloor N^{\mathrm{a}} P_i^{\mathrm{b}} \rfloor \leq n_i^{\mathrm{a,b}} \leq \lceil N^{\mathrm{a}} P_i^{\mathrm{b}} \rceil . \tag{6.65}$$

In the case that the expected value of the occupation number is an integer, the occupation number and the expected value are equal. Thus, in the case that

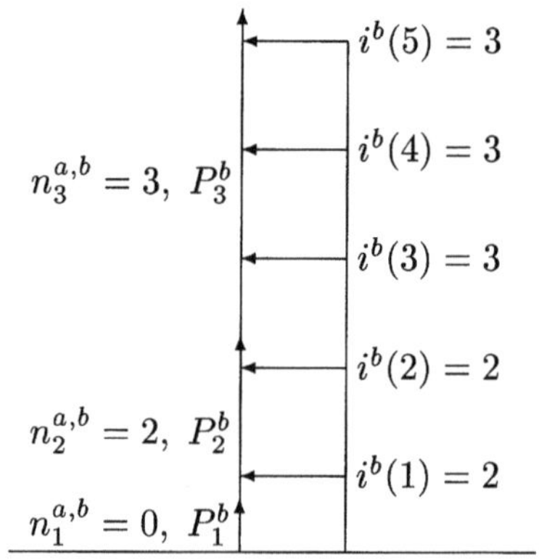

Figure 6.9: A comb for three old particles ($P_1^{\mathrm{b}} = 0.1$, $P_2^{\mathrm{b}} = 0.3$, $P_3^{\mathrm{b}} = 0.6$), five new ones, and $r = 0.7$

the comb is applied to a subensemble with uniformly weighted particles and the desired number of particles ($N^{\mathrm{a}} = N^{\mathrm{b}}$), the subensemble is not changed, because all occupation numbers are equal one.

The comb conserves the expected value of averages over the subensemble. An average before the refresh is given by

$$\tilde{X}^{\mathrm{b}} = \sum_{i=1}^{N^{\mathrm{b}}} X(\zeta_i) P_i^{\mathrm{b}} \tag{6.66}$$

and afterwards by

$$\tilde{X}^{\mathrm{a}} = \sum_{i=1}^{N^{\mathrm{b}}} X(\zeta_i) \frac{n_i^{\mathrm{a,b}}}{N^{\mathrm{a}}} \ . \tag{6.67}$$

The expected value of the latter estimator

$$\mathrm{E}\{\tilde{X}^{\mathrm{a}}\} = \sum_{i=1}^{N^{\mathrm{b}}} X(\zeta_i) \frac{\mathrm{E}\{n_i^{\mathrm{a,b}}\}}{N^{\mathrm{a}}} = \sum_{i=1}^{N^{\mathrm{b}}} X(\zeta_i) P_i^{\mathrm{b}} = \tilde{X}^{\mathrm{b}} \tag{6.68}$$

equals the average over the subensemble before the refresh.

The refresh does not produce any new information. The information contained in the subensemble before the refresh is either the same after the refresh or it is reduced. In the example in Fig. 6.9 one before refresh particle state is lost during the refresh and information is destroyed. In addition, some of the new particles are in the same particle state and are therefore correlated. This correlation is only subsequently broken during the MC simulation by particle scattering and new information is generated.

In order to reduce the CPU time a region is only refreshed if necessary. Two criteria are used to determine whether a region should be refreshed or not. The first criterion controls the particle number in the region

$$N^{\mathrm{b}} < \frac{N^{\mathrm{a}}}{r_{\mathrm{p}}} \quad \vee \quad N^{\mathrm{b}} > r_{\mathrm{p}} N^{\mathrm{a}} \quad \text{with} \quad r_{\mathrm{p}} > 1 \,. \tag{6.69}$$

Whenever the current particle number deviates from the desired number by more than a factor of r_{p} or $1/r_{\mathrm{p}}$ the region is refreshed. As shown in Sec. 3.4 the smallest variance is obtained for an ensemble of independent particles in the case of a uniform particle weight. Therefore, a region is refreshed, whenever the particle weights are too nonuniform

$$\sum_{i=1}^{N^{\mathrm{b}}} (w_i^{\mathrm{b}})^2 > r_{\mathrm{s}} N^{\mathrm{b}} \quad \text{with} \quad r_{\mathrm{s}} > 1 \,, \tag{6.70}$$

where N^{b} is the minimum value of the sum (3.50) obtained for an ensemble with uniform particle weights.

With the comb it is possible to increase or decrease the number of particles in a region to any given value as long as at least one particle is found in that region. This allows to control the number of particles in all regions of the PS. In Fig. 6.10 the electron distribution function is shown for an undoped homogeneous silicon system at room temperature for an electric field of 100kV/cm simulated with and without MR. In order to apply the MR method the KS was partitioned into regions by equienergy surfaces with 0.1eV spacing between 0eV and 3.0eV. To each region 10 simulation particles were assigned and the Refresh was applied every 4fs with $r_{\mathrm{p}} = r_{\mathrm{s}} = 1.5$. The simulations with and without MR were performed for the same CPU time and the data were sampled for bins of 1meV width. The resultant distribution functions with and without MR are the same within the statistical noise. But the standard deviations strongly differ and in the case of the simulation without MR the statistical noise becomes too large for values of the distribution function below 10^{-4}/eV, whereas the simulation with MR yields good results with a low relative standard deviation over the shown range of 10 orders of magnitude. This is due to the distribution of the simulation particles over energy. Without MR the simulation particle distribution is given by the distribution function shown in Fig. 6.10 and the probability density for a simulation particle to have a high energy is very small resulting in a huge relative statistical noise for high energies. With MR the simulation particle density is kept constant over the energy. Thus, more particles are simulated at higher energies with MR than without MR. To obtain the correct distribution function the statistical weight of the simulation particles is accordingly modified by the MR. On the other hand, fewer particles are simulated at low energies and the standard deviation for these energies exceeds

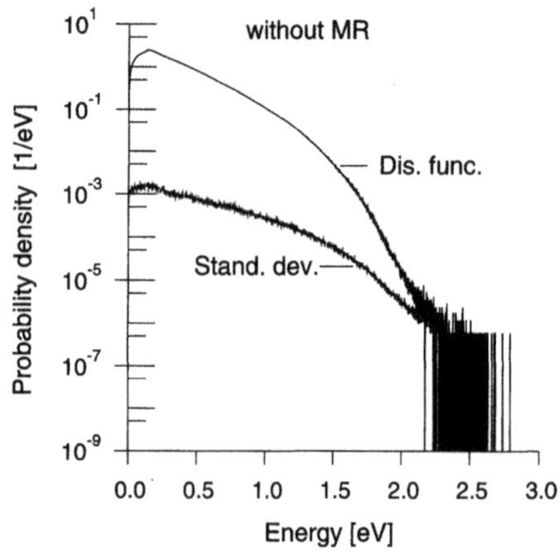

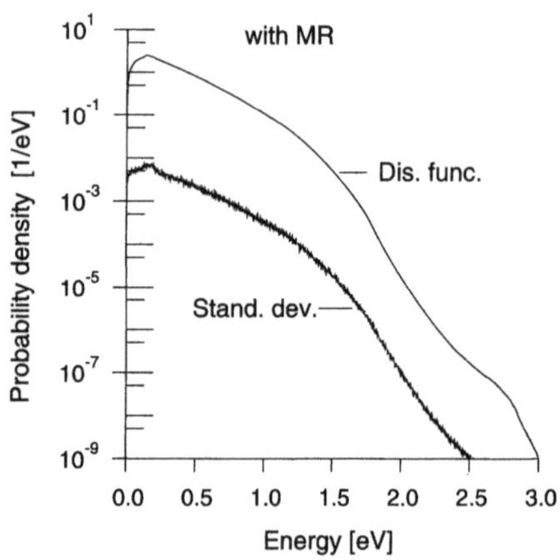

Figure 6.10: One electron distribution function and standard deviation for bulk silicon at room temperature and an electric field of 100kV/cm simulated without (upper graph) and with (lower graph) Multiple Refresh for the same CPU time

the one without MR. This clearly demonstrates the variance redistribution by the MR method. The standard deviation of rare events is decreased at the cost of the common events [6.45]. But the reader should keep in mind that the standard deviation increases only a factor of five at low energies with MR, whereas it is reduced by orders of magnitude at high energies. A similar improvement at high energies without MR would require an increase in CPU time of many orders of magnitude, because the standard deviation is proportional to the square root of the simulation time.

A similar behavior is found, when the MR is applied to the RS. In Fig. 6.11 results are shown for a 1D N^+NN^+ structure biased with 2.0V. The doping concentration and electron density are shown in the upper graph and the relative

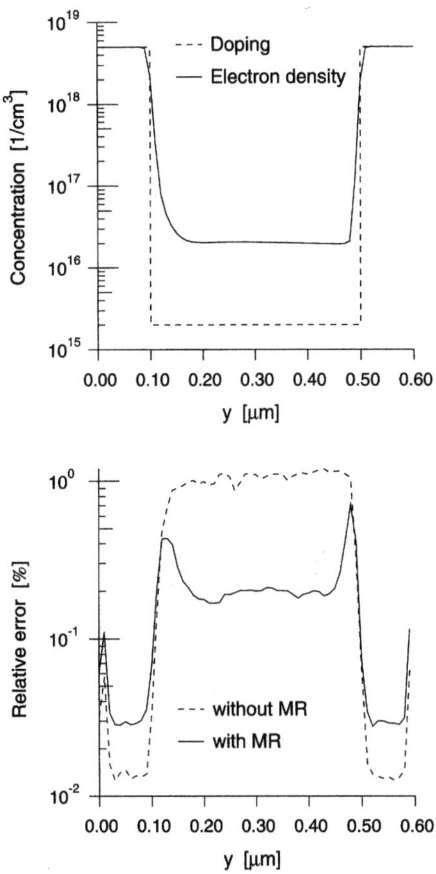

Figure 6.11: Electron density and doping concentration for a 1D N^+NN^+ structure biased at 2.0V (upper graph) and the relative error with and without MR for the same simulation time (lower graph)

error (3.59) with and without MR in the lower one. In both cases the same number of particles was simulated for the same simulation time. Without MR 6000 uniformly weighted particles were used. With MR the simulation particle density was kept constant over the RS with 100 particles per cell of the equidistant RS grid of 10nm spacing, again resulting in a total of 6000 particles. The relative error was calculated by sampling 300 subhistories of 10ps. With MR the variance and thus the relative error are reduced in the low density area at the cost of a higher variance in the high density regions.

In the case of an SC device simulation the Multiple Refresh method has to be applied carefully. Although the Refresh conserves the total particle charge within a region, it might redistribute the charge over the region which in turn might have a spurious impact on the solution of the PE. In order to reduce this effect in the case of an SC device simulation the cells of the RS grid are chosen as the regions for the MR in the RS avoiding an artificial charge transfer from one cell to another. Moreover, the maximum simulation particle charge must be so small that a so-called collisionless plasma is simulated to avoid artifical self-heating [6.1, 6.11] and especially spurious heating of particles with a small charge by particles with a large charge [6.51]. Therefore, in the case of an SC device simulation not only the number of simulation particles per grid cell has to be chosen but also the maximum simulation particle charge. Whenever the charge of a simulation particle exceeds the given maximum, the number of desired particles for the respective cell is increased to reduce the particle charge below the maximum value. Thus, in regions of the device with a high real particle density the number of simulation particles per grid cell might be much larger than in low density areas. Nevertheless the statistics are improved in the low density regions as long as the simulation particle charge is considerably below the maximum value.

6.8 Terminal Current Estimation

Terminal currents are given by the integral of the total current density over the contact area

$$I_l = \int_{A_l} \left(\mathbf{i} + \epsilon \frac{\partial \mathbf{E}}{\partial t} \right)^{\mathrm{T}} d^2\mathbf{r} \ , \tag{6.71}$$

where $\mathbf{i}$ is the expected conduction current density, $\epsilon \partial \mathbf{E}/\partial t$ the displacement current density, I_l the terminal current per unit length, and A_l the area of the lth contact. The latter two quantities are both outwardly facing. Assuming that Neumann-type boundary conditions apply to all surfaces of the device, which are not part of the contacts, the terminal currents must fulfill Kirchhoff's current law

$$\sum_{l=1}^{N_{\mathrm{cont}}} I_l = 0 \ . \tag{6.72}$$

N_{cont} is the number of contacts. A terminal current estimator based on the boundary integral (6.71) which satisfies Eq. (6.72) is given in Ref. [6.52].

In the following, the domain integration technique is used, which is based on test functions [6.53], because it is easier to implement than the boundary integral method of Ref. [6.52] and more general. Test functions are defined for each contact, where the values of the test functions on the contacts are given by

$$h_l(\mathbf{r}) = \delta_{l,m} \quad \text{for} \quad \mathbf{r} \in A_m \ , \tag{6.73}$$

where $\delta_{l,m}$ is the Kronecker symbol. With the test functions the surface integral in Eq. (6.71) can be extended over the whole surface of the device. Application of the theorem of Gauss and using the fact that the total current density is divergenceless and satisfies Neumann-type boundary conditions on all parts of the surface that do not belong to the terminals results in

$$I_l = \oint \left(\mathbf{i} + \epsilon \frac{\partial \mathbf{E}}{\partial t} \right)^{\mathrm{T}} h_l d^2\mathbf{r} = \int \left(\mathbf{i} + \epsilon \frac{\partial \mathbf{E}}{\partial t} \right)^{\mathrm{T}} \nabla_{\mathbf{r}} h_l d^3 r \ . \tag{6.74}$$

Still, the test functions have to be specified within the device domain and on the parts of the surface that do not belong to the contacts.

In the case of the Ramo-Shockley theorem the test functions are used to simplify the estimator [6.54]. To this end, the quasi-static potential is split into two parts related to the space charge (sc) and contact bias (cb)

$$\Psi(\mathbf{r}, t) = \Psi_{\mathrm{sc}}(\mathbf{r}, t) + \Psi_{\mathrm{cb}}(\mathbf{r}, t) \ , \tag{6.75}$$

where the part of the potential related to the space charge must satisfy the Poisson equation (6.11), Dirichlet-type boundary conditions on the contacts with zero bias

$$\Psi_{\mathrm{sc}}(\mathbf{r}, t) = 0 \quad \text{for} \quad \mathbf{r} \in A_m \ , \tag{6.76}$$

and Neumann-type boundary conditions on all other parts of the surface. The part of the potential due to the contact bias must satisfy the Laplace equation

$$\nabla_{\mathbf{r}}^{\mathrm{T}} (\epsilon \nabla_{\mathbf{r}} \Psi_{\mathrm{cb}}) = 0 \ , \tag{6.77}$$

Dirichlet-type boundary conditions on the contacts

$$\Psi_{\mathrm{cb}}(\mathbf{r}, t) = V_m^{\mathrm{a}}(t) \quad \text{for} \quad \mathbf{r} \in A_m \ , \tag{6.78}$$

where V_m^{a} is the voltage applied to the mth contact, and Neumann-type boundary conditions on all other parts of the surface.

The Ramo-Shockley (RS) test functions are defined as the basis functions of the Laplace equation (6.77) [6.19]

$$\nabla_{\mathbf{r}}^{\mathrm{T}} (\epsilon \nabla_{\mathbf{r}} h_l^{\mathrm{RS}}) = 0 \tag{6.79}$$

together with the Dirichlet-type boundary conditions (6.73) on the contacts and Neumann-type boundary conditions on all other parts of the surface

$$\left(-\epsilon \nabla_{\mathbf{r}} h_l^{\mathrm{RS}}\right)^{\mathrm{T}} d^2\mathbf{r} = 0 \quad \text{for} \quad \mathbf{r} \in A \wedge \mathbf{r} \notin A_m (m = 1, \cdots, N_{\mathrm{cont}}) \ , \tag{6.80}$$

such that

$$\Psi_{\mathrm{cb}}(\mathbf{r}, t) = \sum_{m=1}^{N_{\mathrm{cont}}} h_m^{\mathrm{RS}}(\mathbf{r}) V_m^{\mathrm{a}}(t) \ . \tag{6.81}$$

The displacement current at the lth terminal due to the variation of the contact bias is given by

$$\int \epsilon \frac{\partial \left(-\nabla \Psi_{\mathrm{cb}}\right)^{\mathrm{T}}}{\partial t} \nabla_{\mathbf{r}} h_l^{\mathrm{RS}} d^3 r = - \sum_{m=1}^{N_{\mathrm{cont}}} C_{l,m} \frac{\partial V_m^{\mathrm{a}}}{\partial t} \ , \tag{6.82}$$

where the capacitance coefficients $C_{l,m}$ are given by

$$C_{l,m} = \int \epsilon \left(\nabla_{\mathbf{r}} h_l^{\mathrm{RS}}\right)^{\mathrm{T}} \nabla_{\mathbf{r}} h_m^{\mathrm{RS}} d^3 r \ . \tag{6.83}$$

The space charge related part of the displacement current evaluates to

$$\begin{aligned} \int \epsilon \frac{\partial \left(-\nabla \Psi_{\mathrm{sc}}\right)^{\mathrm{T}}}{\partial t} \nabla_{\mathbf{r}} h_l^{\mathrm{RS}} d^3 r &= -\frac{\partial}{\partial t} \int \left(\nabla \Psi_{\mathrm{sc}}\right)^{\mathrm{T}} \epsilon \nabla_{\mathbf{r}} h_l^{\mathrm{RS}} d^3 r \\ &= -\frac{\partial}{\partial t} \oint \Psi_{\mathrm{sc}} \left(\epsilon \nabla_{\mathbf{r}} h_l^{\mathrm{RS}}\right)^{\mathrm{T}} d^2 \mathbf{r} \\ &\quad + \frac{\partial}{\partial t} \int \Psi_{\mathrm{sc}} \nabla_{\mathbf{r}}^{\mathrm{T}} (\epsilon \nabla_{\mathbf{r}} h_l^{\mathrm{RS}}) d^3 r \\ &= 0 \ , \end{aligned} \tag{6.84}$$

where the surface integral on the right-hand side of the second equality vanishes because of the boundary conditions (6.76), (6.80), and the volume integral due to Eq. (6.79).

By substituting Eqs. (6.82) and (6.84) into Eq. (6.74) the Ramo-Shockley theorem is obtained [6.54]

$$I_l = \int \mathbf{i}^{\mathrm{T}} \nabla_{\mathbf{r}} h_l^{\mathrm{RS}} d^3 r - \sum_{m=1}^{N_{\mathrm{cont}}} C_{l,m} \frac{\partial V_m^{\mathrm{a}}}{\partial t} \ . \tag{6.85}$$

Thus, the current due to the charge transport within the device can be calculated without evaluating the displacement current density due to the charge carriers. This makes the evaluation of the terminal currents by MC simulation simpler. The first term on the right-hand side of Eq. (6.85) can be estimated by MC simulation

$$\int \tilde{\mathbf{i}}^{\mathrm{T}} \nabla_{\mathbf{r}} h_l^{\mathrm{RS}} d^3 r = \sum_{k=1}^{N_{\mathrm{sim}}} Q_k^{\mathrm{sim}} \mathbf{v}_k^{\mathrm{T}} \nabla_{\mathbf{r}} h_l^{\mathrm{RS}} (\mathbf{r}_k) \ , \tag{6.86}$$

where Q_k^{sim}, $\mathbf{v}_k$, and $\mathbf{r}_k$ are the charge, velocity, and position of the kth particle, respectively.

The sum of the estimated terminal currents over all contacts satisfies Kirchhoff's current law (6.72) [6.53]

$$\sum_{l=1}^{N_{\mathrm{cont}}} \sum_{k=1}^{N_{\mathrm{sim}}} Q_k^{\mathrm{sim}} \mathbf{v}_k^{\mathrm{T}} \nabla_{\mathbf{r}} h_l^{\mathrm{RS}} (\mathbf{r}_k) = \sum_{k=1}^{N_{\mathrm{sim}}} Q_k^{\mathrm{sim}} \mathbf{v}_k^{\mathrm{T}} \nabla_{\mathbf{r}} \left[\sum_{l=1}^{N_{\mathrm{cont}}} h_l^{\mathrm{RS}} (\mathbf{r}_k) \right] = 0 \ , \tag{6.87}$$

because the sum of all test functions is constant within the device domain

$$\sum_{l=1}^{N_{\mathrm{cont}}} h_l^{\mathrm{RS}} (\mathbf{r}) = 1 \ . \tag{6.88}$$

Due to the principle of superposition the sum of all test functions is the solution of Eq. (6.79) with the boundary condition that the sum must be one on all contacts. Therefore, the sum of all test functions is one within the device domain. The same arguments hold for the contribution of the displacement currents due to the variation of the contact voltages and the total terminal currents also satisfy Kirchhoff's current law.

6.9 Contact Resistance

The area of the device simulated by the MC model is kept as small as possible to increase the efficiency of the simulation (cf. Sec. 8.2). Therefore, a large part of the series resistance of the contact areas and the contact resistance itself are not included in the MC simulation. The neglected parts of the terminal resistances can be included with a linear network of discrete resistors

$$V_l^{\mathrm{a}} = V_l^{\mathrm{cb}} + R_l I_l \ , \tag{6.89}$$

where V_l^{a} is the voltage of the lth terminal, V_l^{cb} the externally applied contact bias, and R_l the contact resistance. Since the terminal current (6.85) depends on the voltages $V_m^{\mathrm{a}'}$ and their derivatives with respect to time, Eq. (6.89) is solved with a forward Euler scheme corresponding to the one shown in Fig. 6.5. To this end the terminal current due to the particles is averaged over the ith time step

$$I_{l,i}^{\mathrm{par}} = \frac{1}{t_{i+1} - t_i} \int_{t_i}^{t_{i+1}} \int \tilde{\mathbf{i}}^{\mathrm{T}}(t) \nabla_{\mathbf{r}} h_l^{\mathrm{RS}} d^3r dt \ . \tag{6.90}$$

The displacement current due to the temporal change in the terminal bias is calculate with finite differences

$$I_{l,i}^{\mathrm{dis}} = - \sum_{m=1}^{N_{\mathrm{cont}}} C_{l,m} \frac{V_{m,i+1}^{\mathrm{a}} - V_{m,i}^{\mathrm{a}}}{t_{i+1} - t_i} \ . \tag{6.91}$$

Inserting these expressions for the particle and displacement currents into Eq. (6.89) yields

$$V_{l,i+1}^{\mathrm{a}} = V_l^{\mathrm{cb}} + R_l I_{l,i}^{\mathrm{par}} - \sum_{m=1}^{N_{\mathrm{cont}}} R_l C_{l,m} \frac{V_{m,i+1}^{\mathrm{a}} - V_{m,i}^{\mathrm{a}}}{t_{i+1} - t_i} \ . \tag{6.92}$$

The solution of this linear system of equations is the terminal voltage $V_{l,i+1}^{\mathrm{a}}$ of the next time step.

Table 6.2: *Fundamental units used within the MC program*

$k_B T_0$	$8.617084 \cdot 10^{-5} T_0$	eV
m_0	$9.109558 \cdot 10^{-31}$	kg
$\hbar$	$6.582183 \cdot 10^{-16}$	eVs
q	$1.602192 \cdot 10^{-19}$	As

6.10 Normalization of Physical Quantities

The formulas describing the MC model contain numerous physical constants of which the consideration consumes CPU time. By using units different from the SI units within the simulation program, these physical constants can be removed. In addition, the normalization homogenizes the magnitude of the internal variables reducing the risk of numerical over- and underflows. Here, the system of units given in Ref. [6.11] is used. It is based on the four fundamental quantities shown in Tab. 6.2. The use of the thermal energy results in a lattice temperature dependent normalization. But since the lattice temperature is assumed to be constant this normalization explicitly removes the lattice temperature dependence from all formulas. With the four fundamental units the normalization factors for other quantities are readily calculated. For example, the normalization factor of the time is $\hbar/k_B T_0$. When a quantity is transfered into the simulator it must be divided by the normalization factor and on output multiplied. Please note that the normalization factor of a dielectric constant is not given by the dielectric constant of the vacuum.

References

[6.1] R.W. Hockney and J.W. Eastwood, *Computer Simulation Using Particles*, Institute of Physics Publishing, Bristol, Philadelphia, 1988.

[6.2] J. D. Cressler, "Re-engineering silicon: Si-Ge heterojunction bipolar technology", *IEEE Spectrum*, vol. 3, pp. 49–55, 1995.

[6.3] A. Schüppen, "SiGe–HBTs for mobile communication", *Solid–State Electron.*, vol. 43, pp. 1373–1381, 1999.

[6.4] C. K. Maiti and G. A. Armstrong, *Applications of Silicon–Germanium Heterostructure Devices*, Series in Optics and Optoelectronics. Institute of Physics Publishing, Bristol, Philadelphia, 2001.

[6.5] L. Risch, H. Schäfer, B. Lustig, F. Hofmann, U. Scheler, M. Franosch, W. Roesner, T. Aeugle, and H. Fischer, "Channel engineering using RP–CVD epitaxy for high performance CMOS transistors", in *Proc. ESSDERC*, Bologna, 1996, vol. 26, pp. 321–324.

[6.6] P. Bouillon, T. Skotnicki, S. Bodnar, C. Morin, J.-L. Regolini, P. Gouagout, and

P. Dollfus, "Experiments with 0.18 μm SiGe channel pMOSFETs and p$^+$ poly-SiGe gate", in *Proc. ESSDERC*, Bologna, 1996, vol. 26, pp. 473–476.
[6.7] T. Takagi, A. Inoue, Y. Hara, Y. Kanzawa, and M. Kubo, "A novel high performance SiGe channel heterostructure dynamic threshold pMOSFET (HDTMOS)", *IEEE Electron Device Lett.*, vol. 22, no. 5, pp. 206–208, 2001.
[6.8] J. L. Hoyt, H. M. Nayfeh, S. Eguchi, I. Aberg, G. Xia, T. Drake, E. A. Fitzgerald, and D. A. Antoniadis, "Strained silicon MOSFET technology", *IEDM Tech. Dig.*, pp. 23–26, 2002.
[6.9] K. Rim, J. L. Hoyt, and J. F. Gibbons, "Fabrication and analysis of deep submicron strained-Si N-MOSFET's", *IEEE Trans. Electron Devices*, vol. 47, no. 7, pp. 1406–1415, 2000.
[6.10] C. Jungemann, S. Keith, and B. Meinerzhagen, "Full-band Monte Carlo device simulation of a Si/SiGe-HBT with a realistic Ge profile", *IEICE Trans. on Electronics*, vol. E83-C, no. 8, pp. 1228–1234, 2000.
[6.11] H.-J. Peifer, "Monte–Carlo Simulation des Hochenergietransports von Elektronen in submikron MOS–Strukturen", Doctor thesis, RWTH Aachen, Aachen, 1992, Augustinus Buchhandlung.
[6.12] F. Schäffler, "High–mobility Si and Ge structures", *Semicond. Sci. Technol.*, vol. 12, pp. 1515–1549, 1997.
[6.13] D. Nuernbergk, "Simulation des Transportverhaltens in Si/Si$_{1-x}$Ge$_x$/Si–Heterobipolartransistoren", Dissertation, Technische Universität Ilmenau, Ilmenau, 1999, (H. Utz Verlag Wissenschaft, München: 1999).
[6.14] J. W. Slotboom and H. C. de Graaf, "Measurements of bandgap narrowing in Si bipolar transistors", *Solid-State Electron.*, vol. 19, pp. 857–862, 1976.
[6.15] D. B. M. Klaassen, J. W. Slotboom, and H. C. de Graaf, "Unified apparent bandgap narrowing in *n*- and *p*-type silicon", *Solid–State Electron.*, vol. 35, pp. 125–129, 1992.
[6.16] R. J. E. Hueting, J. W. Slotboom, A. Pruijmboom, W. B. de Boer, C. E. Timmering, and N. E. B. Cowern, "On the optimization of SiGe-base bipolar transistors", *IEEE Trans. Electron Devices*, vol. 43, no. 9, pp. 1518–1524, 1996.
[6.17] M. A. Green, "Intrinsic concentration, effective densities of states, and effective mass in silicon", *J. Appl. Phys.*, vol. 67, pp. 2944–2954, 1990.
[6.18] H. K. Dirks, "Kapazitätskoeffizienten nichtlinearer dissipativer Systeme", Habilitationsschrift, RWTH Aachen, 1988.
[6.19] J. D. Jackson, *Classical Electrodynamics*, John Wiley & Sons, New York, 2nd edition, 1975.
[6.20] S. M. Sze, *Physics of Semiconductors Devices*, Wiley, New York, 1981.
[6.21] R. S. Varga, *Matrix Iterative Analysis*, Series in Automatic Computation. Prentice-Hall, Englewood Cliffs, New Jersey, 1962.
[6.22] M. V. Fischetti and S. E. Laux, "Monte Carlo analysis of electron transport in small semiconductor devices including band–structure and space–charge effects", *Phys. Rev. B*, vol. 38, pp. 9721–9745, 1988.
[6.23] M. Saraniti, A. Rein, G. Zandler, P. Vogl, and P. Lugli, "An efficient multigrid poisson solver for device simulations", *IEEE Trans. Computer–Aided Des.*, vol. 15, pp. 141–150, 1996.
[6.24] J. Stoer and R. Bulirsch, *Einführung in die Numerische Mathematik*, vol. 2, Springer, Berlin Heidelberg New York, 2nd edition, 1978.

[6.25] P. W. Rambo and J. Denavit, "Time stability of Monte Carlo device simulation", *IEEE Trans. Computer-Aided Des.*, vol. 12, pp. 1734–1741, 1993.

[6.26] A. Ghetti, X. Wang, F. Venturi, and F. A. Leon, "Stability issues in self-consistent Monte Carlo simulations", in *Proc. SISDEP*, Erlangen, 1995, vol. 6, pp. 388–391.

[6.27] P.W. Rambo and J. Denavit, "Time-step stability for self-consistent Monte Carlo device simulations", in *Proc. IWCE*, Portland, Oregon (USA), May 1994, pp. 33–36.

[6.28] F. Liebig, A. Abou Elnour, and K. Schünemann, "An implicit coupling scheme for the use of long time steps in stable self-consistent particle simulation of semiconductor devices with high doping levels", in *Proc. SISPAD*, Tokyo, 1996, pp. 45–46.

[6.29] A. Papoulis, *Probability,Random Variables and Stochastic Processes*, Mc Graw-Hill, 3rd edition, 1991.

[6.30] S. E. Laux, "On particle-mesh coupling in Monte Carlo semiconductor device simulation", *IEEE Trans. Computer-Aided Des.*, vol. 15, pp. 1266–1277, 1996.

[6.31] S. Selberherr, *Analysis and Simulation of Semiconductor Devices*, Springer, Wien, 1984.

[6.32] J. Stoer, *Einführung in die Numerische Mathematik*, vol. 1, Springer, Berlin Heidelberg New York, 2nd edition, 1978.

[6.33] E. Sangiorgi, B. Riccò, and F. Venturi, "MOS^2: An efficient Monte Carlo simulator for MOS devices", *IEEE Trans. Computer-Aided Des.*, vol. 7, no. 2, pp. 259–271, 1988.

[6.34] J. M. Higman, K. Hess, C. G. Hwang, and R. W. Dutton, "Coupled Monte Carlo-drift diffusion analysis of hot-electron effects in MOSFET's", *IEEE Trans. Electron Devices*, vol. 36, no. 5, pp. 930–937, 1989.

[6.35] D. Y. Cheng, C. G. Hwang, and R. W. Dutton, "PISCES-MC: A multiwindow and multimethod 2-D device simulator", *IEEE Trans. Computer-Aided Des.*, vol. 7, pp. 1017–1026, 1988.

[6.36] J. D. Bude and M. Mastrapasqua, "Impact ionization and distribution functions in sub-micron nMOSFET technologies", *IEEE Electron Device Lett.*, vol. 16, no. 10, pp. 439–441, 1995.

[6.37] C. Jungemann, S. Yamaguchi, and H. Goto, "On the accuracy and efficiency of substrate current calculations for sub-μm n-MOSFET's", *IEEE Electron Device Lett.*, vol. 17, no. 10, pp. 464–466, 1996.

[6.38] J. D. Bude, "Monte Carlo simulation of impact ionization feedback in sub-micron MOSFET technologies", in *Exte. Abst. of the 1995 Inte. Conf. on Solid State Devices and Materials*, Osaka, 1995.

[6.39] J. Bude, T. Iizuka, and Y. Kamakura, "Determination of threshold energy for hot electron interface state generation", *IEDM Tech. Dig.*, pp. 865–868, 1996.

[6.40] C. Jungemann, S. Yamaguchi, and H. Goto, "Accurate prediction of hot-carrier effects for a deep sub-μm CMOS technology based on inverse modeling and full band Monte Carlo device simulation", in *Proc. SISPAD*, 1996, vol. 1, pp. 59–60.

[6.41] C. Jungemann, S. Yamaguchi, and H. Goto, "Investigation of the influence of impact ionization feedback on the spatial distribution of hot carriers in an NMOSFET", in *Proc. ESSDERC*, Stuttgart, 1997, vol. 27, pp. 336–339.

[6.42] R. Y. Rubinstein, *Simulation and the Monte Carlo method*, Wiley series in probability and mathematical statistics. John Wiley & Sons, New York, 1981.

[6.43] H.-J. Schlebusch, "Importance Sampling Techniken zur effizienten Simulation von Kommunikationssystemen", Doctor thesis, RWTH Aachen, 1991.

[6.44] C. Jungemann, S. Decker, R. Thoma, W.-L. Engl, and H. Goto, "Phase space multiple refresh: A general purpose statistical enhancement technique for Monte Carlo device simulation", *IEEE J. Tech. Comp. Aided Design*, , no. 2, 1997.

[6.45] M. G. Gray, T. E. Booth, T. J. T. Kwan, and C. M. Snell, "A multi-comb variance reduction scheme for Monte Carlo semiconductor simulators", *IEEE Trans. Electron Devices*, vol. 45, no. 4, pp. 918–924, 1998.

[6.46] A. M. Kriman and R. P. Joshi, "Scaled ensemble Monte Carlo studies of impact ionization", in *Proc. IWCE*, Portland, Oregon, May 1994, pp. 57–60.

[6.47] A. Phillips Jr. and P. J. Price, "Monte Carlo calculations on hot electron energy tails", *Appl. Phys. Lett.*, vol. 30, pp. 528–530, 1977.

[6.48] S. E. Laux, M. V. Fischetti, and D. J. Frank, "Monte Carlo analysis of semiconductor devices: The DAMOCLES program ", *IBM J. Res. Develop.*, vol. 34, pp. 466–494, 1990.

[6.49] F. Venturi, R. K. Smith, E. C. Sangiorgi, M. R. Pinto, and B. Riccò, "A general purpose device simulation coupling Poisson and Monte Carlo transport with application to deep submicron MOSFET's", *IEEE Trans. Computer-Aided Des.*, vol. 8, no. 4, pp. 360–369, 1989.

[6.50] R. Thoma, H. J. Peifer, W. L. Engl, W. Quade, R. Brunetti, and C. Jacoboni, "An improved impact–ionization model for high–energy electron transport in Si with Monte Carlo simulation", *J. Appl. Phys.*, vol. 69, pp. 2300–2311, 1991.

[6.51] A. Pacelli and U. Ravaioli, "Analysis of variance-reduction schemes for ensemble Monte-Carlo simulation of semiconductor devices", *Solid–State Electron.*, vol. 41, pp. 599–605, 1997.

[6.52] P. Graf, S. Keith, and B. Meinerzhagen, "Evaluation of solenoidal and statistically enhanced total current densities", *Proc. SISPAD*, Cambridge, MA (USA), 1997.

[6.53] P. D. Yoder, K. Gärtner, and W. Fichtner, "A generalized Ramo-Shockley theorem for classical to quantum transport at arbitrary frequencies", *J. Appl. Phys.*, vol. 79, pp. 1951–1954, 1996.

[6.54] H. Kim, H. S. Min, T. W. Tang, and Y. J. Park, "An extended proof of the Ramo-Shockley theorem", *Solid–State Electron.*, vol. 34, pp. 1251–1253, 1991.

7 Momentum-Based Transport Models

The MC model is very CPU intensive and simpler but more CPU efficient models based on balance equations have been developed. The most widely used momentum-based models are the drift-diffusion (DD) and the hydrodynamic (HD) models [7.1–7.13]. Both models are derived by applying different degrees of approximation to balance equations of the type (2.39). The DD model is the simplest momentum-based model and consists of balance equations for the particle and current densities. Thus, only the first two moments of the distribution function are calculated instead of the full distribution function and a large fraction of the information contained in the distribution function is lost. On the other hand, the dimensionality of the problem is reduced by the integration of the **k**-space and the CPU efficiency is improved by orders of magnitude. In the case of the HD model the first four moments are considered including the particle density, current density, particle gas temperature, and the energy current density. This already enables the simulation of nonlocal effects, like the velocity overshoot, which have a strong impact on the device behavior of modern deep sub-micron devices.

In the following, the derivation of the HD model for large-signal (Sec. 7.1), small-signal (Sec. 7.2), and noise simulations (Sec. 7.3) is discussed in detail. Since the DD model is a straight forward approximation of the HD model, its derivation is described in less detail (Sec. 7.4). For the DD model on the other hand the proof of the Nyquist theorem is given. Transport and noise parameters for both models are discussed in the last section.

7.1 The Hydrodynamic Model

The generalized HD model of Refs. [7.12, 7.13] extended to the case of a position-dependent band structure is used in this work, because it is applicable in the case of a nonparabolic band structure. First the HD model for holes is derived and then the results are given for the electron model.

In the following the notation for the hole density is $p(\mathbf{r}, t)$ and for the electron density $n(\mathbf{r},t)$. Hole quantities are labeled by the subscript p and electron quantities by n. The HD model is based on balance equations of the type (2.39) for the four microscopic quantities $X = 1$, $\mathbf{v}$, ε, and $\varepsilon\mathbf{v}$. The corresponding macroscopic densities are for holes

$$p = p\,\langle 1\rangle_{\mathbf{k}} \,, \tag{7.1}$$
$$\mathbf{j}_p = p\,\langle \mathbf{v}\rangle_{\mathbf{k}} \,, \tag{7.2}$$
$$w_p = p\,\langle \varepsilon\rangle_{\mathbf{k}} \,, \tag{7.3}$$
$$\mathbf{s}_p = p\,\langle \varepsilon\mathbf{v}\rangle_{\mathbf{k}} \,. \tag{7.4}$$

With Eq. (2.39) and $X = 1$ the hole continuity equation is obtained without approximations

$$\nabla_{\mathbf{r}}^{\mathrm{T}}\mathbf{j}_p + \frac{\partial p}{\partial t} = G^{\mathrm{II}} \,, \tag{7.5}$$

where G^{II} is the generation rate due to II.

The balance equation for the current density (2.41) is derived with the macroscopic relaxation time approximation (2.42) for $\mathbf{X} = \mathbf{v}$

$$\frac{\partial \mathbf{j}_p}{\partial t} - p\left\langle \hat{m}^{-1}\right\rangle_{\mathbf{k}} \nabla_{\mathbf{r}} \left(E_{\mathrm{V}} - q\Psi\right) + p\left\langle \hat{m}^{-1}\nabla_{\mathbf{r}}\varepsilon\right\rangle_{\mathbf{k}} + \left(\nabla_{\mathbf{r}}^{\mathrm{T}}\left[p\left\langle \mathbf{v}\mathbf{v}^{\mathrm{T}}\right\rangle_{\mathbf{k}}\right]\right)^{\mathrm{T}} - p\left\langle \mathbf{v}^{\mathrm{T}}\nabla_{\mathbf{r}}\mathbf{v}^{\mathrm{T}}\right\rangle_{\mathbf{k}}^{\mathrm{T}} = -\frac{\mathbf{j}_p}{\tau_{\mathbf{j}_p}} \,. \tag{7.6}$$

In order to simplify this equation rather stringent approximations are used. The nearly negligible forces due to the position-dependent band structure are calculated with the assumption that the velocity can be separated into position- and wave vector dependent factors $\mathbf{v}_n(\mathbf{r},\mathbf{k}) = g(\mathbf{r})\mathbf{v}_n(\mathbf{k})$ [7.14], so that

$$\nabla_{\mathbf{r}}\mathbf{v}^{\mathrm{T}} = \frac{\nabla_{\mathbf{r}} g}{g}\,\mathbf{v}^{\mathrm{T}} \,. \tag{7.7}$$

and

$$\left\langle \mathbf{v}^{\mathrm{T}}\nabla_{\mathbf{r}}\mathbf{v}^{\mathrm{T}}\right\rangle_{\mathbf{k}}^{\mathrm{T}} = \left\langle \mathbf{v}\mathbf{v}^{\mathrm{T}}\right\rangle_{\mathbf{k}} \frac{\nabla_{\mathbf{r}} g}{g} \,. \tag{7.8}$$

Matrix-valued quantities are replaced by scalars

$$\left\langle \hat{m}^{-1}\right\rangle_{\mathbf{k}} \approx \frac{\mathrm{Tr}\left\langle \hat{m}^{-1}\right\rangle_{\mathbf{k}}}{3}\,\hat{I} = \left\langle \frac{1}{m}\right\rangle_{\mathbf{k}} \hat{I} \,, \tag{7.9}$$
$$\left\langle \hat{m}^{-1}\nabla_{\mathbf{r}}\varepsilon\right\rangle_{\mathbf{k}} \approx \left\langle \frac{\nabla_{\mathbf{r}}\varepsilon}{m}\right\rangle_{\mathbf{k}} \,, \tag{7.10}$$
$$\left\langle \mathbf{v}\mathbf{v}^{\mathrm{T}}\right\rangle_{\mathbf{k}} \approx \frac{\left\langle \mathbf{v}^2\right\rangle_{\mathbf{k}}}{3}\,\hat{I} \,, \tag{7.11}$$

and the dynamic hole temperature $T_p^\star$ is defined by

$$\frac{3}{2}k_\mathrm{B}T_p^\star = \frac{1}{2}m_p^\star \left\langle \mathbf{v}^2 \right\rangle_\mathbf{k} , \tag{7.12}$$

where the mass $m_p^\star$ is calculated under equilibrium conditions

$$\frac{1}{m_p^\star} = \left\langle \frac{1}{m} \right\rangle_\mathbf{k}^\mathrm{eq} = \frac{\left\langle \mathbf{v}^2 \right\rangle_\mathbf{k}^\mathrm{eq}}{k_\mathrm{B}T_0} . \tag{7.13}$$

Thus, under equilibrium conditions the dynamic temperature equals the lattice temperature as can be shown with integration by parts. The expected value of the inverse mass is eliminated by the introduction of a new relaxation time $\tau_{\mathbf{j}_p}^\star$

$$\left\langle \frac{1}{m} \right\rangle_\mathbf{k} \tau_{\mathbf{j}_p} = \frac{\tau_{\mathbf{j}_p}^\star}{m_p^\star} . \tag{7.14}$$

Under equilibrium conditions the following relation holds for the gradient of the energy

$$\langle \nabla_\mathbf{r} \varepsilon \rangle_\mathbf{k}^\mathrm{eq} = -k_\mathrm{B}T_0 \frac{\nabla_\mathbf{r} N_\mathrm{V}}{N_\mathrm{V}} \tag{7.15}$$

due to the definition of the effective DOS (5.40). With this relation the position-dependent force is further simplified in analogy to Ref. [7.4]

$$\left\langle \frac{\nabla_\mathbf{r} \varepsilon}{m} \right\rangle_\mathbf{k} \approx -\frac{k_\mathrm{B}T_p^\star}{m_p^\star} \frac{\nabla_\mathbf{r} N_\mathrm{V}}{N_\mathrm{V}} . \tag{7.16}$$

With these approximations the final current balance equation is obtained

$$\mathbf{j}_p + \tau_{\mathbf{j}_p} \frac{\partial \mathbf{j}_p}{\partial t} = \frac{\tau_{\mathbf{j}_p}^\star}{m_p^\star} \nabla_\mathbf{r}(E_\mathrm{V} - q\Psi) - \frac{\tau_{\mathbf{j}_p}}{m_p^\star} \left(\nabla_\mathbf{r} \left[p k_\mathrm{B} T_p^\star \right] - p k_\mathrm{B} T_p^\star \frac{\nabla_\mathbf{r} N_\mathrm{V}}{N_\mathrm{V}} \right) . \tag{7.17}$$

Thus, the diffusion constant is given by a generalized Einstein relation [7.12]

$$D = \frac{\tau_{\mathbf{j}_p} k_\mathrm{B} T_p^\star}{m_p^\star} = \underbrace{\frac{q\tau_{\mathbf{j}_p}^\star}{m_p^\star}}_{=\mu} \frac{\tau_{\mathbf{j}_p}}{\tau_{\mathbf{j}_p}^\star} \underbrace{\frac{k_\mathrm{B} T_p^\star}{q}}_{=V_{T_p^\star}} = \mu V_{T_p^\star} \frac{\tau_{\mathbf{j}_p}}{\tau_{\mathbf{j}_p}^\star} , \tag{7.18}$$

where the thermal voltage is evaluated with the dynamic temperature of the particle gas instead of the lattice temperature.

The balance equation for the energy is obtained with $X = \varepsilon$

$$\frac{\partial w_p}{\partial t} - \mathbf{j}_p^\mathrm{T} \nabla_\mathbf{r}(E_\mathrm{V} - q\Psi) + p \left\langle \mathbf{v}^\mathrm{T} \nabla_\mathbf{r} \varepsilon \right\rangle_\mathbf{k} + \nabla_\mathbf{r}^\mathrm{T} \mathbf{s}_p - p \left\langle \mathbf{v}^\mathrm{T} \nabla_\mathbf{r} \varepsilon \right\rangle_\mathbf{k} = -p \frac{\langle \varepsilon \rangle - \langle \varepsilon \rangle^\mathrm{eq}}{\tau_{w_p}} , \tag{7.19}$$

where it has been assumed that the energy generation by II is negligible. With the new energy relaxation time $\tau^\star_{w_p}$

$$\frac{\langle\varepsilon\rangle - \langle\varepsilon\rangle^{\mathrm{eq}}}{\tau_{w_p}} = \frac{3k_\mathrm{B}}{2}\frac{T^\star_p - T_0}{\tau^\star_{w_p}} \tag{7.20}$$

the final equation reads

$$\nabla^\mathrm{T}_\mathbf{r}\mathbf{s}_p + \frac{\partial w_p}{\partial t} = \mathbf{j}^\mathrm{T}_p\nabla_\mathbf{r}(E_\mathrm{v} - q\Psi) - p\frac{3k_\mathrm{B}}{2}\frac{T^\star_p - T_0}{\tau^\star_{w_p}} . \tag{7.21}$$

The balance equation for the energy current density is derived with $\mathbf{X} = \varepsilon\mathbf{v}$

$$\frac{\partial\mathbf{s}_p}{\partial t} - p\left\langle\hat{m}^{-1}\varepsilon + \mathbf{v}\mathbf{v}^\mathrm{T}\right\rangle_\mathbf{k}\nabla_\mathbf{r}\left(E_\mathrm{v} - q\Psi\right) + p\left\langle\hat{m}^{-1}\varepsilon\nabla_\mathbf{r}\varepsilon\right\rangle_\mathbf{k}$$
$$+ \left(\nabla^\mathrm{T}_\mathbf{r}\left[p\left\langle\varepsilon\mathbf{v}\mathbf{v}^\mathrm{T}\right\rangle_\mathbf{k}\right]\right)^\mathrm{T} - p\left\langle\mathbf{v}^\mathrm{T}\nabla_\mathbf{r}\varepsilon\mathbf{v}^\mathrm{T}\right\rangle^\mathrm{T}_\mathbf{k} = -\frac{\mathbf{s}_p}{\tau_{\mathbf{s}_p}} . \tag{7.22}$$

The term involving the position-dependent velocity is simplified in the same manner as above

$$\left\langle\mathbf{v}^\mathrm{T}\nabla_\mathbf{r}\varepsilon\mathbf{v}^\mathrm{T}\right\rangle^\mathrm{T}_\mathbf{k} = \left\langle\varepsilon\mathbf{v}\mathbf{v}^\mathrm{T}\right\rangle_\mathbf{k} 2\frac{\nabla_\mathbf{r}g}{g} . \tag{7.23}$$

The matrix-valued quantities are approximated by

$$\left\langle\hat{m}^{-1}\varepsilon + \mathbf{v}\mathbf{v}^\mathrm{T}\right\rangle_\mathbf{k} \approx \left[\left\langle\frac{\varepsilon}{m}\right\rangle_\mathbf{k} + \frac{\left\langle\mathbf{v}^2\right\rangle_\mathbf{k}}{3}\right]\hat{I} , \tag{7.24}$$

$$\left\langle\hat{m}^{-1}\varepsilon\nabla_\mathbf{r}\varepsilon\right\rangle_\mathbf{k} \approx \left\langle\frac{\varepsilon\nabla_\mathbf{r}\varepsilon}{m}\right\rangle_\mathbf{k} , \tag{7.25}$$

and

$$\left\langle\varepsilon\mathbf{v}\mathbf{v}^\mathrm{T}\right\rangle_\mathbf{k} \approx \frac{5\left\langle\mathbf{v}^2\right\rangle^2_\mathbf{k}}{6\mathrm{Tr}\left\langle\hat{m}^{-1}\right\rangle_\mathbf{k}}\hat{I} , \tag{7.26}$$

where the last approximation is used to close the set of equations and is exact for a Maxwell-Boltzmann distribution function in conjunction with a parabolic band structure. The position-dependent force is further simplified[1]

$$\left\langle\frac{\varepsilon\nabla_\mathbf{r}\varepsilon}{m}\right\rangle_\mathbf{k} \approx -\frac{15}{2\mathrm{Tr}\left\langle\hat{m}^{-1}\right\rangle_\mathbf{k}}\left(\frac{k_\mathrm{B}T^\star_p}{m^\star_p}\right)^2\frac{\nabla_\mathbf{r}N_\mathrm{v}}{N_\mathrm{v}} , \tag{7.27}$$

and a new relaxation time is defined

$$\left\langle\frac{\varepsilon}{m} + \frac{\mathbf{v}^2}{3}\right\rangle_\mathbf{k}\tau_{\mathbf{s}_p} = \frac{5}{6}\left\langle\mathbf{v}^2\right\rangle_\mathbf{k}\tau^\star_{\mathbf{s}_p} . \tag{7.28}$$

[1]The approximations (7.16) and (7.27) are formulated in such a way that in conjunction with the boundary conditions the current density and the energy current density vanish under equilibrium conditions for arbitrary band structures.

It turns out that the difference between these two relaxation times is negligible $\tau^\star_{\mathbf{s}_p} \approx \tau_{\mathbf{s}_p}$. In addition, the gradient of the expected value of the inverse mass is neglected

$$\nabla_{\mathbf{r}} \frac{3}{\mathrm{Tr}\,\langle \hat{m}^{-1}\rangle_{\mathbf{k}}} \approx \mathbf{0}\,. \tag{7.29}$$

These approximations yield the final equation for the energy current density

$$\begin{aligned}\mathbf{s}_p + \tau_{\mathbf{s}_p} \frac{\partial \mathbf{s}_p}{\partial t} = p \frac{5 k_{\mathrm{B}} T^\star_p \tau^\star_{\mathbf{s}_p}}{2 m^\star_p} \nabla_{\mathbf{r}} \left(E_{\mathrm{v}} - q\Psi\right) \\ - \frac{5 \tau^\star_{\mathbf{s}_p} \tau_{\mathbf{j}_p}}{2 m^\star_p \tau^\star_{\mathbf{j}_p}} \left(\nabla_{\mathbf{r}} \left[p \left(k_{\mathrm{B}} T^\star_p\right)^2 \right] - p (k_{\mathrm{B}} T^\star_p)^2 \frac{\nabla_{\mathbf{r}} N_{\mathrm{v}}}{N_{\mathrm{v}}} \right)\,.\end{aligned} \tag{7.30}$$

Eqs. (7.17), (7.21), and (7.30) contain first order derivatives with respect to time. These derivatives lead in conjunction with the derivative in Eq. (7.5) and the PE (6.11) to plasma oscillations, which cause numerical problems, and the three derivatives are therefore usually omitted. Since the relaxation times are smaller than one picosecond, this limits the HD model to frequencies below 100GHz which is sufficient for nearly all technical applications. The final HD model reads for holes

$$\nabla^{\mathrm{T}}_{\mathbf{r}} \mathbf{j}_p + \frac{\partial p}{\partial t} = G^{\mathrm{II}}\,, \tag{7.31}$$

$$\begin{aligned}\mathbf{j}_p &= p \frac{\tau^\star_{\mathbf{j}_p}}{m^\star_p} \nabla_{\mathbf{r}} (E_{\mathrm{v}} - q\Psi) \\ &\quad - \frac{\tau_{\mathbf{j}_p}}{m^\star_p} \left(\nabla^{\mathrm{T}}_{\mathbf{r}} \left[p k_{\mathrm{B}} T^\star_p \right] - p k_{\mathrm{B}} T^\star_p \frac{\nabla_{\mathbf{r}} N_{\mathrm{v}}}{N_{\mathrm{v}}} \right)\,,\end{aligned} \tag{7.32}$$

$$\nabla^{\mathrm{T}}_{\mathbf{r}} \mathbf{s}_p = \mathbf{j}^{\mathrm{T}}_p \nabla_{\mathbf{r}} (E_{\mathrm{v}} - q\Psi) - p \frac{3 k_{\mathrm{B}}}{2} \frac{T^\star_p - T_0}{\tau^\star_{w_p}}\,, \tag{7.33}$$

$$\begin{aligned}\mathbf{s}_p &= p \frac{5 k_{\mathrm{B}} T^\star_p \tau^\star_{\mathbf{s}_p}}{2 m^\star_p} \nabla_{\mathbf{r}} \left(E_{\mathrm{v}} - q\Psi\right) \\ &\quad - \frac{5 \tau^\star_{\mathbf{s}_p} \tau_{\mathbf{j}_p}}{2 m^\star_p \tau^\star_{\mathbf{j}_p}} \left(\nabla_{\mathbf{r}} \left[p \left(k_{\mathrm{B}} T^\star_p\right)^2 \right] - p (k_{\mathrm{B}} T^\star_p)^2 \frac{\nabla_{\mathbf{r}} N_{\mathrm{v}}}{N_{\mathrm{v}}} \right)\,,\end{aligned} \tag{7.34}$$

and electrons

$$\nabla^{\mathrm{T}}_{\mathbf{r}} \mathbf{j}_n + \frac{\partial n}{\partial t} = G^{\mathrm{II}}\,, \tag{7.35}$$

$$\mathbf{j}_n = -n \frac{\tau^\star_{\mathbf{j}_n}}{m^\star_n} \nabla_{\mathbf{r}} (E_{\mathrm{c}} - q\Psi)$$

$$-\frac{\tau_{\mathbf{j}_n}}{m_n^\star}\left(\nabla_{\mathbf{r}}^{\mathrm{T}}\left[nk_{\mathrm{B}}T_n^\star\right]-nk_{\mathrm{B}}T_n^\star\frac{\nabla_{\mathbf{r}}N_{\mathrm{c}}}{N_{\mathrm{c}}}\right), \tag{7.36}$$

$$\nabla_{\mathbf{r}}^{\mathrm{T}}\mathbf{s}_n = -\mathbf{j}_n^{\mathrm{T}}\nabla_{\mathbf{r}}(E_{\mathrm{c}}-q\Psi)-n\frac{3k_{\mathrm{B}}}{2}\frac{T_n^\star-T_0}{\tau_{w_n}^\star}, \tag{7.37}$$

$$\mathbf{s}_n = -n\frac{5k_{\mathrm{B}}T_n^\star\tau_{\mathbf{s}_n}^\star}{2m_n^\star}\nabla_{\mathbf{r}}\left(E_{\mathrm{c}}-q\Psi\right)$$
$$-\frac{5\tau_{\mathbf{s}_n}^\star\tau_{\mathbf{j}_n}}{2m_n^\star\tau_{\mathbf{j}_n}^\star}\left(\nabla_{\mathbf{r}}\left[n\left(k_{\mathrm{B}}T_n^\star\right)^2\right]-n(k_{\mathrm{B}}T_n^\star)^2\frac{\nabla_{\mathbf{r}}N_{\mathrm{c}}}{N_{\mathrm{c}}}\right). \tag{7.38}$$

The (energy) current densities are eliminated from the HD model by inserting Eqs. (7.32), (7.34), (7.36), and (7.38) into Eqs. (7.31), (7.33), (7.35), and (7.37). The remaining five equations including the PE (6.11) for the unknown variables p, $T_p^\star$, n, $T_n^\star$, and Ψ are

$$(7.31)+(7.32) \rightarrow F_{\dot{p}}(p,T_p^\star,n,T_n^\star,\Psi)+\frac{\partial p}{\partial t}=0, \tag{7.39}$$

$$(7.33)+(7.32)+(7.34) \rightarrow F_{\dot{w}_p}(p,T_p^\star,n,T_n^\star,\Psi)=0, \tag{7.40}$$

$$(7.35)+(7.36) \rightarrow F_{\dot{n}}(p,T_p^\star,n,T_n^\star,\Psi)+\frac{\partial n}{\partial t}=0, \tag{7.41}$$

$$(7.37)+(7.36)+(7.38) \rightarrow F_{\dot{w}_n}(p,T_p^\star,n,T_n^\star,\Psi)=0, \tag{7.42}$$

$$(6.11) \rightarrow F_{\Psi}(p,T_p^\star,n,T_n^\star,\Psi)=0, \tag{7.43}$$

where F_α with $\alpha=\dot{p}$, $\dot{w}_p$, $\dot{n}$, $\dot{w}_n$, Ψ is an abbreviation including all the resultant terms besides the derivatives in time. The abbreviations for the right-hand sides of Eqs. (7.32) and (7.36) are

$$\mathbf{j}_p = \mathbf{F}_{\mathbf{j}_p}(p,T_p^\star,n,T_n^\star,\Psi), \tag{7.44}$$

$$\mathbf{j}_n = \mathbf{F}_{\mathbf{j}_n}(p,T_p^\star,n,T_n^\star,\Psi). \tag{7.45}$$

The final set of Eqs. (7.39)-(7.43) contains four macroscopic relaxation times for holes ($\tau_{\mathbf{j}_p}^\star$, $\tau_{\mathbf{j}_p}$, $\tau_{w_p}^\star$, $\tau_{\mathbf{s}_p}^\star$) and four for electrons ($\tau_{\mathbf{j}_n}^\star$, $\tau_{\mathbf{j}_n}$, $\tau_{w_n}^\star$, $\tau_{\mathbf{s}_n}^\star$), which are assumed to depend only on the hole or electron dynamic temperature, respectively, for a given lattice temperature, doping, and germanium content. They are determined under homogeneous bulk conditions by single-particle MC simulations for a constant electric field $\mathbf{E}$

$$\tau^\star_{\mathbf{j}_p} = \frac{m^\star_p \left\langle \mathbf{v}^\mathrm{T} \right\rangle_\mathrm{hom} \mathbf{E}}{q\mathbf{E}^2} , \tag{7.46}$$

$$\tau_{\mathbf{j}_p} = \frac{3\tau^\star_{\mathbf{j}_p}}{m^\star_p \mathrm{Tr} \left\langle \hat{m}^{-1} \right\rangle_\mathrm{hom}} , \tag{7.47}$$

$$\tau^\star_{w_p} = \frac{3qk_\mathrm{B}(T^\star_p - T_0)}{2 \left\langle \mathbf{v}^\mathrm{T} \right\rangle_\mathrm{hom} \mathbf{E}} , \tag{7.48}$$

$$\tau^\star_{\mathbf{s}_p} = \frac{2m^\star_p \left\langle \varepsilon \mathbf{v}^\mathrm{T} \right\rangle_\mathrm{hom} \mathbf{E}}{5qk_\mathrm{B}T^\star_p \mathbf{E}^2} . \tag{7.49}$$

The corresponding dynamic temperature is given by

$$T^\star_p = \frac{m^\star_p}{3k_\mathrm{B}} \left\langle \mathbf{v}^2 \right\rangle_\mathrm{hom} . \tag{7.50}$$

Similar equations can be derived for electrons. For small electric fields the efficiency of the estimators for the odd moments $\langle \mathbf{v} \rangle_\mathrm{hom}$ and $\langle \varepsilon \mathbf{v} \rangle_\mathrm{hom}$ is rather low and the moments are estimated based on Eqs. (2.58) and (2.61)

$$\langle \mathbf{v} \rangle_\mathrm{hom} = q \left\langle \tau \hat{m}^{-1} + \frac{d\tau}{d\varepsilon} \mathbf{v}\mathbf{v}^\mathrm{T} \right\rangle_\mathrm{hom} \mathbf{E} , \tag{7.51}$$

$$\langle \varepsilon \mathbf{v} \rangle_\mathrm{hom} = q \left\langle \tau \varepsilon \hat{m}^{-1} + \tau \mathbf{v}\mathbf{v}^\mathrm{T} + \frac{d\tau}{d\varepsilon} \varepsilon \mathbf{v}\mathbf{v}^\mathrm{T} \right\rangle_\mathrm{hom} \mathbf{E} . \tag{7.52}$$

The estimator (7.48) yields an infinitely large energy relaxation time for a vanishing electric field under bulk conditions. To circumvent this artefact due to the homogeneous bulk conditions, the maximum value of the relaxation time is limited to the equilibrium value given by

$$\tau^{\star\ \mathrm{eq}}_{w_p} = \frac{m^\star_p \left(\langle \varepsilon v^2 \rangle^\mathrm{eq}_\mathrm{hom} - \langle \varepsilon \rangle^\mathrm{eq}_\mathrm{hom} \langle v^2 \rangle^\mathrm{eq}_\mathrm{hom} \right)}{2 \left\langle \varepsilon S\{\varepsilon\} \right\rangle^\mathrm{eq}_\mathrm{hom}} , \tag{7.53}$$

which is derived by assuming a Maxwell-Boltzmann distribution function for the particle gas and calculating the derivative with respect to the particle gas temperature of the approximation and the exact scattering integral under the assumption of a constant relaxation time.

The transport parameters are stored together with the dynamic temperature in lookup tables as a function of the electric field, doping, germanium content, and lattice temperature. For an HD device simulation they are interpolated with appropriate methods [7.15] and the CPU efficiency of the HD model is not impaired compared to models based on analytical expressions for the transport parameters. Under device conditions the dynamic temperature might be lower than the lattice temperature. This state is not accessible under homogeneous bulk conditions and the values of the transport parameters for equilibrium are used.

The boundary conditions on ideal ohmic contacts are for the particle densities

$$p = -\frac{N_{\mathrm{don}} - N_{\mathrm{acc}}}{2} + \sqrt{\left(\frac{N_{\mathrm{don}} - N_{\mathrm{acc}}}{2}\right)^2 + n_{\mathrm{i}}^2} , \tag{7.54}$$

$$n = \frac{N_{\mathrm{don}} - N_{\mathrm{acc}}}{2} + \sqrt{\left(\frac{N_{\mathrm{don}} - N_{\mathrm{acc}}}{2}\right)^2 + n_{\mathrm{i}}^2} , \tag{7.55}$$

and temperatures

$$T_p^\star = T_0 , \tag{7.56}$$

$$T_n^\star = T_0 . \tag{7.57}$$

On all other surfaces Neumann-type boundary conditions are applied for the particle current and energy current densities. The boundary conditions for the electrostatic potential are given in Sec. 6.3.

The total current at the lth terminal is obtained by integrating the displacement and conduction current densities over the area of the contact

$$I_l(t) = \int_{A_l} \left(q\mathbf{j}_p(\mathbf{r},t) - q\mathbf{j}_n(\mathbf{r},t) - \varepsilon \frac{\partial \nabla_{\mathbf{r}} \Psi(\mathbf{r},t)}{\partial t} \right)^{\mathrm{T}} d^2\mathbf{r} . \tag{7.58}$$

7.2 Small-Signal Analysis

With small-signal analysis the linear transient response of the terminal currents to small voltage changes under otherwise stationary boundary conditions is investigated. To this end the terminal currents and voltages are expanded up to the first order

$$I_l(t) = I_{l,\mathrm{s}} + \delta I_l(t) , \tag{7.59}$$

$$V_l(t) = V_{l,\mathrm{s}} + \delta V_l(t) , \tag{7.60}$$

where the subscript s indicates the stationary solution and δ the linear term. The corresponding expansion of the variables of the HD model is given by

$$p(\mathbf{r},t) = p_{\mathrm{s}}(\mathbf{r}) + \delta p(\mathbf{r},t) , \tag{7.61}$$

$$T_p^\star(\mathbf{r},t) = T_{p,\mathrm{s}}^\star(\mathbf{r}) + \delta T_p^\star(\mathbf{r},t) , \tag{7.62}$$

$$n(\mathbf{r},t) = n_{\mathrm{s}}(\mathbf{r}) + \delta n(\mathbf{r},t) , \tag{7.63}$$

$$T_n^\star(\mathbf{r},t) = T_{n,\mathrm{s}}^\star(\mathbf{r}) + \delta T_n^\star(\mathbf{r},t) , \tag{7.64}$$

$$\Psi(\mathbf{r},t) = \Psi_{\mathrm{s}}(\mathbf{r}) + \delta \Psi(\mathbf{r},t) . \tag{7.65}$$

The stationary solution is obtained with

$$F_\alpha(p_s, T^\star_{p,s}, n_s, T^\star_{n,s}, \Psi_s) = 0 \quad \text{for} \quad \alpha = \dot{p}, \dot{w}_p, \dot{n}, \dot{w}_n, \Psi \tag{7.66}$$

together with the appropriate boundary conditions for constant voltage operation. The Jacobian of the stationary HD model is defined as

$$\Lambda_\alpha(\delta p, \delta T^\star_p, \delta n, \delta T^\star_n, \delta\Psi) = \frac{\partial F_\alpha}{\partial p}\delta p + \frac{\partial F_\alpha}{\partial T^\star_p}\delta T^\star_p + \frac{\partial F_\alpha}{\partial n}\delta n + \frac{\partial F_\alpha}{\partial T^\star_n}\delta T^\star_n + \frac{\partial F_\alpha}{\partial \Psi}\delta\Psi \ , \tag{7.67}$$

where the derivatives are evaluated for the steady-state. The corresponding linearized HD model reads

$$\Lambda_\alpha(\delta p, \delta T^\star_p, \delta n, \delta T^\star_n, \delta\Psi) + \delta_{\alpha,\dot{p}}\frac{\partial \delta p}{\partial t} + \delta_{\alpha,\dot{n}}\frac{\partial \delta n}{\partial t} = 0 \ , \tag{7.68}$$

where $\delta_{\alpha,\beta}$ is the Kronecker symbol. Since the small-signal HD model is a time invariant, linear PDE system, the Fourier transform is more convenient

$$\Lambda_\alpha(\delta \underline{p}, \delta \underline{T}^\star_p, \delta \underline{n}, \delta \underline{T}^\star_n, \delta\underline{\Psi}) + \delta_{\alpha,\dot{p}}\mathrm{i}\omega\delta\underline{p} + \delta_{\alpha,\dot{n}}\mathrm{i}\omega\delta\underline{n} = 0 \ , \tag{7.69}$$

together with the boundary conditions for the lth contact

$$\left.\begin{array}{rcl} \delta\underline{p}(\mathbf{r}) &=& 0 \\ \delta\underline{T}^\star_p(\mathbf{r}) &=& 0 \\ \delta\underline{n}(\mathbf{r}) &=& 0 \\ \delta\underline{T}^\star_n(\mathbf{r}) &=& 0 \\ \delta\underline{\Psi}(\mathbf{r}) &=& \delta\underline{V}_l \end{array}\right\} \quad \text{for} \quad \mathbf{r} \in \mathbf{A}_l \ . \tag{7.70}$$

All other boundary conditions are the same as for the large-signal HD model.

The small-signal currents are defined in the same manner as the Jacobian of the HD model

$$\begin{aligned} \delta\underline{\mathbf{j}}_p &= \mathbf{\Lambda}_{\mathbf{j}_p}(\delta \underline{p}, \delta \underline{T}^\star_p, \delta \underline{n}, \delta \underline{T}^\star_n, \delta\underline{\Psi}) \\ &= \frac{\partial \mathbf{F}_{\mathbf{j}_p}}{\partial p}\delta\underline{p} + \frac{\partial \mathbf{F}_{\mathbf{j}_p}}{\partial T^\star_p}\delta\underline{T}^\star_p + \frac{\partial \mathbf{F}_{\mathbf{j}_p}}{\partial n}\delta\underline{n} + \frac{\partial \mathbf{F}_{\mathbf{j}_p}}{\partial T^\star_n}\delta\underline{T}^\star_n + \frac{\partial \mathbf{F}_{\mathbf{j}_p}}{\partial \Psi}\delta\underline{\Psi} \ , \end{aligned} \tag{7.71}$$

$$\begin{aligned} \delta\underline{\mathbf{j}}_n &= \mathbf{\Lambda}_{\mathbf{j}_n}(\delta \underline{p}, \delta \underline{T}^\star_p, \delta \underline{n}, \delta \underline{T}^\star_n, \delta\underline{\Psi}) \\ &= \frac{\partial \mathbf{F}_{\mathbf{j}_n}}{\partial p}\delta\underline{p} + \frac{\partial \mathbf{F}_{\mathbf{j}_n}}{\partial T^\star_p}\delta\underline{T}^\star_p + \frac{\partial \mathbf{F}_{\mathbf{j}_n}}{\partial n}\delta\underline{n} + \frac{\partial \mathbf{F}_{\mathbf{j}_n}}{\partial T^\star_n}\delta\underline{T}^\star_n + \frac{\partial \mathbf{F}_{\mathbf{j}_n}}{\partial \Psi}\delta\underline{\Psi} \ , \end{aligned} \tag{7.72}$$

and the small-signal current of the kth terminal reads

$$\delta\underline{I}_k(\omega) = \int_{A_k} \left(q\delta\underline{\mathbf{j}}_p(\mathbf{r},\omega) - q\delta\underline{\mathbf{j}}_n(\mathbf{r},\omega) - \mathrm{i}\omega\varepsilon\nabla_{\mathbf{r}}\delta\underline{\Psi}(\mathbf{r},\omega)\right)^{\mathrm{T}} d^2\mathbf{r} \ . \tag{7.73}$$

Due to the linear nature of the small-signal HD model the terminal currents are linear in the terminal voltages

$$\delta \underline{I}_k(\omega) = \sum_{l=1}^{N_{\text{cont}}} \underline{Y}_{k,l}(\omega) \delta \underline{V}_l(\omega) \ , \tag{7.74}$$

where $\underline{Y}_{k,l}$ the corresponding Y-parameter [7.16].

7.3 Noise Analysis

The Langevin-type HD model for fluctuations under stationary bias conditions is derived from the LBTE (2.68) along the same lines as the HD model from the BTE (cf. Sec. 7.1). It reads for holes

$$\nabla_{\mathbf{r}}^{\mathrm{T}} \mathbf{j}_p + \frac{\partial p}{\partial t} = G^{\mathrm{II}} \qquad + \xi_{\dot{p}} \ , \tag{7.75}$$

$$\begin{aligned} \mathbf{j}_p =& p \frac{\tau^\star_{\mathbf{j}_p}}{m^\star_p} \nabla_{\mathbf{r}} (E_{\mathrm{v}} - q\Psi) \\ & - \frac{\tau_{\mathbf{j}_p}}{m^\star_p} \left(\nabla_{\mathbf{r}}^{\mathrm{T}} \left[p k_{\mathrm{B}} T^\star_p \right] - p k_{\mathrm{B}} T^\star_p \frac{\nabla_{\mathbf{r}} N_{\mathrm{v}}}{N_{\mathrm{v}}} \right) \qquad + \boldsymbol{\xi}_{\mathbf{j}_p} \ , \end{aligned} \tag{7.76}$$

$$\nabla_{\mathbf{r}}^{\mathrm{T}} \mathbf{s}_p = \mathbf{j}_p^{\mathrm{T}} \nabla_{\mathbf{r}} (E_{\mathrm{v}} - q\Psi) - p \frac{3 k_{\mathrm{B}}}{2} \frac{T^\star_p - T_0}{\tau^\star_{w_p}} \qquad + \xi_{\dot{w}_p} \ , \tag{7.77}$$

$$\begin{aligned} \mathbf{s}_p =& p \frac{5 k_{\mathrm{B}} T^\star_p \tau^\star_{\mathbf{s}_p}}{2 m^\star_p} \nabla_{\mathbf{r}} (E_{\mathrm{v}} - q\Psi) \\ & - \frac{5 \tau^\star_{\mathbf{s}_p} \tau_{\mathbf{j}_p}}{2 m^\star_p \tau^\star_{\mathbf{j}_p}} \left(\nabla_{\mathbf{r}} \left[p \left(k_{\mathrm{B}} T^\star_p \right)^2 \right] - p (k_{\mathrm{B}} T^\star_p)^2 \frac{\nabla_{\mathbf{r}} N_{\mathrm{v}}}{N_{\mathrm{v}}} \right) + \boldsymbol{\xi}_{\mathbf{s}_p} \ , \end{aligned} \tag{7.78}$$

and electrons

$$\nabla_{\mathbf{r}}^{\mathrm{T}} \mathbf{j}_n + \frac{\partial n}{\partial t} = G^{\mathrm{II}} \qquad + \xi_{\dot{n}} \ , \tag{7.79}$$

$$\begin{aligned} \mathbf{j}_n =& - n \frac{\tau^\star_{\mathbf{j}_n}}{m^\star_n} \nabla_{\mathbf{r}} (E_{\mathrm{c}} - q\Psi) \\ & - \frac{\tau_{\mathbf{j}_n}}{m^\star_n} \left(\nabla_{\mathbf{r}}^{\mathrm{T}} \left[n k_{\mathrm{B}} T^\star_n \right] - n k_{\mathrm{B}} T^\star_n \frac{\nabla_{\mathbf{r}} N_{\mathrm{c}}}{N_{\mathrm{c}}} \right) \qquad + \boldsymbol{\xi}_{\mathbf{j}_n} \ , \end{aligned} \tag{7.80}$$

$$\nabla_{\mathbf{r}}^{\mathrm{T}} \mathbf{s}_n = - \mathbf{j}_n^{\mathrm{T}} \nabla_{\mathbf{r}} (E_{\mathrm{c}} - q\Psi) - n \frac{3 k_{\mathrm{B}}}{2} \frac{T^\star_n - T_0}{\tau^\star_{w_n}} \qquad + \xi_{\dot{w}_n} \ , \tag{7.81}$$

$$\begin{aligned} \mathbf{s}_n =& - n \frac{5 k_{\mathrm{B}} T^\star_n \tau^\star_{\mathbf{s}_n}}{2 m^\star_n} \nabla_{\mathbf{r}} (E_{\mathrm{c}} - q\Psi) \\ & - \frac{5 \tau^\star_{\mathbf{s}_n} \tau_{\mathbf{j}_n}}{2 m^\star_n \tau^\star_{\mathbf{j}_n}} \left(\nabla_{\mathbf{r}} \left[n \left(k_{\mathrm{B}} T^\star_n \right)^2 \right] - n (k_{\mathrm{B}} T^\star_n)^2 \frac{\nabla_{\mathbf{r}} N_{\mathrm{c}}}{N_{\mathrm{c}}} \right) + \boldsymbol{\xi}_{\mathbf{s}_n} \ . \end{aligned} \tag{7.82}$$

The only new terms compared to Eqs. (7.31)-(7.38) are the Langevin forces of which the expected values vanish. The Langevin forces can be viewed as moments of the Langevin force of the LBTE (2.72). But since the Langevin forces based on the macroscopic relaxation time approximation do not satisfy the Einstein relation (2.63) (cf. Sec. 2.4), the concept of modified Langevin forces is used here which is described further below.

The fluctuations induced by the Langevin forces are assumed to be so small that the response is linear. The corresponding linearized Langevin-type HD model is derived in the frequency domain in analogy to the small-signal HD model (7.69)

$$\Lambda_{\dot p}(\delta \underline{p}, \delta \underline{T}_p^\star, \delta \underline{n}, \delta \underline{T}_n^\star, \delta \underline{\Psi}) + \mathrm{i}\omega \delta \underline{p} = \underline{R}_{\dot p} = \underline{\xi}_{\dot p} - \nabla_{\mathbf{r}}^{\mathrm{T}} \underline{\boldsymbol{\xi}}_{\mathbf{j}_p} , \tag{7.83}$$

$$\Lambda_{\dot w_p}(\delta \underline{p}, \delta \underline{T}_p^\star, \delta \underline{n}, \delta \underline{T}_n^\star, \delta \underline{\Psi}) = \underline{R}_{\dot w_p} = \underline{\xi}_{\dot w_p} + \underline{\boldsymbol{\xi}}_{\mathbf{j}_p}^{\mathrm{T}} \nabla_{\mathbf{r}}(E_{\mathrm{V}} - q\Psi) - \nabla_{\mathbf{r}}^{\mathrm{T}} \underline{\boldsymbol{\xi}}_{\mathbf{s}_p} , \tag{7.84}$$

$$\Lambda_{\dot n}(\delta \underline{n}, \delta \underline{T}_p^\star, \delta \underline{n}, \delta \underline{T}_n^\star, \delta \underline{\Psi}) + \mathrm{i}\omega \delta \underline{n} = \underline{R}_{\dot n} = \underline{\xi}_{\dot n} - \nabla_{\mathbf{r}}^{\mathrm{T}} \underline{\boldsymbol{\xi}}_{\mathbf{j}_n} , \tag{7.85}$$

$$\Lambda_{\dot w_n}(\delta \underline{p}, \delta \underline{T}_p^\star, \delta \underline{n}, \delta \underline{T}_n^\star, \delta \underline{\Psi}) = \underline{R}_{\dot w_n} = \underline{\xi}_{\dot w_n} - \underline{\boldsymbol{\xi}}_{\mathbf{j}_n}^{\mathrm{T}} \nabla_{\mathbf{r}}(E_{\mathrm{C}} - q\Psi) - \nabla_{\mathbf{r}}^{\mathrm{T}} \underline{\boldsymbol{\xi}}_{\mathbf{s}_p} , \tag{7.86}$$

$$\Lambda_{\Psi}(\delta \underline{p}, \delta \underline{T}_p^\star, \delta \underline{n}, \delta \underline{T}_n^\star, \delta \underline{\Psi}) = \underline{R}_{\Psi} = 0 , \tag{7.87}$$

where all terms involving Langevin forces have been moved to the right-hand side and are abbreviated by $\underline{R}_\alpha$. Due to Joule's heat terms in Eqs. (7.77) and (7.81) the Langevin forces of the current densities also appear in the respective conservation equations for the energy (7.84), (7.86). The Neumann boundary conditions are the same as for the stationary case, whereas the Dirichlet boundary conditions of the ideal ohmic contacts are set to zero for each terminal in accordance with the constant voltage operation

$$\left.\begin{aligned} \delta \underline{p}(\mathbf{r}) &= 0 \\ \delta \underline{T}_p^\star(\mathbf{r}) &= 0 \\ \delta \underline{n}(\mathbf{r}) &= 0 \\ \delta \underline{T}_n^\star(\mathbf{r}) &= 0 \\ \delta \underline{\Psi}(\mathbf{r}) &= 0 \end{aligned}\right\} \quad \text{for} \quad \mathbf{r} \in \mathbf{A}_l . \tag{7.88}$$

In order to calculate the terminal current fluctuations, Green's functions are defined [7.17–7.21]

$$\Lambda_\alpha(\underline{G}_{\delta p}^\beta, \underline{G}_{\delta T_p^\star}^\beta, \underline{G}_{\delta n}^\beta, \underline{G}_{\delta T_n^\star}^\beta, \underline{G}_{\delta \Psi}^\beta) + \delta_{\alpha,\dot p}\mathrm{i}\omega \underline{G}_{\delta p}^\beta + \delta_{\alpha,\dot n}\mathrm{i}\omega \underline{G}_{\delta n}^\beta = \delta_{\alpha,\beta}\delta(\mathbf{r} - \mathbf{r}') , \tag{7.89}$$

with α, β = $\dot p$, $\dot w_p$, $\dot n$, $\dot w_n$, Ψ and where $\delta(\mathbf{r} - \mathbf{r}')$ is the Dirac function. The boundary conditions for the terminals are of the Dirichlet type

$$\left.\begin{array}{rcl} \underline{G}^{\beta}_{\delta p}(\mathbf{r},\mathbf{r}',\omega) & = & 0 \\ \underline{G}^{\beta}_{\delta T^{\star}_{p}}(\mathbf{r},\mathbf{r}',\omega) & = & 0 \\ \underline{G}^{\beta}_{\delta n}(\mathbf{r},\mathbf{r}',\omega) & = & 0 \\ \underline{G}^{\beta}_{\delta T^{\star}_{n}}(\mathbf{r},\mathbf{r}',\omega) & = & 0 \\ \underline{G}^{\beta}_{\delta \Psi}(\mathbf{r},\mathbf{r}',\omega) & = & 0 \end{array}\right\} \quad \text{for} \quad \mathbf{r} \in \mathbf{A}_l \ , \tag{7.90}$$

and on all other surfaces of the Neumann type. With these Green's functions and Eqs. (7.71) and (7.72) the Green's functions of the current densities can be calculated

$$\underline{\mathbf{G}}^{\beta}_{\delta \mathbf{j}_p} = \mathbf{\Lambda}_{\mathbf{j}_p}(\underline{G}^{\beta}_{\delta p}, \underline{G}^{\beta}_{\delta T^{\star}_{p}}, \underline{G}^{\beta}_{\delta n}, \underline{G}^{\beta}_{\delta T^{\star}_{n}}, \underline{G}^{\beta}_{\delta \Psi}) \ , \tag{7.91}$$

$$\underline{\mathbf{G}}^{\beta}_{\delta \mathbf{j}_n} = \mathbf{\Lambda}_{\mathbf{j}_n}(\underline{G}^{\beta}_{\delta p}, \underline{G}^{\beta}_{\delta T^{\star}_{p}}, \underline{G}^{\beta}_{\delta n}, \underline{G}^{\beta}_{\delta T^{\star}_{n}}, \underline{G}^{\beta}_{\delta \Psi}) \ . \tag{7.92}$$

The resultant current density fluctuations read

$$\delta \underline{\mathbf{j}}_p(\mathbf{r},\omega) = \sum_{\beta} \int \underline{\mathbf{G}}^{\beta}_{\delta \mathbf{j}_p}(\mathbf{r},\mathbf{r}',\omega) \underline{R}_{\beta}(\mathbf{r}',\omega) d^3r' + \underline{\boldsymbol{\xi}}_{\mathbf{j}_p}(\mathbf{r},\omega) \ , \tag{7.93}$$

$$\delta \underline{\mathbf{j}}_n(\mathbf{r},\omega) = \sum_{\beta} \int \underline{\mathbf{G}}^{\beta}_{\delta \mathbf{j}_n}(\mathbf{r},\mathbf{r}',\omega) \underline{R}_{\beta}(\mathbf{r}',\omega) d^3r' + \underline{\boldsymbol{\xi}}_{\mathbf{j}_n}(\mathbf{r},\omega) \ , \tag{7.94}$$

where the Langevin forces of the current densities have been explicitly added, because Eqs. (7.71) and (7.72) are based on the linearizations of Eqs. (7.32) and (7.36), which in contrast to Eqs. (7.76) and (7.80) do not include the Langevin forces of the current densities.

The Green's function for the terminal current is defined as

$$\underline{G}^{\beta}_{\delta I_l}(\mathbf{r}',\omega) = \int_{A_l} \left(q\underline{\mathbf{G}}^{\beta}_{\delta \mathbf{j}_p}(\mathbf{r},\mathbf{r}',\omega) - q\underline{\mathbf{G}}^{\beta}_{\delta \mathbf{j}_n}(\mathbf{r},\mathbf{r}',\omega) - \mathrm{i}\omega\epsilon(\mathbf{r})\nabla_{\mathbf{r}}\underline{G}^{\beta}_{\delta \Psi}(\mathbf{r},\mathbf{r}',\omega) \right)^{\mathrm{T}} d^2\mathbf{r} \ . \tag{7.95}$$

It yields the current for the lth terminal when all terminals are grounded and a unit sources of the type β is placed at the location $\mathbf{r}'$. The sum over all terminals yields

$$\sum_{l=1}^{N_{\mathrm{cont}}} \underline{G}_{\delta I_l}^{\dot{p}}(\mathbf{r}',\omega) = q\,, \tag{7.96}$$

$$\sum_{l=1}^{N_{\mathrm{cont}}} \underline{G}_{\delta I_l}^{\dot{n}}(\mathbf{r}',\omega) = -q\,, \tag{7.97}$$

$$\sum_{l=1}^{N_{\mathrm{cont}}} \underline{G}_{\delta I_l}^{\dot{w}_p}(\mathbf{r}',\omega) = 0\,, \tag{7.98}$$

$$\sum_{l=1}^{N_{\mathrm{cont}}} \underline{G}_{\delta I_l}^{\dot{w}_n}(\mathbf{r}',\omega) = 0\,. \tag{7.99}$$

With Eq. (7.95) the Fourier transform of the current fluctuations at the lth terminal reads

$$\begin{aligned}\delta \underline{I}_l = \int & \underline{G}_{\delta I_l}^{\dot{p}} \underline{\xi}_{\dot{p}} - \underline{G}_{\delta I_l}^{\dot{p}} \nabla_{\mathbf{r}'}^{\mathrm{T}} \underline{\boldsymbol{\xi}}_{\mathbf{j}_p} \\ & + \underline{G}_{\delta I_l}^{\dot{w}_p} \underline{\xi}_{\dot{w}_p} + \underline{G}_{\delta I_l}^{\dot{w}_p} \underline{\boldsymbol{\xi}}_{\mathbf{j}_p}^{\mathrm{T}} \nabla_{\mathbf{r}'} \left(E_{\mathrm{v}} - q\Psi_{\mathrm{s}}\right) - \underline{G}_{\delta I_l}^{\dot{w}_p} \nabla_{\mathbf{r}'}^{\mathrm{T}} \underline{\boldsymbol{\xi}}_{\mathbf{s}_p} \\ & + \underline{G}_{\delta I_l}^{\dot{n}} \underline{\xi}_{\dot{n}} - \underline{G}_{\delta I_l}^{\dot{n}} \nabla_{\mathbf{r}'}^{\mathrm{T}} \underline{\boldsymbol{\xi}}_{\mathbf{j}_n} \\ & + \underline{G}_{\delta I_l}^{\dot{w}_n} \underline{\xi}_{\dot{w}_n} - \underline{G}_{\delta I_l}^{\dot{w}_n} \underline{\boldsymbol{\xi}}_{\mathbf{j}_n}^{\mathrm{T}} \nabla_{\mathbf{r}'} \left(E_{\mathrm{c}} - q\Psi_{\mathrm{s}}\right) - \underline{G}_{\delta I_l}^{\dot{w}_n} \nabla_{\mathbf{r}'}^{\mathrm{T}} \underline{\boldsymbol{\xi}}_{\mathbf{s}_n} \, d^3 r' \\ + & \int_{A_l} q \left(\underline{\boldsymbol{\xi}}_{\mathbf{j}_p} - \underline{\boldsymbol{\xi}}_{\mathbf{j}_n}\right)^{\mathrm{T}} d^2\mathbf{r}\,. \end{aligned} \tag{7.100}$$

Integration by parts is used to shift the ∇-operator in the case of the Langevin forces of the (energy) current densities [7.22] and the surface integrals are eliminated by neglecting surface scattering on all surfaces that do not belong to the ideal ohmic contacts

$$\begin{aligned}\delta \underline{I}_l = \int & \underline{G}_{\delta I_l}^{\dot{p}} \underline{\xi}_{\dot{p}} + \left(\nabla_{\mathbf{r}'} \underline{G}_{\delta I_l}^{\dot{p}} + \underline{G}_{\delta I_l}^{\dot{w}_p} \nabla_{\mathbf{r}'} \left(E_{\mathrm{v}} - q\Psi_{\mathrm{s}}\right)\right)^{\mathrm{T}} \underline{\boldsymbol{\xi}}_{\mathbf{j}_p} + \\ & \underline{G}_{\delta I_l}^{\dot{w}_p} \underline{\xi}_{\dot{w}_p} + \left(\nabla_{\mathbf{r}'} \underline{G}_{\delta I_l}^{\dot{w}_p}\right)^{\mathrm{T}} \underline{\boldsymbol{\xi}}_{\mathbf{s}_p} + \\ & \underline{G}_{\delta I_l}^{\dot{n}} \underline{\xi}_{\dot{n}} + \left(\nabla_{\mathbf{r}'} \underline{G}_{\delta I_l}^{\dot{n}} - \underline{G}_{\delta I_l}^{\dot{w}_n} \nabla_{\mathbf{r}'} \left(E_{\mathrm{c}} - q\Psi_{\mathrm{s}}\right)\right)^{\mathrm{T}} \underline{\boldsymbol{\xi}}_{\mathbf{j}_n} + \\ & \underline{G}_{\delta I_l}^{\dot{w}_n} \underline{\xi}_{\dot{w}_n} + \left(\nabla_{\mathbf{r}'} \underline{G}_{\delta I_l}^{\dot{w}_n}\right)^{\mathrm{T}} \underline{\boldsymbol{\xi}}_{\mathbf{s}_n} \, d^3 r'\,. \end{aligned} \tag{7.101}$$

In the following noise due to particle scattering within the conduction or valence bands (diffusion noise) and noise due to particle generation by II (generation noise) will be treated separately.

7.3.1 Diffusion Noise

In the case of particle scattering exclusively within the conduction or valence bands the Langevin forces of the respective particle densities $\underline{\xi}_{\dot{p}}$, $\underline{\xi}_{\dot{n}}$ are zero. With the abbreviations for holes

$$\underline{\mathbf{G}}_{\delta I_l,p} = \begin{pmatrix} \nabla_{\mathbf{r}'} \underline{G}^{\dot{p}}_{\delta I_l} + \underline{G}^{\dot{w}_p}_{\delta I_l} \nabla_{\mathbf{r}'} (E_\mathrm{v} - q\Psi_\mathrm{s}) \\ \underline{G}^{\dot{w}_p}_{\delta I_l} \\ \nabla_{\mathbf{r}'} \underline{G}^{\dot{w}_p}_{\delta I_l} \end{pmatrix} , \qquad \underline{\boldsymbol{\xi}}_p = \begin{pmatrix} \underline{\boldsymbol{\xi}}_{\mathbf{j}_p} \\ \underline{\xi}_{\dot{w}_p} \\ \underline{\boldsymbol{\xi}}_{\mathbf{s}_p} \end{pmatrix} \tag{7.102}$$

and for electrons

$$\underline{\mathbf{G}}_{\delta I_l,n} = \begin{pmatrix} \nabla_{\mathbf{r}'} \underline{G}^{\dot{n}}_{\delta I_l} - \underline{G}^{\dot{w}_n}_{\delta I_l} \nabla_{\mathbf{r}'} (E_\mathrm{c} - q\Psi_\mathrm{s}) \\ \underline{G}^{\dot{w}_n}_{\delta I_l} \\ \nabla_{\mathbf{r}'} \underline{G}^{\dot{w}_n}_{\delta I_l} \end{pmatrix} , \qquad \underline{\boldsymbol{\xi}}_n = \begin{pmatrix} \underline{\boldsymbol{\xi}}_{\mathbf{j}_n} \\ \underline{\xi}_{\dot{w}_n} \\ \underline{\boldsymbol{\xi}}_{\mathbf{s}_n} \end{pmatrix} \tag{7.103}$$

the current fluctuations (7.101) due to particle scattering are given by

$$\delta \underline{I}_l = \int \underline{\mathbf{G}}^\mathrm{T}_{\delta I_l,p} \underline{\boldsymbol{\xi}}_p + \underline{\mathbf{G}}^\mathrm{T}_{\delta I_l,n} \underline{\boldsymbol{\xi}}_n d^3 r' \; . \tag{7.104}$$

The spectrum of the cross-correlation function of the lth and kth terminal currents reads for diffusion noise

$$\begin{aligned} \underline{S}^\mathrm{diff}_{I_l I_k}(\omega) = \int\int & \underline{\mathbf{G}}^\mathrm{T}_{\delta I_l,p}(\mathbf{r},\omega) \hat{S}_{\boldsymbol{\xi}_p \boldsymbol{\xi}_p}(\mathbf{r},\mathbf{r}') \underline{\mathbf{G}}^*_{\delta I_k,p}(\mathbf{r}',\omega) \\ & + \underline{\mathbf{G}}^\mathrm{T}_{\delta I_l,p}(\mathbf{r},\omega) \hat{S}_{\boldsymbol{\xi}_p \boldsymbol{\xi}_n}(\mathbf{r},\mathbf{r}') \underline{\mathbf{G}}^*_{\delta I_k,n}(\mathbf{r}',\omega) \\ & + \underline{\mathbf{G}}^\mathrm{T}_{\delta I_l,n}(\mathbf{r},\omega) \hat{S}_{\boldsymbol{\xi}_n \boldsymbol{\xi}_p}(\mathbf{r},\mathbf{r}') \underline{\mathbf{G}}^*_{\delta I_k,p}(\mathbf{r}',\omega) \\ & + \underline{\mathbf{G}}^\mathrm{T}_{\delta I_l,n}(\mathbf{r},\omega) \hat{S}_{\boldsymbol{\xi}_n \boldsymbol{\xi}_n}(\mathbf{r},\mathbf{r}') \underline{\mathbf{G}}^*_{\delta I_k,n}(\mathbf{r}',\omega) \, d^3 r' d^3 r \; , \end{aligned} \tag{7.105}$$

where * indicates the complex conjugate. For scattering events involving only holes or electrons the hole and electron Langevin forces are uncorrelated

$$\hat{S}_{\boldsymbol{\xi}_n \boldsymbol{\xi}_p}(\mathbf{r},\mathbf{r}') = \hat{S}^\dagger_{\boldsymbol{\xi}_p \boldsymbol{\xi}_n}(\mathbf{r}',\mathbf{r}) = \hat{0} \; , \tag{7.106}$$

where † labels the complex conjugate of the transpose. Since the scattering events in the MC simulator are instantaneous, independent and do not change the position of the particle, the correlations of the Langevin forces can be written as (cf. Eq. 2.69)

$$\hat{S}_{\boldsymbol{\xi}_p \boldsymbol{\xi}_p}(\mathbf{r},\mathbf{r}') = p_\mathrm{s}(\mathbf{r}) \hat{S}_{\boldsymbol{\xi}'_p \boldsymbol{\xi}'_p}(\mathbf{r}) \delta(\mathbf{r}-\mathbf{r}') \; , \tag{7.107}$$

$$\hat{S}_{\boldsymbol{\xi}_n \boldsymbol{\xi}_n}(\mathbf{r},\mathbf{r}') = n_\mathrm{s}(\mathbf{r}) \hat{S}_{\boldsymbol{\xi}'_n \boldsymbol{\xi}'_n}(\mathbf{r}) \delta(\mathbf{r}-\mathbf{r}') \; , \tag{7.108}$$

where the matrix-valued local noise sources $\hat{S}_{\boldsymbol{\xi}'_p \boldsymbol{\xi}'_p}$ and $\hat{S}_{\boldsymbol{\xi}'_n \boldsymbol{\xi}'_n}$ are given by the spectral intensities of the single-particle Langevin forces [7.23–7.27]. Inserting

Eqs. (7.107) and (7.108) into Eq. (7.105) yields

$$\underline{S}^{\text{diff}}_{I_l I_k} = \int p_{\text{s}}(\mathbf{r})\underline{\mathbf{G}}^{\text{T}}_{\delta I_l,p}(\mathbf{r},\omega)\hat{S}_{\boldsymbol{\xi}'_p\boldsymbol{\xi}'_p}(\mathbf{r})\underline{\mathbf{G}}^{*}_{\delta I_k,p}(\mathbf{r},\omega)$$
$$+n_{\text{s}}(\mathbf{r})\underline{\mathbf{G}}^{\text{T}}_{\delta I_l,n}(\mathbf{r},\omega)\hat{S}_{\boldsymbol{\xi}'_n\boldsymbol{\xi}'_n}(\mathbf{r})\underline{\mathbf{G}}^{*}_{\delta I_k,n}(\mathbf{r},\omega)\, d^3r \ . \tag{7.109}$$

Thus, the uncorrelated electron and hole contribution to diffusion noise can be treated separately. Since the methods for the evaluation of the spectral intensities of the single-particle Langevin forces are similar for electrons and holes, only the hole case is investigated below.

In accordance with the HD approach it is assumed that the local noise source $\hat{S}_{\boldsymbol{\xi}'_p\boldsymbol{\xi}'_p}$ can be modeled as a function of the local hole temperature and can be evaluated under homogeneous bulk conditions neglecting the real space [7.12,7.28, 7.29]. Since under these conditions the particle density is constant, the current density is replaced by the velocity $\delta\mathbf{v}_p$, the energy density by the energy $\delta\varepsilon_p$, and the energy current density by the energy current $\delta\varepsilon\mathbf{v}_p$ for a single particle. With the definitions

$$\delta\underline{\mathbf{X}}_p = \begin{pmatrix} \delta\underline{\mathbf{v}}_p \\ \delta\underline{T}^{\star}_p \\ \delta\underline{\varepsilon\mathbf{v}}_p \end{pmatrix} , \qquad \underline{\xi}'_p = \begin{pmatrix} \underline{\xi}_{\mathbf{v}_p} \\ \underline{\xi}_{\dot{\varepsilon}_p} \\ \underline{\xi}_{\varepsilon\mathbf{v}_p} \end{pmatrix} \tag{7.110}$$

the linearized single-particle HD model under homogeneous bulk conditions reads

$$\hat{\Lambda}_p\delta\underline{\mathbf{X}}_p = \underline{\xi}'_p \ . \tag{7.111}$$

The Jacobian is given by

$$\hat{\Lambda}_p = \begin{pmatrix} \hat{I} & -\frac{q}{m^{\star}_p}\frac{\partial\tau^{\star}_{\mathrm{j}_p}}{\partial T^{\star}_p}\mathbf{E}_{\text{s}} & \hat{0} \\ -q\mathbf{E}^{\text{T}}_{\text{s}} & \frac{3}{2}\frac{k_{\text{B}}}{\tau^{\star}_{w_p}}\left(1-(T^{\star}_{p,\text{s}}-T_0)\frac{\partial\ln\tau^{\star}_{w_p}}{\partial T^{\star}_p}\right) & \mathbf{0}^{\text{T}} \\ \hat{0} & -\frac{5qk_{\text{B}}\tau^{\star}_{s_p}}{2m^{\star}_p}\mathbf{E}_{\text{s}}\left(1+\frac{\partial\ln\tau^{\star}_{s_p}}{\partial T^{\star}_p}T^{\star}_{p,\text{s}}\right) & \hat{I} \end{pmatrix} , \tag{7.112}$$

where $\hat{I}$ is the identity matrix, $\mathbf{E}_{\text{s}}$ the homogeneous electric field, and the particle conservation equation (7.31) has been dropped due to the assumption of a constant particle number and the omission of the position dependence. With Eq. (7.111) the spectral intensity of the single-particle Langevin forces is given by

$$\hat{S}_{\boldsymbol{\xi}'_p\boldsymbol{\xi}'_p} = \hat{\Lambda}_p\hat{S}_{\mathbf{X}_p\mathbf{X}_p}\hat{\Lambda}^{\text{T}}_p \ . \tag{7.113}$$

Since first order derivatives in time have been neglected in Eqs. (7.32)-(7.34), the linearized HD bulk model (7.112) does not dependent on the frequency. Together with the white spectra of the Langevin forces this means that the spectral intensities of the fluctuations must also be white. This is not correct for very high frequencies [7.28, 7.30], but even the spectral intensity of the rather slow temperature fluctuations is white below 100GHz (cf. Sec. 7.5), above which our momentum-based models are not applied so that the approximation holds.

The spectral intensity of the fluctuations is the Fourier transform of the corresponding autocorrelation function (2.66) [7.31]. Since the spectrum is only required below 100GHz, where it is constant, it can be calculated with (2.67) at zero frequency

$$\hat{S}_{\mathbf{X}_p \mathbf{X}_p} = 2 \int_{-\infty}^{\infty} \langle \delta \mathbf{X}_p, \delta \mathbf{X}_p^{\mathrm{T}} \rangle_{\mathrm{hom}}(t) dt \tag{7.114}$$

simplifying the evaluation by MC simulation.

In contrast to [7.28] the spectral intensities of the single-particle Langevin forces are not calculated as averages of the scattering processes of the MC simulation which would result in the **exact** spectral intensities[2] [7.23]. Instead, modified spectral intensities of the Langevin forces are calculated in such a way that the HD model **exactly** reproduces the low-frequency MC autocorrelation and cross-correlation spectra of the fluctuations of the velocity, carrier temperature, and energy current under bulk conditions (cf. Eqs. (7.113) and (7.114)) [7.29]. This has the advantage that the simple dynamics of the HD model, which are an approximation of the MC model, do not distort the bulk fluctuations. It is worth noting that because of the inexact dynamics of the HD model it is in general not possible to have exact Langevin forces and fluctuations at the same time. Since the objective is the evaluation of the terminal current fluctuations, which correspond to the velocity fluctuations under homogeneous bulk conditions, the concept of the modified Langevin forces is used instead of the exact ones.

In the case of equilibrium ($\mathbf{E}_s = \mathbf{0}$) the Jacobian (7.112) reduces to a diagonal matrix and the modified Langevin forces of the (energy times) velocity equal the exact ones. Therefore, the Einstein relation (2.63) holds for the velocity and energy times velocity, and the problems of the macroscopic relaxation time approximation described in Sec. 2.4 are avoided.

The HD model is based on the assumption of an isotropic particle distribution function in **k**-space (cf. Sec. 7.1) and the homogeneous bulk simulations are performed for fixed directions of the electric field, whereas the driving force in a device might have any direction. In order to account for these effects in a device simulation, the spectral intensity of the Langevin forces (7.113) is projected onto the field direction of the bulk simulation via the longitudinal component of the

[2]With Eq. (8) of [7.28] the exact spectral intensities of the Langevin forces can be estimated by MC simulation. In order to be consistent with this HD model and to avoid the macroscopic relaxation time approximation used in [7.28], the instantaneous variations $\Delta\alpha_i, \Delta\beta_i$ of that scheme must be calculated for the variables $\tau\mathbf{v}$, ε, and $\tau\varepsilon\mathbf{v}$, where τ is the microscopic relaxation time (2.46) [7.26] (see also discussion in Sec. 2.4).

vector-valued Langevin forces

$$\underline{\xi}_{v_l} = \frac{\mathbf{E}_s^T \underline{\xi}_\mathbf{v}}{|\mathbf{E}_s|} \quad \text{and} \quad \underline{\xi}_{\varepsilon v_l} = \frac{\mathbf{E}_s^T \underline{\xi}_{\varepsilon\mathbf{v}}}{|\mathbf{E}_s|} , \tag{7.115}$$

where the particle type index has been dropped. On the device level matrix-valued components of $\hat{S}_{\boldsymbol{\xi}'\boldsymbol{\xi}'}$ are approximated by isotropic quantities, e.g.

$$\hat{S}_{\boldsymbol{\xi}_\mathbf{v}\boldsymbol{\xi}_\mathbf{v}} \approx S_{\xi_{v_l}\xi_{v_l}} \hat{I} , \tag{7.116}$$

whereas vector-valued components are aligned along the direction of the local force, which here is assumed to be parallel to the current density, e.g.

$$\mathbf{S}_{\boldsymbol{\xi}_\mathbf{v}\xi_{\dot{\varepsilon}}} \approx \pm \frac{\mathbf{j}_s}{|\mathbf{j}_s|} S_{\xi_{v_l}\xi_{\dot{\varepsilon}}} , \tag{7.117}$$

where the upper sign refers to holes and the lower to electrons. The approximation of the complete matrix $\hat{S}_{\boldsymbol{\xi}'\boldsymbol{\xi}'}$ on the device level is given by

$$\hat{S}_{\boldsymbol{\xi}'\boldsymbol{\xi}'} = \begin{pmatrix} \hat{S}_{\boldsymbol{\xi}_\mathbf{v}\boldsymbol{\xi}_\mathbf{v}} & \mathbf{S}_{\boldsymbol{\xi}_\mathbf{v}\xi_{\dot{\varepsilon}}} & \hat{S}_{\boldsymbol{\xi}_\mathbf{v}\boldsymbol{\xi}_{\varepsilon\mathbf{v}}} \\ \mathbf{S}^T_{\xi_{\dot{\varepsilon}}\boldsymbol{\xi}_\mathbf{v}} & S_{\xi_{\dot{\varepsilon}}\xi_{\dot{\varepsilon}}} & \mathbf{S}^T_{\xi_{\dot{\varepsilon}}\boldsymbol{\xi}_{\varepsilon\mathbf{v}}} \\ \hat{S}_{\boldsymbol{\xi}_{\varepsilon\mathbf{v}}\boldsymbol{\xi}_\mathbf{v}} & \mathbf{S}_{\boldsymbol{\xi}_{\varepsilon\mathbf{v}}\xi_{\dot{\varepsilon}}} & \hat{S}_{\boldsymbol{\xi}_{\varepsilon\mathbf{v}}\boldsymbol{\xi}_{\varepsilon\mathbf{v}}} \end{pmatrix}$$

$$\approx \begin{pmatrix} S_{\xi_{v_l}\xi_{v_l}}\hat{I} & \pm\frac{\mathbf{j}_s}{|\mathbf{j}_s|}S_{\xi_{v_l}\xi_{\dot{\varepsilon}}} & S_{\xi_{v_l}\xi_{\varepsilon v_l}}\hat{I} \\ \pm S_{\xi_{\dot{\varepsilon}}\xi_{v_l}}\frac{\mathbf{j}_s^T}{|\mathbf{j}_s|} & S_{\xi_{\dot{\varepsilon}}\xi_{\dot{\varepsilon}}} & \pm S_{\xi_{\dot{\varepsilon}}\xi_{\varepsilon v_l}}\frac{\mathbf{j}_s^T}{|\mathbf{j}_s|} \\ S_{\xi_{\varepsilon v_l}\xi_{v_l}}\hat{I} & \pm\frac{\mathbf{j}_s}{|\mathbf{j}_s|}S_{\xi_{\varepsilon v_l}\xi_{\dot{\varepsilon}}} & S_{\xi_{\varepsilon v_l}\xi_{\varepsilon v_l}}\hat{I} \end{pmatrix} \tag{7.118}$$

and depends on the direction of the respective particle current.

7.3.2 Generation Noise

During an II event a secondary electron and hole are generated and in contrast to the scattering processes investigated above, the Langevin forces of the density fluctuations $\xi_{\dot{p}}, \xi_{\dot{n}}$ are not zero. On the other hand, the other Langevin forces $\boldsymbol{\xi}_{\mathbf{j}_p}$, $\boldsymbol{\xi}_{\mathbf{j}_n}$, $\xi_{\dot{w}_p}$, $\xi_{\dot{w}_n}$, $\boldsymbol{\xi}_{\mathbf{s}_p}$, and $\boldsymbol{\xi}_{\mathbf{s}_n}$ can be neglected for II, because the average II scattering rate is much smaller than the average scattering rate of the other scattering processes [7.32].

In the case of an II event the secondary electron is transfered from the valence into the conduction band generating a secondary hole in the valence band. The generation of the hole is therefore correlated with the generation of the electron and the spectral intensities of the Langevin forces are given by [7.19]

$$S_{\xi_{\dot{p}}\xi_{\dot{p}}} = S_{\xi_{\dot{p}}\xi_{\dot{n}}} = S_{\xi_{\dot{n}}\xi_{\dot{p}}} = S_{\xi_{\dot{n}}\xi_{\dot{n}}} = 2G^{\mathrm{II}} , \tag{7.119}$$

where the last relation holds, because the generation process is a Poisson process [7.31]. Thus, the terminal current noise due to generation by II reads

$$\underline{S}^{\mathrm{II}}_{I_l I_k} = \int (\underline{G}^{\dot{p}}_{\delta I_l} + \underline{G}^{\dot{n}}_{\delta I_l}) 2G^{\mathrm{II}} (\underline{G}^{\dot{p}}_{\delta I_k} + \underline{G}^{\dot{n}}_{\delta I_k})^* d^3 r \,. \tag{7.120}$$

Since the generation noise is uncorrelated with the diffusion noise, the total noise is obtained by adding Eqs. (7.109) and (7.120).

In the case of trap-assisted generation/recombination noise the same arguments as above can be used to eliminate all Langevin forces besides the ones for the density fluctuations and the fundamental scheme presented by F. Bonani and G. Ghione for the Shockley-Read-Hall process can be used in the HD model [7.19].

7.3.3 Monte-Carlo Evaluation of the Spectral Intensities

The spectral intensities (7.114) of the bulk fluctuations are evaluated by stationary MC simulations of homogeneous systems. The symmetry of the correlation functions for stationary processes allows to limit the integral to negative times [7.31]

$$\int_{-\infty}^{\infty} \langle \delta \mathbf{X}(t) \delta \mathbf{X}^{\mathrm{T}}(0) \rangle_{\mathrm{hom}} dt = \int_{-\infty}^{0} \langle \delta \mathbf{X}(t) \delta \mathbf{X}^{\mathrm{T}}(0) + \delta \mathbf{X}(0) \delta \mathbf{X}^{\mathrm{T}}(t) \rangle_{\mathrm{hom}} dt \,, \tag{7.121}$$

and the expectation is estimated by single-particle MC simulation

$$\int_{-\infty}^{0} \langle \delta \mathbf{X}(0) \delta \mathbf{X}^{\mathrm{T}}(t) \rangle_{\mathrm{hom}} dt \approx \frac{1}{N - M + 1} \sum_{i=M}^{N} \delta \mathbf{X}(i\Delta t) \delta \mathbf{Y}^{\mathrm{T}}(i\Delta t) \tag{7.122}$$

with

$$\delta \mathbf{Y}(i\Delta t) = \sum_{j=0}^{M-1} \delta \mathbf{Y}_{i-j} \quad \text{and} \quad \delta \mathbf{Y}_k = \int_{(k-1)\Delta t}^{k\Delta t} \delta \mathbf{X}(t') dt' \,, \tag{7.123}$$

where the above integral is exactly solved by path integration. During the MC simulation the last M realizations of $\delta \mathbf{Y}_k$ are stored and for each time interval i the sum over j is evaluated by

$$\delta \mathbf{Y}(i\Delta t) = \delta \mathbf{Y}((i-1)\Delta t) - \delta \mathbf{Y}_{i-M} + \delta \mathbf{Y}_i \,. \tag{7.124}$$

Thus, the CPU intensive summation of all M realizations of $\delta \mathbf{Y}_k$ for each time interval i is avoided and the CPU efficiency of the MC bulk simulation is increased by more than one order of magnitude for $M = 1000$, $N = 20000M$, and $\Delta t = 10$fs. The small value of Δt ensures that all information contained in the MC simulation is sampled. The large value of M is necessary to completely capture the rather slow decay of the energy related correlation functions. The value of N must be large, because Eq. (7.122) is only an asymptotically unbiased estimator and to ensure that the stochastic errors of the estimated noise parameters are sufficiently small (cf. Sec. 3.5).

7.4 The Drift-Diffusion Model

The DD model is obtained by applying two further approximations to the HD model

$$\tau_{\mathbf{j}_p} \approx \tau^\star_{\mathbf{j}_p} , \tag{7.125}$$

$$T^\star_p \approx T_0 , \tag{7.126}$$

which imply the Einstein relation (2.63). Eq. (7.17) now reads

$$\mathbf{j}_p + \tau^\star_{\mathbf{j}_p} \frac{\partial \mathbf{j}_p}{\partial t} = p \frac{\tau^\star_{\mathbf{j}_p}}{m^\star_p} \nabla_{\mathbf{r}} (E_{\mathrm{v}} - q\Psi) - \frac{k_{\mathrm{B}} T_0 \tau^\star_{\mathbf{j}_p}}{m^\star_p} \left(\nabla_{\mathbf{r}} p - p \frac{\nabla_{\mathbf{r}} N_{\mathrm{v}}}{N_{\mathrm{v}}} \right) , \tag{7.127}$$

and Eqs. (7.21) and (7.30) can be dropped, because the particle gas temperature is assumed to be equal to the lattice temperature. The final DD model for holes and electrons is obtained neglecting again the derivative in time in the balance equations for the current densities

$$\nabla^{\mathrm{T}}_{\mathbf{r}} \mathbf{j}_p + \frac{\partial p}{\partial t} = G^{\mathrm{II}} , \tag{7.128}$$

$$\mathbf{j}_p = p \frac{\tau^\star_{\mathbf{j}_p}}{m^\star_p} \nabla_{\mathbf{r}} (E_{\mathrm{v}} - q\Psi) - \frac{k_{\mathrm{B}} T_0 \tau^\star_{\mathbf{j}_p}}{m^\star_p} \left(\nabla_{\mathbf{r}} p - p \frac{\nabla_{\mathbf{r}} N_{\mathrm{v}}}{N_{\mathrm{v}}} \right) , \tag{7.129}$$

$$\nabla^{\mathrm{T}}_{\mathbf{r}} \mathbf{j}_n + \frac{\partial n}{\partial t} = G^{\mathrm{II}} , \tag{7.130}$$

$$\mathbf{j}_n = -n \frac{\tau^\star_{\mathbf{j}_n}}{m^\star_n} \nabla_{\mathbf{r}} (E_{\mathrm{c}} - q\Psi) - \frac{k_{\mathrm{B}} T_0 \tau^\star_{\mathbf{j}_n}}{m^\star_n} \left(\nabla_{\mathbf{r}} n - n \frac{\nabla_{\mathbf{r}} N_{\mathrm{c}}}{N_{\mathrm{c}}} \right) . \tag{7.131}$$

The boundary conditions for the particle and current densities are the same as for the HD model.

Quasifermi potentials Φ_p and Φ_n for holes and electrons can be defined by the relations

$$p = n_{\mathrm{i}} \exp\left(\frac{\Phi_p - \Psi - \Psi_h}{V_{\mathrm{T}}} \right) \tag{7.132}$$

and

$$n = n_{\mathrm{i}} \exp\left(\frac{\Psi + \Psi_h - \Phi_n}{V_{\mathrm{T}}} \right) . \tag{7.133}$$

where Ψ_h is the workfunction difference relative to relaxed silicon (6.17) and n_{i} the intrinsic particle density (5.41). With this transformation the current continuity equations read

$$\mathbf{j}_p = -p\frac{q\tau^\star_{\mathbf{j}_p}}{m^\star_p}\nabla_{\mathbf{r}}\Phi_p = -p\mu_p\nabla_{\mathbf{r}}\Phi_p \tag{7.134}$$

and

$$\mathbf{j}_n = n\frac{q\tau^\star_{\mathbf{j}_n}}{m^\star_n}\nabla_{\mathbf{r}}\Phi_n = n\mu_n\nabla_{\mathbf{r}}\Phi_n \ . \tag{7.135}$$

Thus, the driving force is given by the gradient of the quasifermi potential of which the modulus $|\nabla_{\mathbf{r}}\Phi|$ is used to parameterize the mobility in order to account for nonequilibrium effects in the DD model [7.5, 7.6].

The balance equations of the current densities (7.129) and (7.131) are eliminated from the DD model similar to the HD case and the final DD model for the variables p, n, and Ψ reads

$$(7.128) + (7.129) \rightarrow \quad F_{\dot p}(p,n,\Psi) + \frac{\partial p}{\partial t} = 0 \ , \tag{7.136}$$

$$(7.130) + (7.131) \rightarrow \quad F_{\dot n}(p,n,\Psi) + \frac{\partial n}{\partial t} = 0 \ , \tag{7.137}$$

$$(6.11) \rightarrow \quad F_{\Psi}(p,n,\Psi) = 0 \ . \tag{7.138}$$

The right-hand sides of the balance equations for the current densities (7.129) and (7.131) are abbreviated by

$$\mathbf{j}_p = \mathbf{F}_{\mathbf{j}_p}(p,n,\Psi) \ , \tag{7.139}$$

$$\mathbf{j}_n = \mathbf{F}_{\mathbf{j}_n}(p,n,\Psi) \ . \tag{7.140}$$

The Jacobians are given by

$$\Lambda_\alpha(\delta p,\delta n,\delta\Psi) = \frac{\partial F_\alpha}{\partial p}\delta p + \frac{\partial F_\alpha}{\partial n}\delta n + \frac{\partial F_\alpha}{\partial \Psi}\delta\Psi \ , \tag{7.141}$$

and

$$\begin{aligned}\delta\underline{\mathbf{j}}_p &= \mathbf{\Lambda}_{\mathbf{j}_p}(\delta\underline{p},\delta\underline{n},\delta\underline{\Psi}) \\ &= \frac{\partial \mathbf{F}_{\mathbf{j}_p}}{\partial p}\delta\underline{p} + \frac{\partial \mathbf{F}_{\mathbf{j}_p}}{\partial n}\delta\underline{n} + \frac{\partial \mathbf{F}_{\mathbf{j}_p}}{\partial \Psi}\delta\underline{\Psi} \ , \end{aligned} \tag{7.142}$$

$$\begin{aligned}\delta\underline{\mathbf{j}}_n &= \mathbf{\Lambda}_{\mathbf{j}_n}(\delta\underline{p},\delta\underline{n},\delta\underline{\Psi}) \\ &= \frac{\partial \mathbf{F}_{\mathbf{j}_n}}{\partial p}\delta\underline{p} + \frac{\partial \mathbf{F}_{\mathbf{j}_n}}{\partial n}\delta\underline{n} + \frac{\partial \mathbf{F}_{\mathbf{j}_n}}{\partial \Psi}\delta\underline{\Psi} \ . \end{aligned} \tag{7.143}$$

With these definitions small-signal analysis can be performed in the same manner as described in Sec. 7.2 for the HD model.

The linearized Langevin-type DD model in the frequency domain evaluates to

$$\Lambda_{\dot p}(\delta\underline{p},\delta\underline{n},\delta\underline{\Psi}) + \mathrm{i}\omega\delta\underline{p} = \underline{\xi}_{\dot p} - \nabla^{\mathrm{T}}_{\mathbf{r}}\underline{\boldsymbol{\xi}}_{\mathbf{j}_p} \ , \tag{7.144}$$

$$\Lambda_{\dot n}(\delta\underline{n},\delta\underline{n},\delta\underline{\Psi}) + \mathrm{i}\omega\delta\underline{n} = \underline{\xi}_{\dot n} - \nabla^{\mathrm{T}}_{\mathbf{r}}\underline{\boldsymbol{\xi}}_{\mathbf{j}_n} \ , \tag{7.145}$$

$$\Lambda_{\Psi}(\delta\underline{p},\delta\underline{n},\delta\underline{\Psi}) = 0 \ , \tag{7.146}$$

where the boundary conditions for the particles densities and potential are the same as in the HD case. Again, Green's functions are defined for the small-signal variables $\delta\underline{p}$, $\delta\underline{n}$, and $\delta\underline{\Psi}$, which must satisfy the same boundary conditions as the linearized Langevin-type DD model and

$$\Lambda_\alpha(\underline{G}^\beta_{\delta p},\underline{G}^\beta_{\delta n},\underline{G}^\beta_{\delta\Psi})+\delta_{\alpha,\dot{p}}\mathrm{i}\omega\underline{G}^\beta_{\delta p}+\delta_{\alpha,\dot{n}}\mathrm{i}\omega\underline{G}^\beta_{\delta n}=\delta_{\alpha,\beta}\delta(\mathbf{r}-\mathbf{r}')\ . \tag{7.147}$$

The Green's functions of the current densities are given by

$$\underline{\mathbf{G}}^\beta_{\delta\mathbf{j}_p} = \mathbf{\Lambda}_{\mathbf{j}_p}(\underline{G}^\beta_{\delta p},\underline{G}^\beta_{\delta n},\underline{G}^\beta_{\delta\Psi})\ , \tag{7.148}$$
$$\underline{\mathbf{G}}^\beta_{\delta\mathbf{j}_n} = \mathbf{\Lambda}_{\mathbf{j}_n}(\underline{G}^\beta_{\delta p},\underline{G}^\beta_{\delta n},\underline{G}^\beta_{\delta\Psi})\ . \tag{7.149}$$

With Eq. (7.95) the Fourier transform of the current fluctuations of the lth terminal is obtained [7.19]

$$\delta\underline{I}_l=\int \underline{G}^{\dot{p}}_{\delta I_l}\underline{\xi}_{\dot{p}}+\left(\nabla_{\mathbf{r}'}\underline{G}^{\dot{p}}_{\delta I_l}\right)^{\mathrm{T}}\underline{\boldsymbol{\xi}}_{\mathbf{j}_p}+\underline{G}^{\dot{n}}_{\delta I_l}\underline{\xi}_{\dot{n}}+\left(\nabla_{\mathbf{r}'}\underline{G}^{\dot{n}}_{\delta I_l}\right)^{\mathrm{T}}\underline{\boldsymbol{\xi}}_{\mathbf{j}_n}\,d^3r'\ , \tag{7.150}$$

where the Green's functions $\underline{G}^{\dot{p}}_{\delta I_l},\underline{G}^{\dot{n}}_{\delta I_l}$ (also called transfer functions) are similar to Shockley's impedance fields [7.17] with the difference that they are used to calculate the terminal current fluctuations instead of the voltage fluctuations [7.21].

Comparison of Eqs. (7.101) and (7.150) reveals a significant difference in the Green's functions for the Langevin force of the current density (shown only for holes)

$$\begin{aligned}\mathrm{HD}:&\quad \nabla\underline{G}^{\dot{p}}_{\delta I_l}+\underline{G}^{\dot{w}_p}_{\delta I_l}\nabla\left(E_\mathrm{v}-q\Psi_\mathrm{s}\right)\ ,\\ \mathrm{DD}:&\quad \nabla\underline{G}^{\dot{p}}_{\delta I_l}\ .\end{aligned}$$

This difference means that the intuitive argumentation applied by Shockley et al. in the framework of the DD approximation [7.17] is not easily extended to the case of the HD model, because this idea is based on the fact that the Green's function of the current fluctuations must be given by the gradient of the Green's function for the density fluctuations. This is not the case for the HD model, where additional contributions due to Joule's heat term appear. If these additional terms are neglected, noise can be largely overestimated in devices with strong internal electric fields (e.g. MOSFETs).

In the case of the DD model diffusion noise is only due to the Langevin force for the current density of which the corresponding single-particle Langevin force is the one for the particle velocity and the following relations hold for the respective spectral intensities

$$\begin{aligned}\hat{S}_{\boldsymbol{\xi}_{\mathbf{j}_p}\boldsymbol{\xi}_{\mathbf{j}_p}}(\mathbf{r},\mathbf{r}') &= p_\mathrm{s}(\mathbf{r})\hat{S}_{\boldsymbol{\xi}_{\mathbf{v}_p}\boldsymbol{\xi}_{\mathbf{v}_p}}(\mathbf{r})\delta(\mathbf{r}-\mathbf{r}')\ ,\\ \hat{S}_{\boldsymbol{\xi}_{\mathbf{j}_n}\boldsymbol{\xi}_{\mathbf{j}_n}}(\mathbf{r},\mathbf{r}') &= n_\mathrm{s}(\mathbf{r})\hat{S}_{\boldsymbol{\xi}_{\mathbf{v}_n}\boldsymbol{\xi}_{\mathbf{v}_n}}(\mathbf{r})\delta(\mathbf{r}-\mathbf{r}')\ .\end{aligned} \tag{7.151}$$

With this the spectral intensity of current noise reads

$$\underline{S}^{\text{diff}}_{I_l I_k}(\omega) = \int p_{\text{s}} \left(\nabla_{\mathbf{r}} \underline{G}^{\dot{p}}_{\delta I_l}\right)^{\text{T}} \hat{S}_{\boldsymbol{\xi}_{\mathbf{v}_p} \boldsymbol{\xi}_{\mathbf{v}_p}} \left(\nabla_{\mathbf{r}} \underline{G}^{\dot{p}}_{\delta I_k}\right)^{*} + n_{\text{s}} \left(\nabla_{\mathbf{r}} \underline{G}^{\dot{n}}_{\delta I_l}\right)^{\text{T}} \hat{S}_{\boldsymbol{\xi}_{\mathbf{v}_n} \boldsymbol{\xi}_{\mathbf{v}_n}} \left(\nabla_{\mathbf{r}} \underline{G}^{\dot{n}}_{\delta I_k}\right)^{*} d^3 r \ , \tag{7.152}$$

where the spectral intensities of the Langevin forces for the particle velocities are evaluated under homogeneous bulk conditions.

In the following the discussion is limited to the hole case because the electron case is similar. Under homogeneous bulk conditions the Jacobian of the DD model is given by the identity matrix so that

$$\hat{S}_{\boldsymbol{\xi}_{\mathbf{v}_p} \boldsymbol{\xi}_{\mathbf{v}_p}} = \hat{S}_{\mathbf{v}_p \mathbf{v}_p} \ . \tag{7.153}$$

It is worth noting that this relation is a rather stringent approximation due to the very simple dynamics of the DD model. Only in the case of equilibrium it is exact. The spectral intensity of the velocity fluctuations is given by the Fourier transform of the velocity autocorrelation function at zero frequency

$$\hat{S}_{\mathbf{v}_p \mathbf{v}_p} = 2 \int_{-\infty}^{\infty} \langle \delta \mathbf{v}_p(t) \delta \mathbf{v}_p^{\text{T}}(0) \rangle_{\text{hom}} dt \ . \tag{7.154}$$

Under equilibrium conditions this relation can be simplified with Eqs. (2.98) and (2.63)

$$\hat{S}_{\mathbf{v}_p \mathbf{v}_p} = 4 \hat{D}_p = 4 V_{\text{T}} \hat{\mu}_p \ . \tag{7.155}$$

It is worth noting that this relation only holds under equilibrium conditions. Under nonequilibrium conditions the Einstein relation (2.63) and its various generalizations fail to give the correct local noise source [7.33]. The same holds for the fluctuation-dissipation relations corresponding to the other correlations covered by (7.114), because these as well are generally valid only under equilibrium conditions [7.23]. Moreover, the correlation functions must be calculated under bulk conditions without considering the real space in order to obtain the correct noise sources (cf. Sec. 2.4) in contrast to the methods proposed P. Shiktorov et al. [7.28, 7.34].

In the case of generation noise due to II Eq. (7.120) can be used without any modifications in conjunction with the DD model.

7.4.1 Proof of the Nyquist Theorem for zero frequency

Under equilibrium conditions the terminal current noise can be calculated with the Nyquist theorem for zero frequency [7.17, 7.35, 7.36]

$$S_{I_l I_K} = 4 k_{\text{B}} T_0 G_{l,k} \ , \tag{7.156}$$

which is proved here for diffusion noise [7.17, 7.37]. The conductance $G_{l,k}$ is evaluated with the small-signal DD model based on the quasifermi potentials

$$\nabla_{\mathbf{r}}^{\mathrm{T}}\delta\mathbf{j}_p = 0\,, \tag{7.157}$$

$$\delta\mathbf{j}_p = -p_{\mathrm{s}}\mu_p\nabla_{\mathbf{r}}\delta\Phi_p\,, \tag{7.158}$$

$$\nabla_{\mathbf{r}}^{\mathrm{T}}\delta\mathbf{j}_n = 0\,, \tag{7.159}$$

$$\delta\mathbf{j}_n = n_{\mathrm{s}}\mu_n\nabla_{\mathbf{r}}\delta\Phi_n\,, \tag{7.160}$$

which takes this simple form because of $\nabla_{\mathbf{r}}\Phi_{p,\mathrm{s}} = \nabla_{\mathbf{r}}\Phi_{n,\mathrm{s}} = \mathbf{0}$ for equilibrium, and the transport equations are decoupled from the PE. The boundary conditions are of the Dirichlet type for the contacts

$$\delta\Phi_p(\mathbf{r}) = \delta\Phi_n(\mathbf{r}) = \delta V_l \quad \text{for} \quad \mathbf{r}\in\mathbf{A}_l \tag{7.161}$$

and of the Neumann type on all other surfaces. Since the hole and electron DD model are also decoupled, in the following only the hole model is derived. The small-signal quasifermi potential for the holes is expressed with its basis functions

$$\delta\Phi_p(\mathbf{r}) = \sum_{l=1}^{N_{\mathrm{cont}}}\phi_p^l(\mathbf{r})\delta V_l \tag{7.162}$$

with

$$\nabla_{\mathbf{r}}^{\mathrm{T}}\left(-p_{\mathrm{s}}\mu_p\nabla_{\mathbf{r}}\phi_p^l\right) = 0 \tag{7.163}$$

and

$$\phi_p^l(\mathbf{r}) = \delta_{l,k} \quad \text{for} \quad \mathbf{r}\in\mathbf{A}_k\,. \tag{7.164}$$

With these basis functions the current flowing out of the lth terminal is given by

$$\delta I_l = \sum_{k=1}^{N_{\mathrm{cont}}}\int_{A_l}\left(-qp_{\mathrm{s}}\mu_p\nabla_{\mathbf{r}}\phi_p^k\delta V_k\right)^{\mathrm{T}}d^2\mathbf{r} = -\sum_{k=1}^{N_{\mathrm{cont}}}G_{l,k}^p\delta V_k\,. \tag{7.165}$$

Since the basis functions have the properties of test functions (6.73), the surface integral can be extended to the whole surface by multiplying the integrand with the basis function of the lth contact, and a positive-definite symmetric expression is obtained for the hole conductance [7.37]

$$\begin{aligned} G_{l,k}^p &= \int_{A_l}\left(qp_{\mathrm{s}}\mu_p\nabla_{\mathbf{r}}\phi_p^k\right)^{\mathrm{T}}d^2\mathbf{r} \\ &= \oint\phi_p^l\left(qp_{\mathrm{s}}\mu_p\nabla_{\mathbf{r}}\phi_p^k\right)^{\mathrm{T}}d^2\mathbf{r} \\ &= \int\left(\nabla_{\mathbf{r}}\phi_p^l\right)^{\mathrm{T}}qp_{\mathrm{s}}\mu_p\nabla_{\mathbf{r}}\phi_p^k d^3r \\ &= G_{k,l}^p\,. \end{aligned} \tag{7.166}$$

Under equilibrium conditions the Einstein relation (2.63) holds and the terminal current noise due to diffusion noise (7.152) is given for holes by

$$\underline{S}_{I_l I_k}^{\text{diff},p} = \int p_{\text{s}} \left(\nabla_{\mathbf{r}} \underline{G}_{\delta I_l}^{\dot{p}}\right)^{\text{T}} 4V_{\text{T}}\mu_p \left(\nabla_{\mathbf{r}} \underline{G}_{\delta I_k}^{\dot{p}}\right)^{*} d^3r \ . \tag{7.167}$$

The Green's function of the terminal current is calculated for zero frequency by

$$\underline{G}_{\delta I_l}^{\dot{p}}(\mathbf{r}',0) = \int\limits_{A_l} \left(-qp_{\text{s}}(\mathbf{r})\mu_p(\mathbf{r})\nabla_{\mathbf{r}} G_{\delta\Phi_p}^{\dot{p}}(\mathbf{r},\mathbf{r}')\right)^{\text{T}} d^2\mathbf{r} \ , \tag{7.168}$$

where the electron contribution to the terminal current is zero, because the electron and hole small-signal DD models are decoupled. The Green's function of the small-signal quasifermi potential is real valued, because zero frequency is assumed, and it is determined with a relation similar to Eq. (7.147)

$$\nabla_{\mathbf{r}}^{\text{T}} \left(-p_{\text{s}}(\mathbf{r})\mu_p(\mathbf{r})\nabla_{\mathbf{r}} G_{\delta\Phi_p}^{\dot{p}}(\mathbf{r},\mathbf{r}')\right) = \delta(\mathbf{r}-\mathbf{r}') \ . \tag{7.169}$$

Multiplication of this relation with the basis function of the lth contact and integration over $\mathbf{r}$ yields

$$\begin{aligned} \int \phi_p^l(\mathbf{r})\nabla_{\mathbf{r}}^{\text{T}} \left(-p_{\text{s}}(\mathbf{r})\mu_p(\mathbf{r})\nabla_{\mathbf{r}} G_{\delta\Phi_p}^{\dot{p}}(\mathbf{r},\mathbf{r}')\right) d^3r &= \int \phi_p^l(\mathbf{r})\delta(\mathbf{r}-\mathbf{r}')d^3r \\ &= \phi_p^l(\mathbf{r}') \ . \end{aligned} \tag{7.170}$$

Integration by parts on the left-hand side results in

$$\begin{aligned} &\oint \phi_p^l(\mathbf{r}) \left(-p_{\text{s}}(\mathbf{r})\mu_p(\mathbf{r})\nabla_{\mathbf{r}} G_{\delta\Phi_p}^{\dot{p}}(\mathbf{r},\mathbf{r}')\right)^{\text{T}} d^2\mathbf{r} \\ &+ \int \left(\nabla_{\mathbf{r}}\phi_p^l(\mathbf{r})\right)^{\text{T}} p_{\text{s}}(\mathbf{r})\mu_p(\mathbf{r})\nabla_{\mathbf{r}} G_{\delta\Phi_p}^{\dot{p}}(\mathbf{r},\mathbf{r}')d^3r = \phi_p^l(\mathbf{r}') \ . \end{aligned} \tag{7.171}$$

Since the basis function has the properties of a test function the surface integral can be limited to the area of the lth contact

$$\begin{aligned} \oint \phi_p^l(\mathbf{r}) \left(-p_{\text{s}}(\mathbf{r})\mu_p(\mathbf{r})\nabla_{\mathbf{r}} G_{\delta\Phi_p}^{\dot{p}}(\mathbf{r},\mathbf{r}')\right)^{\text{T}} d^2\mathbf{r} &= \\ \int\limits_{A_l} \left(-p_{\text{s}}(\mathbf{r})\mu_p(\mathbf{r})\nabla_{\mathbf{r}} G_{\delta\Phi_p}^{\dot{p}}(\mathbf{r},\mathbf{r}')\right)^{\text{T}} d^2\mathbf{r} &= \frac{1}{q}\underline{G}_{\delta I_l}^{\dot{p}}(\mathbf{r}',0) \ . \end{aligned} \tag{7.172}$$

The volume integral in Eq. (7.171) can be evaluated with further integration by parts

$$\begin{aligned} \int \left(p_{\text{s}}(\mathbf{r})\mu_p(\mathbf{r})\nabla_{\mathbf{r}}\phi_p^l(\mathbf{r})\right)^{\text{T}} \nabla_{\mathbf{r}} G_{\delta\Phi_p}^{\dot{p}}(\mathbf{r},\mathbf{r}')d^3r &= \\ \oint \left(p_{\text{s}}(\mathbf{r})\mu_p(\mathbf{r})\nabla_{\mathbf{r}}\phi_p^l(\mathbf{r})\right)^{\text{T}} G_{\delta\Phi_p}^{\dot{p}}(\mathbf{r},\mathbf{r}')d^2\mathbf{r} & \end{aligned}$$

$$- \int \nabla_{\mathbf{r}}^{\mathrm{T}} \left(p_{\mathrm{s}}(\mathbf{r}) \mu_p(\mathbf{r}) \nabla_{\mathbf{r}} \phi_p^l(\mathbf{r}) \right) G_{\delta\Phi_p}^{\dot{p}}(\mathbf{r},\mathbf{r}') d^3r \;=\; 0 \,. \qquad (7.173)$$

The surface integral vanishes on the contacts due to the Dirichlet boundary conditions for the Green's function and on all other surfaces due to the Neumann boundary condition for the small-signal quasifermi potential. The second volume integral is zero because of Eq. (7.163). These results yield

$$\underline{G}_{\delta I_l}^{\dot{p}}(\mathbf{r}',0) = q\phi_p^l(\mathbf{r}') \,, \qquad (7.174)$$

and the hole diffusion noise (7.167) reads

$$\underline{S}_{I_l I_k}^{\mathrm{diff},p} = \int p_{\mathrm{s}} \left(q \nabla_{\mathbf{r}} \phi_p^l \right)^{\mathrm{T}} 4V_{\mathrm{T}} \mu_p q \nabla_{\mathbf{r}} \phi_p^k \, d^3r = 4k_{\mathrm{B}} T_0 G_{l,k}^p \,. \qquad (7.175)$$

A similar expression can be obtained for electrons and the sum of both yields the Nyquist theorem (7.156) [7.17, 7.37].

7.5 Transport and Noise Parameter Simulation

MC bulk simulations are used to generate the transport and noise parameters for the DD and HD models to achieve consistency of the momentum-based models and the BTE, and because these parameters are either experimentally inaccessible or have not been measured in the necessary range of experimental conditions. An example of an experimentally almost inaccessible quantity is the inverse mass tensor[3] which is needed for the calculation of hydrodynamic transport parameters (cf. Sec. 7.1) [7.12, 7.13, 7.39]. In Fig. 7.1 the expected value of the inverse mass tensor based on Eq. (5.48) is shown for electrons and holes in undoped silicon at room temperature as a function of the electric field. The anisotropy and dependence on the electric field are more pronounced in the case of electrons, although the electron band structure is simpler than the one of holes especially at low energies (cf. Fig. 5.3).

In Fig. 7.2 the four macroscopic relaxation times of the GHDM are shown for electrons. The relaxation times are always smaller than 0.3ps and the first order derivatives with respect to time can be safely neglected in Eqs. (7.17), (7.21), and (7.30) for frequencies below 100GHz.

In Fig. 7.3 different correlation functions of the single-particle fluctuations (δv_l, $\delta T^{\star}$, $\delta\varepsilon v_l$) evaluated by MC bulk simulations are shown for electrons. The slowest decay is found in the case of the correlation functions involving temperature fluctuations. But even these correlations have almost vanished after 1ps and the limits of the integral in (7.114) can be safely reduced to $\pm$10ps with negligible error. The corresponding spectral intensities of the single-particle fluctuations are

[3]The effective mass tensor at the conduction band minimum can be determined by low-temperature cyclotron resonance measurements [7.38].

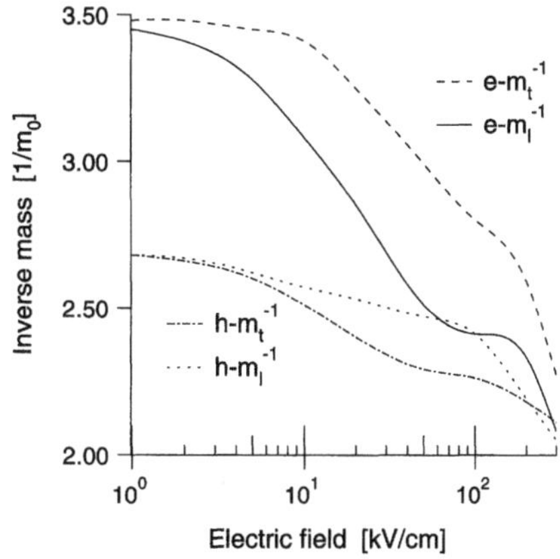

Figure 7.1: Expected values of the longitudinal and transverse inverse mass for electrons and holes in undoped silicon at room temperature as a function of the electric field in ⟨100⟩-direction

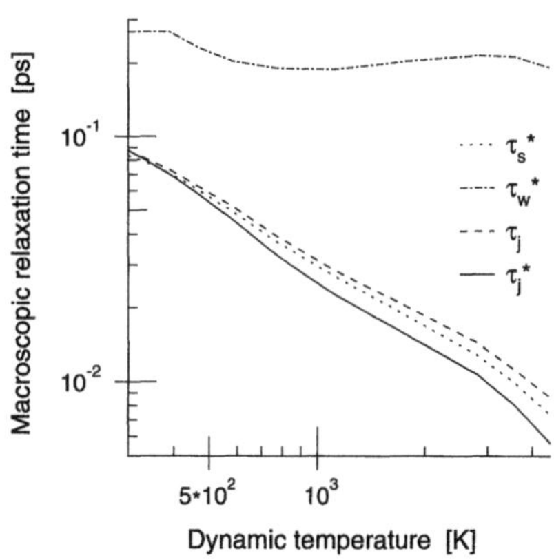

Figure 7.2: Macroscopic relaxation times for relaxed silicon with a donor doping of $2 \cdot 10^{17}/\mathrm{cm}^3$ at room temperature under homogeneous bulk conditions as a function of the dynamic temperature where the electric field is in ⟨100⟩-direction

shown in Fig. 7.4 as a function of the frequency. The spectral intensities are nearly white (constant) below 100GHz, and the differences between the results for 0 and 100GHz are less than 5%. Thus, the assumption of white noise in (7.113) is well justified in the frequency range for which our new models have been developed.

In Fig. 7.5 the low-frequency spectral intensities of the velocity fluctuations in the direction of the electric field $S_{v_l v_l}$ are shown for electrons in a homogeneous bulk system as directly evaluated by MC simulations (i.e. integral over the velocity

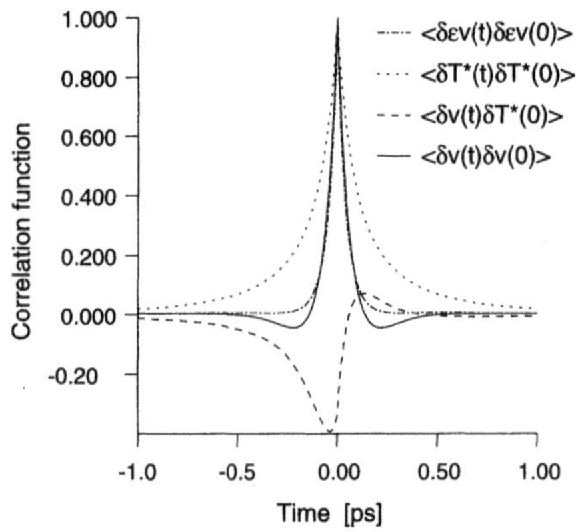

Figure 7.3: Normalized velocity/velocity, velocity/temperature, temperature/temperature, and energy-current/energy-current correlation functions for relaxed silicon with a donor doping of $2 \cdot 10^{17}/\mathrm{cm}^3$ at room temperature under homogeneous bulk conditions for an electric field of 30kV/cm

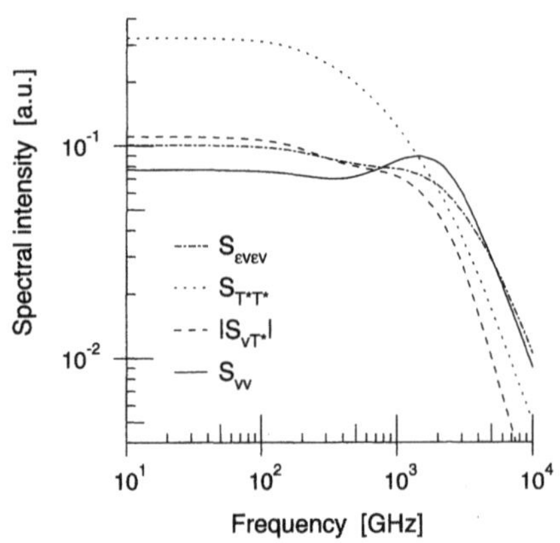

Figure 7.4: Velocity/velocity, temperature/temperature, and energy-current/energy-current correlation spectra for electrons in relaxed silicon with a donor doping of $2 \cdot 10^{17}/\mathrm{cm}^3$ at room temperature under homogeneous bulk conditions for an electric field of 30kV/cm in $\langle 100 \rangle$ direction and the absolute value of the velocity/temperature cross correlation spectrum

autocorrelation function (7.154)), based on the Einstein relation (7.155), and based on a generalized Einstein relation $S_{v_l v_l} = 4k_\mathrm{B}T_n^\star \mu_n$. As already shown in Ref. [7.33] the Einstein relation and its generalization fail to give the correct local noise source. In addition, the Einstein relation does not capture the strong

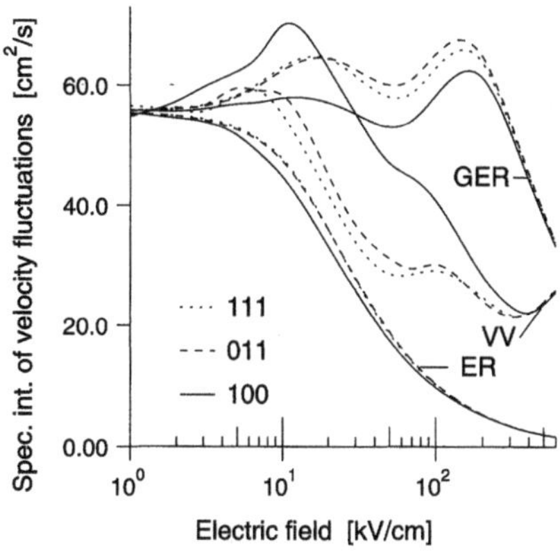

Figure 7.5: Spectral intensity of the electron velocity fluctuations in the direction of the electric field for three field directions (⟨100⟩, ⟨011⟩, ⟨111⟩) evaluated directly with the velocity autocorrelation function (VV), based on the Einstein relation (ER) and on the generalized Einstein relation (GER) for relaxed silicon with a donor doping of $2 \cdot 10^{17}/\mathrm{cm}^3$ at room temperature under homogeneous bulk conditions for 0Hz

dependence of the velocity fluctuations on the direction of the electric field [7.33], where ⟨100⟩ is the main direction of the current flow in an HBT and ⟨011⟩ in a MOSFET. Thus, models based on Einstein relations (e.g. [7.18, 7.40–7.44]) are bound to fail in cases, where the noise sources are modified by nonequilibrium effects. In [7.33] this problem is circumvented by deducing a new hot-carrier Einstein relation based on the differential mobility and the noise temperature. But as pointed out in [7.27] the noise temperature cannot be calculated based on transport parameters and MC simulations are still required to determine the local noise source.

Based on the LBTE it is possible to derive exact spectral intensities of the Langevin forces for the momentum-based models [7.23–7.26, 7.28]. But due to the simplified kinetics of the DD and HD models the exact Langevin forces do not yield the exact fluctuations (cf. Sec. 7.3.1). This is depicted in Fig. 7.6, where the spectral intensities of the velocity fluctuations are shown calculated by the DD and HD models based on the exact Langevin forces together with the exact MC results. Under equilibrium conditions all three models yield the same result, because fluctuation-dissipation relations hold [7.23]. With increasing electric field differences occur and the accuracy of the DD and HD models based on exact Langevin forces degrades, where the HD model yields the better overall agreement. These results clearly demonstrate the advantage of the modified Langevin forces, which yield the exact results in both cases, DD and HD by definition. It is worth noting that the differences between our simplified and the full HD model

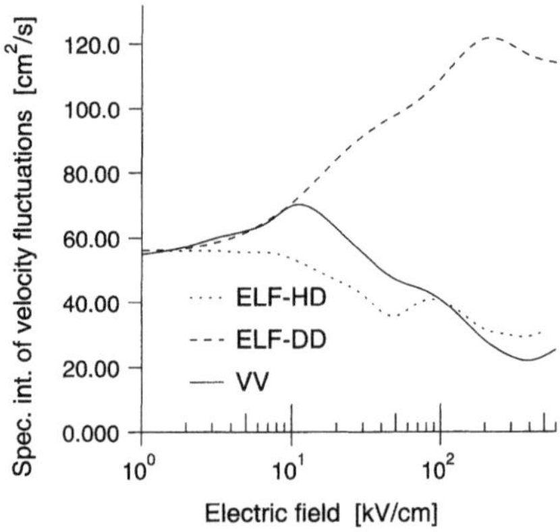

Figure 7.6: Spectral intensity of the electron velocity fluctuations in the direction of the electric field for the ⟨100⟩ direction evaluated directly with the velocity autocorrelation function (VV), based on the exact spectral intensities of the Langevin forces within the DD framework (ELF-DD) and in the framework of the HD model (ELF-HD) for relaxed silicon with a donor doping of $2 \cdot 10^{17}/\mathrm{cm}^3$ at room temperature under homogeneous bulk conditions for 0Hz

of [7.28] vanish under bulk conditions in the limit of zero frequency, which are the conditions used in the above comparison.

References

[7.1] W. V. Roosbroeck, "Theory of the flow of electrons and holes in germanium and other semiconductors", *Bell System Technical Journal*, pp. 561–607, 1950.

[7.2] W. Shockley, *Electrons and Holes in Semiconductors*, van Nostrand, Princeton, New Jersey, 1950.

[7.3] R. Stratton, "Diffusion of hot and cold electrons in semiconductor barriers", *Phys. Rev.*, vol. 126, pp. 2002–2013, 1962.

[7.4] A. H. Marshak and K. M. van Vliet, "Electrical current in solids with position–dependent band structure", *Solid–State Electron.*, vol. 21, pp. 417–427, 1978.

[7.5] W. L. Engl, H. K. Dirks, and B. Meinerzhagen, "Device modeling", *Proc. IEEE*, vol. 71, pp. 10–33, 1983.

[7.6] S. Selberherr, *Analysis and Simulation of Semiconductor Devices*, Springer, Wien, 1984.

[7.7] K. Bløtekjær, "Transport equations for electrons in two-valley semiconductors", *IEEE Trans. Electron Devices*, vol. 17, no. 1, pp. 38–47, 1970.

[7.8] R. K. Cook and J. Frey, "An efficient technique for two-dimensional simulation of velocity overshoot effects in Si and GaAs devices", *COMPEL Int. J. Comput. Math. Electr. Electron. Eng.*, vol. 1, pp. 65–87, 1982.

[7.9] M. Fukuma and R. H. Uebbing, "Two-dimensional MOSFET simulation with energy transport phenomena", in *IEDM*, 1984, pp. 621–624.

[7.10] W. Hänsch and M. Miura-Mattausch, "A new current relation for hot electron transport", in *Proc. NASECODE IV*, 1985, pp. 311–314.

[7.11] B. Meinerzhagen and W. L. Engl, "The influence of the thermal equilibrium approximation on the accuracy of classical two-dimensional numerical modeling of silicon submicrometer MOS transistors", *IEEE Trans. Electron Devices*, vol. 35, no. 5, pp. 689–697, 1988.

[7.12] R. Thoma, A. Emunds, B. Meinerzhagen, H. J. Peifer, and W. L. Engl, "Hydrodynamic equations for semiconductors with nonparabolic bandstructures", *IEEE Trans. Electron Devices*, vol. 38, no. 6, pp. 1343–1352, 1991.

[7.13] R. Thoma, "Entwicklung eines hydrodynamischen Gleichungssystems zur Beschreibung von Halbleiterbauelementen", Doktorarbeit, RWTH Aachen, 1991, Aachen.

[7.14] C. Jungemann, B. Neinhüs, and B. Meinerzhagen, "Investigation of the local force approximation in numerical device simulation by full-band Monte Carlo simulation", *VLSI Design*, vol. 13, no. 1-4, pp. 281–285, 2001.

[7.15] B. Neinhüs, S. Decker, P. Graf, F. M. Bufler, and B. Meinerzhagen, "Consistent hydrodynamic and Monte-Carlo simulation of SiGe HBTs based on table models for the relaxation times", *VLSI Design*, vol. 8, pp. 387–391, 1998.

[7.16] H. Beneking, *High Speed Semiconductor Devices*, Chapman & Hall, London, 1994.

[7.17] W. Shockley, John A. Copeland, and R. P. James, "The impedance field method of noise calculation in active semiconductor devices", in *Quantum theory of atoms, molecules and solid state*, P. O. Lowdin, Ed., pp. 537–563. Academic Press, 1966.

[7.18] F. Bonani, G. Ghione, M. R. Pinto, and R. K. Smith, "An efficient approach to noise analysis through multidimentional physics-based models", *IEEE Trans. Electron Devices*, vol. 45, no. 1, pp. 261–269, 1998.

[7.19] F. Bonani and G. Ghione, "Generation–recombination noise modelling in semiconductor devices through population or approximate equivalent current density fluctuations", *Solid–State Electron.*, vol. 43, pp. 285–295, 1999.

[7.20] F. Bonani and G. Ghione, "Noise modeling for PDE based device simulations", in *Proc. ICNF*, Gainesville, FL, 2001, vol. 16, pp. 631–636.

[7.21] P. Shiktorov, E. Starikov, V. Gruzinskis, L. Varani, J.-C. Vaissiere, J.-P. Nougier, and L. Reggiani, "Admittance field method for the calculation of the spectral density of current fluctuations", *Fluctuation and Noise Letters*, vol. 1, no. 1, pp. R1–R11, 2001.

[7.22] K. M. van Vliet, "Markov approach to density fluctuations due to transport and scattering. I. Mathematical formalism*", *J. Math. Phys.*, vol. 12, pp. 1981–1998, 1971.

[7.23] Sh. Kogan, *Electronic Noise and Fluctuations in Solids*, Cambridge University Press, Cambridge, New York, Melbourne, 1996.

[7.24] S. M. Kogan, "Equations for the correlation functions using a generalized Keldysh technique", *Phys. Rev. A*, vol. 44, pp. 8072–8082, 1991.

[7.25] H. S. Min, "A unified theory of noise in nondegenerate semiconductors", *J. Appl. Phys.*, vol. 61, pp. 4549–4565, 1987.

[7.26] C. Jungemann and B. Meinerzhagen, "Analysis of the stochastic error of stationary Monte Carlo device simulations", *IEEE Trans. Electron Devices*, vol. 48, no. 5, pp. 985–992, 2001.

[7.27] F. Bonani and G. Ghione, *Noise in Semiconductor Devices, Modeling and Simulation*, Advanced Microelectronics. Springer, Berlin, Heidelberg, New York, 2001.

[7.28] P. Shiktorov, E. Starikov, V. Gruzinskis, T. Gonzalez, J. Mateos, D. Pardo, L. Reggiani, L. Varani, and J. C. Vaissere, "Langevin forces and generalized transfer fields for noise modeling in deep submicron devices", *IEEE Trans. Electron Devices*, vol. 47, no. 10, pp. 1992–1998, 2000.

[7.29] C. Jungemann, B. Neinhüs, S. Decker, and B. Meinerzhagen, "Hierarchical 2D RF noise simulation of Si and SiGe devices by Langevin-type DD and HD models based on MC generated noise parameters", in *IEDM Tech. Dig.*, Washington (USA), 2001, pp. 481–484.

[7.30] L. Reggiani, E. Starikov, P. Shiktorov, V. Gružinskis, and L. Varani, "Modelling of small-signal response and electronic noise in semiconductor high-field transport", *Semicond. Sci. Technol.*, vol. 12, pp. 141–156, 1997.

[7.31] A. Papoulis, *Probability,Random Variables and Stochastic Processes*, Mc Graw-Hill, 3rd edition, 1991.

[7.32] M. V. Fischetti, N. Sano, S. E. Laux, and K. Natori, "Full-band-structure theory of high-field transport and impact ionization of electrons and holes in Ge, Si, and GaAs", *IEEE J. Tech. Comp. Aided Design*, , no. 3, 1997.

[7.33] J.-P. Nougier, "Fluctuations and noise of hot carriers in semiconducor materials and devices", *IEEE Trans. Electron Devices*, vol. ED-41, no. 11, pp. 2034–2049, 1994.

[7.34] P. Shiktorov, V. Gružinskis, E. Starikov, T. González, J. Mateos, D. Pardo, L. Reggiani, and L.Varani, "On the spectral strength of the noise source entering the transfer impedance method", *Appl. Phys. Lett.*, vol. 71, pp. 3093–3095, 1997.

[7.35] G. Röpke, *Statistische Mechanik für das Nichtgleichgewicht*, Physik-Verlag, Weinheim, 1987.

[7.36] A. Blum, *Elektronisches Rauschen*, B. G. Teubner, Stuttgart, 1996.

[7.37] E. V. Sukhorukov and D. Loss, "Noise in multiterminal diffuse conductors: Universality, nonlocality, and exchange effects", *Phys. Rev. B*, vol. 59, pp. 13054–13066, 1999.

[7.38] J. C. Hensel, H. Hasegawa, and M. Nakayama, "Cyclotron resonance in uniaxially stressed silicon. II. Nature of the covalent bond", *Phys. Rev. A*, vol. 138, pp. 225–238, 1965.

[7.39] D. Nuernbergk, "Simulation des Transportverhaltens in $Si/Si_{1-x}Ge_x/Si$-Heterobipolartransistoren", Dissertation, Technische Universität Ilmenau, Ilmenau, 1999, (H. Utz Verlag Wissenschaft, München: 1999).

[7.40] A. van der Ziel, *Noise in Solid State Devices and Circuits*, John Wiley and Sons, Cansada, 1986.

[7.41] F. M. Klaassen and J. Prins, "Thermal noise of MOS transistors", *Philips Res. Repts*, pp. 505–514, 1967.

[7.42] P. Klein, "An analytical thermal noise model of deep submicron MOSFET's", *IEEE Electron Device Lett.*, vol. 20, no. 8, pp. 399–401, 1999.

[7.43] A. J. Scholten, H. J. Tromp, L. F. Tiemeijer, R. van Langevelde, R. J. Havens, P. W. H. de Vreede, R. F. M. Roes, P. H. Woerlee, A. H. Montree, , and D. B. M.

Klaassen, "Accurate thermal noise model for deep-submicron CMOS", *IEDM Tech. Dig.*, pp. 155–158, 1999.

[7.44] J.-S. Goo, C. Choi, F. Danneville, E. Morifuji, H. Sasaki Momose, Z. Yu, H. Iwai, T. H. Lee, and R. W. Dutton, "An accurate and efficient high frequency noise simulation technique for deep submicron MOSFETs", *IEEE Trans. Electron Devices*, vol. 47, no. 12, pp. 2410–2419, 2000.

8 Stochastic Properties of Monte-Carlo Device Simulations

The relative stochastic error of a quantity evaluated by stationary MC simulation is proportional to its variance, which can be evaluated with the corresponding autocorrelation function (cf. Sec. 3.4). Since the spectral intensity is the Fourier transform of the autocorrelation function (2.66), the variance of a quantity X averaged over the simulation time T_{sim} (Eq. (3.60)) can be evaluated with the spectral intensity similar to MacDonald's function

$$\begin{aligned}
\sigma^2\{\overline{X}\} &= \frac{\sigma^2\{X\}}{T_{\text{sim}}} \int_{-T_{\text{sim}}}^{T_{\text{sim}}} \left(1 - \frac{|\tau|}{T_{\text{sim}}}\right) C\{X\}(\tau) d\tau \\
&= \frac{1}{\pi T_{\text{sim}}} \int_{0}^{\infty} S_{XX}\left(\frac{\nu}{T_{\text{sim}}}\right) \frac{1-\cos(\nu)}{\nu^2} d\nu \qquad (8.1) \\
&\approx \frac{S_{XX}(0)}{2T_{\text{sim}}}, \qquad (8.2)
\end{aligned}$$

where the approximation (8.2) holds for sufficiently large simulation times. Without statistical enhancement the spectral intensity can be calculated exactly with the methods of Sec. 2.4 or approximately with the methods of Chap. 7. The latter approach is used here in a modified form to investigate in Sec. 8.1 the properties of terminal current estimators. In Sec. 8.2 it is shown how to estimate the CPU time of MC simulations in-advance with the momentum-based noise models of Chap. 7.

8.1 Stochastic Properties of Terminal Current Estimators

In this section the stochastic noise of stationary MC device simulations is discussed for a generalization of the Ramo-Shockley estimator for terminal currents. Under stationary bias conditions the expected value of the displacement current density within the device is zero and Eq. (6.74) reduces to [8.1]

$$\tilde{I}_l = \int \tilde{\mathbf{i}}^{\mathrm{T}} \nabla_{\mathbf{r}} h_l \; d^3 r \; . \tag{8.3}$$

The stochastic noise of this estimator is characterized by the spectral intensity of its fluctuations (cf. Eq. (8.1)), which can be calculated with sufficient precision with the methods described in Chap. 7. Only in the case of very high frequencies (> 100GHz) a modified approach must be used, which includes the first order derivatives with respect to time in Eqs. (7.17)-(7.30). But as already pointed out in Chap. 7 this causes numerical problems. To circumvent these problems in the following the DD approach including all first order derivate with respect to time is used only in the case of equilibrium for which these instabilities do not occur.

The linearized Langevin-type DD model in the frequency domain is derived in this case by calculating balance equations for the microscopic quantities [8.2]

$$X = 1 \; , \tag{8.4}$$

$$\mathbf{X} = \frac{\tau \mathbf{v}}{1 + \mathrm{i}\omega\tau} \tag{8.5}$$

avoiding the macroscopic relaxation time approximation (cf. Sec. 2.4). The resultant small-signal DD model for the fluctuations reads under equilibrium conditions ($\nabla_{\mathbf{r}} \Phi_{p,\mathrm{s}} = \nabla_{\mathbf{r}} \Phi_{n,\mathrm{s}} = \mathbf{0}$)

$$\nabla_{\mathbf{r}}^{\mathrm{T}} \delta \underline{\mathbf{j}}_p + \mathrm{i}\omega \delta \underline{p} = 0 \; , \tag{8.6}$$

$$\delta \underline{\mathbf{j}}_p = -p_{\mathrm{s}} \underline{\hat{\mu}}_p \nabla_{\mathbf{r}} \delta \underline{\Phi}_p + \underline{\boldsymbol{\xi}}_{\mathbf{j}_p} \; , \tag{8.7}$$

$$\nabla_{\mathbf{r}}^{\mathrm{T}} \delta \underline{\mathbf{j}}_n + \mathrm{i}\omega \delta \underline{n} = 0 \; , \tag{8.8}$$

$$\delta \underline{\mathbf{j}}_n = n_{\mathrm{s}} \underline{\hat{\mu}}_n \nabla_{\mathbf{r}} \delta \underline{\Phi}_n + \underline{\boldsymbol{\xi}}_{\mathbf{j}_n} \; , \tag{8.9}$$

where generation/recombination phenomena have been neglected and the mobility is exactly given by (cf. Sec. 2.4)

$$\underline{\hat{\mu}} = \left\langle \frac{q}{\hbar} \nabla_{\mathbf{k}} \frac{\tau \mathbf{v}^{\mathrm{T}}}{1 + \mathrm{i}\omega\tau} \right\rangle_{\mathbf{k}}^{\mathrm{eq}} \; . \tag{8.10}$$

The only difference of the above DD model compared to the one in Sec. 7.4 is the frequency-dependent mobility. Therefore, the same definitions and methods can be used as in Sec. 7.4.

The spectral intensities of the Langevin forces can be exactly evaluated with Eqs. (2.73), (2.74), and (2.94)

$$\hat{S}_{\xi_{\mathbf{j}_p}\xi_{\mathbf{j}_p}}(\mathbf{r},\mathbf{r}',\omega) = 4p_s(\mathbf{r})\Re\left\{\hat{\underline{D}}_p(\mathbf{r},\omega)\right\}\delta(\mathbf{r}-\mathbf{r}') , \tag{8.11}$$

$$\hat{S}_{\xi_{\mathbf{j}_n}\xi_{\mathbf{j}_n}}(\mathbf{r},\mathbf{r}',\omega) = 4n_s(\mathbf{r})\Re\left\{\hat{\underline{D}}_n(\mathbf{r},\omega)\right\}\delta(\mathbf{r}-\mathbf{r}') , \tag{8.12}$$

where the diffusion constant is given by the Einstein relation (2.95). In contrast to Chap. 7 the spectral intensities of the Langevin forces are here not white, because they are based on a frequency-dependent microscopic quantity (8.5).

The fluctuations of the current at the lth terminal estimated with Eq. (8.3) are due to the fluctuations of the conduction current density within the device

$$\delta \underline{I}_l = \int \delta \underline{\mathbf{i}}^{\mathrm{T}} \nabla_{\mathbf{r}} h_l \, d^3 r , \tag{8.13}$$

where the conduction current density fluctuations are given by

$$\delta\underline{\mathbf{i}}(\mathbf{r},\omega) = \int \hat{\underline{G}}_{\delta\mathbf{i}}^{\mathbf{j}_p}(\mathbf{r},\mathbf{r}',\omega)\underline{\xi}_{\mathbf{j}_p}(\mathbf{r}',\omega) + \hat{\underline{G}}_{\delta\mathbf{i}}^{\mathbf{j}_n}(\mathbf{r},\mathbf{r}',\omega)\underline{\xi}_{\mathbf{j}_n}(\mathbf{r}',\omega) d^3 r' \tag{8.14}$$

with

$$\left(\hat{\underline{G}}_{\delta\mathbf{i}}^{\mathbf{j}_p}(\mathbf{r},\mathbf{r}',\omega)\right)^{\mathrm{T}} = q\nabla_{\mathbf{r}'}\left(\underline{\mathbf{G}}_{\delta\mathbf{j}_p}^{\dot{p}}(\mathbf{r},\mathbf{r}',\omega) - \underline{\mathbf{G}}_{\delta\mathbf{j}_n}^{\dot{p}}(\mathbf{r},\mathbf{r}',\omega)\right)^{\mathrm{T}} + q\hat{I}\delta(\mathbf{r}-\mathbf{r}') , \tag{8.15}$$

$$\left(\hat{\underline{G}}_{\delta\mathbf{i}}^{\mathbf{j}_n}(\mathbf{r},\mathbf{r}',\omega)\right)^{\mathrm{T}} = q\nabla_{\mathbf{r}'}\left(\underline{\mathbf{G}}_{\delta\mathbf{j}_p}^{\dot{n}}(\mathbf{r},\mathbf{r}',\omega) - \underline{\mathbf{G}}_{\delta\mathbf{j}_n}^{\dot{n}}(\mathbf{r},\mathbf{r}',\omega)\right)^{\mathrm{T}} - q\hat{I}\delta(\mathbf{r}-\mathbf{r}') . \tag{8.16}$$

The spectral intensity of the current density evaluates to

$$\begin{aligned}\hat{\underline{S}}_{\mathbf{ii}}(\mathbf{r}_1,\mathbf{r}_2,\omega) = 4\int & \left(\hat{\underline{G}}_{\delta\mathbf{i}}^{\mathbf{j}_p}(\mathbf{r}_1,\mathbf{r}',\omega)\right)^{\mathrm{T}} p_s(\mathbf{r}')\Re\left\{\hat{\underline{D}}_p(\mathbf{r}',\omega)\right\} \\ & \hat{\underline{G}}_{\delta\mathbf{i}}^{\mathbf{j}_p}(\mathbf{r}_2,\mathbf{r}',\omega) \\ & + \left(\hat{\underline{G}}_{\delta\mathbf{i}}^{\mathbf{j}_n}(\mathbf{r}_1,\mathbf{r}',\omega)\right)^{\mathrm{T}} n_s(\mathbf{r}')\Re\left\{\hat{\underline{D}}_n(\mathbf{r}',\omega)\right\} \\ & \hat{\underline{G}}_{\delta\mathbf{i}}^{\mathbf{j}_n}(\mathbf{r}_2,\mathbf{r}',\omega)\, d^3 r' .\end{aligned} \tag{8.17}$$

Although the noise sources (8.11) and (8.12) are local, these spectral intensities are nonlocal due to the Green's functions.

The corresponding spectral intensity of the terminal current estimator (8.3) is given by

$$S_{I_l I_l}(\omega) = \int\int \left(\nabla_{\mathbf{r}_1} h_l(\mathbf{r}_1)\right)^{\mathrm{T}} \hat{\underline{S}}_{\mathbf{ii}}(\mathbf{r}_1,\mathbf{r}_2,\omega)\nabla_{\mathbf{r}_2} h_l(\mathbf{r}_2) d^3 r_1 d^3 r_2 . \tag{8.18}$$

Thus, the spectral intensity of the current estimator depends on the choice of the test functions. Only in the case of the Ramo-Shockley test functions the estimator is an unbiased estimator for the terminal current noise. In all other cases the noise is biased and only the estimated stationary terminal current is a physical quantity. On the other hand, the variance of the estimated terminal current depends on the spectral intensity and a reduction of this quantity would improve the efficiency of the estimator. Based on this idea P. D. Yoder et al. proposed an optimized terminal current estimator for stationary simulations which supposedly reduces the variance of the time average of the terminal current [8.1]. But instead of minimizing the variance of the terminal current estimator based on Eq. (8.18), which includes a triple integral over the RS, a simplified expression was minimized, which is obtained by approximating the Green's functions in Eq. (8.17) by Dirac functions

$$\begin{aligned} S_{I_l I_l}^{\mathrm{App}}(\omega) &= 4q^2 \int \left(\nabla_{\mathbf{r}'} h_{l,p}(\mathbf{r}')\right)^{\mathrm{T}} p_{\mathrm{s}}(\mathbf{r}') \Re\{\underline{\hat{D}}_p(\mathbf{r}',\omega)\} \nabla_{\mathbf{r}'} h_{l,p}(\mathbf{r}') \\ &\quad + \left(\nabla_{\mathbf{r}'} h_{l,n}(\mathbf{r}')\right)^{\mathrm{T}} n_{\mathrm{s}}(\mathbf{r}') \Re\{\underline{\hat{D}}_n(\mathbf{r}',\omega)\} \nabla_{\mathbf{r}'} h_{l,n}(\mathbf{r}') d^3 r' \ . \end{aligned} \quad (8.19)$$

By minimizing this expression for zero frequency P. D. Yoder et al. obtained the Euler-Lagrange equations of the test functions for electrons and holes [8.1]

$$\nabla_{\mathbf{r}}^{\mathrm{T}} \left(p_{\mathrm{s}} \hat{D}_p \nabla_{\mathbf{r}} h_{l,p}^{\mathrm{Y}}\right) = 0 \ , \quad (8.20)$$

$$\nabla_{\mathbf{r}}^{\mathrm{T}} \left(n_{\mathrm{s}} \hat{D}_n \nabla_{\mathbf{r}} h_{l,n}^{\mathrm{Y}}\right) = 0 \ . \quad (8.21)$$

Assuming that electron diffusion noise does not induce hole current fluctuations and vice versa ($\nabla_{\mathbf{r}'}(\underline{\mathbf{G}}_{\delta \mathbf{j}_n}^{\dot{p}})^{\mathrm{T}} = \nabla_{\mathbf{r}'}(\underline{\mathbf{G}}_{\delta \mathbf{j}_p}^{\dot{n}})^{\mathrm{T}} = \hat{0}$) different test functions are used to estimate the electron and hole part of the terminal current

$$\tilde{I}_l = \int q \left(\tilde{\mathbf{j}}_p^{\mathrm{T}} \nabla_{\mathbf{r}} h_{l,p} - \tilde{\mathbf{j}}_n^{\mathrm{T}} \nabla_{\mathbf{r}} h_{l,n}\right) d^3 r \ . \quad (8.22)$$

In Fig. 8.1 the terminal current is shown evaluated with Eq. (8.22) based on the Ramo-Shockley test functions and the optimized ones for an MC simulation of the 1D N^+NN^+ structure of Fig. 6.11. The fluctuations of the terminal current are considerably reduced by the optimized test functions, because the optimized test functions restrict the sampling to the low-density areas in the device due to their much lower gradient in the high-density areas compared to the Ramo-Shockley test functions (Fig. 8.2). The gradient of the Ramo-Shockley test functions on the other hand is constant throughout the 1D structure and the current is also sampled in the high density regions with their much higher fluctuations leading to the larger terminal current fluctuations in Fig. 8.1. Whether this is also the case for the time-average of the terminal current remains to be shown.

In Fig. 8.3 the spectral intensity of the terminal current as a function of the frequency is shown as calculated with Eq. (8.18) based on the Ramo-Shockley test

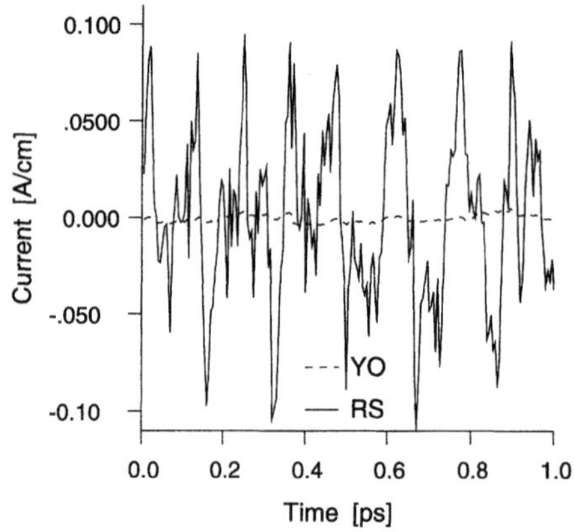

Figure 8.1: Terminal current of the 1D N^+NN^+ structure biased at 0.0V calculated with Eq. (8.22) based on the Ramo-Shockley test functions (RS) and based on the optimized test functions by P. D. Yoder et al. (YO)

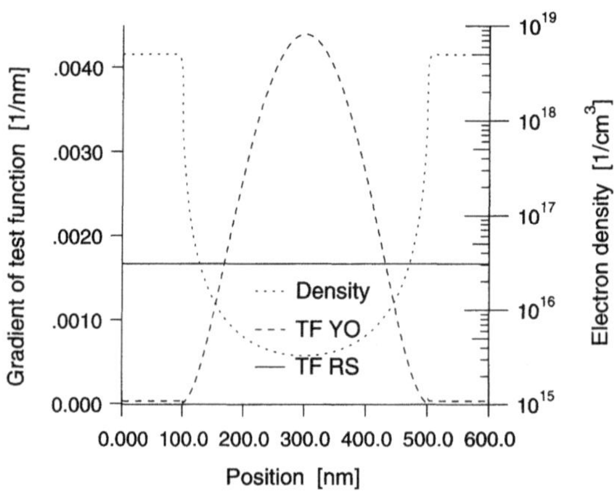

Figure 8.2: Gradient of the Ramo-Shockley (RS) and optimized test functions by P. D. Yoder et al. (YO) and the electron density for the 1D N^+NN^+ structure biased at 0.0V

functions and the optimized ones, and based on the approximation (8.19), again for both sets of test functions. The spectral intensities are normalized to the value given by the Nyquist theorem (7.156) at zero frequency. The result of Eq. (8.18) based on the Ramo-Shockley test functions shows the well-known behavior. A resonance peak is found at the plasma frequency (about 10THz for the maximum

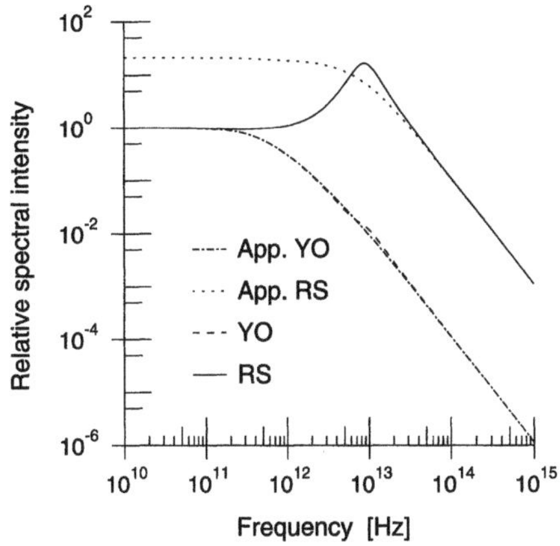

Figure 8.3: Relative spectral intensity of the terminal current for the 1D N^+NN^+ structure biased at 0.0V calculated with Eq. (8.18) based on the Ramo-Shockley test functions (RS) and based on the optimized test functions by Yoder et al. (YO), and calculated with the approximation (8.19) (App. RS) and (App. YO)

electron concentration of $5 \cdot 10^{18}/\mathrm{cm}^3$) and at higher frequencies the spectral intensity is inversely proportional to the square of the frequency [8.3]. In the case of the optimized test functions the resonance is strongly suppressed and the spectral intensity decays monotonically. The results based on the approximative spectral intensity (8.19) reproduce the exact results based on Eq. (8.18) very well at high frequencies, whereas the approximation fails in the case of low frequencies and the Ramo-Shockley test functions. It yields a much larger noise than the exact expression for the Ramo-Shockley test functions. This spuriously enhanced noise is reduced by optimizing the test functions. But this is in contrast to the exact results, were both sets of test functions yield the same spectral intensity at low frequencies. This has a strong impact on the variance of the time-averaged current, which is strongly influenced by the spectral intensity at low frequencies (cf. Eq. (8.2)). In Fig. 8.4 the variance is shown as a function of the simulated time. For very short simulation times the variance of the optimized estimator is smaller than the one of the Ramo-Shockley estimator. But simulation times that short are not reasonable, because the results are still correlated with the initial data. For simulation times much longer than the correlation time of the terminal current both estimators yield the same variance, which is inversely proportional to the simulated time (cf. Eq. (8.2)). This means that the variance of the terminal current estimator, which is proportional to the low-frequency spectral intensity, is *not* improved by the so-called optimized test functions. This result is due to the failure of the approximations used to derive Eq. (8.19). For zero frequency this is easily shown, because in this case the fluctuations of the particle current

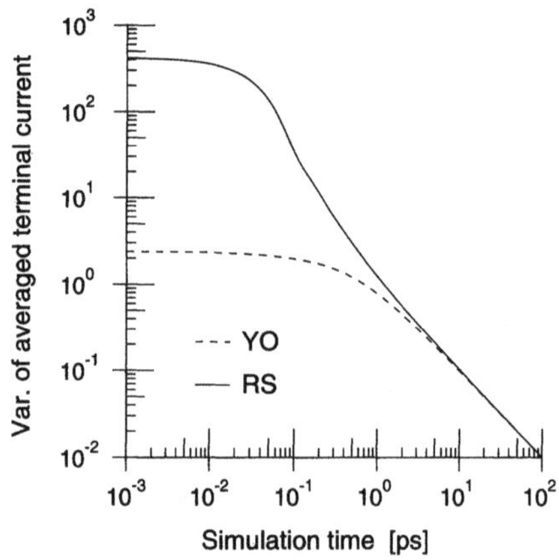

Figure 8.4: Variance of the terminal current averaged over the simulation time for the 1D N^+NN^+ structure biased at 0.0V based on the Ramo-Shockley test functions (RS) and based on the optimized test functions by Yoder et al. (YO)

densities are divergenceless (cf. Eqs. (8.6), (8.8)). Therefore, in the case of zero frequency the terminal current is independent of the test functions and therewith its variance, which thus can not be reduced by the so-called optimized test functions. It is worth noting that this result also holds in the case of nonequilibrium, because Eqs. (8.6), (8.8) remain valid.

The huge differences in the fluctuations of the terminal current depending on the test functions (Fig. 8.1) are due to high-frequency fluctuations (Fig. 8.3). In Fig. 8.5 the spatial origin of the terminal current noise (the integrand of Eq. (8.18) with respect to $\mathbf{r}'$ of Eq. (8.17)) is shown for different frequencies based on the Ramo-Shockley test functions. In the case of low frequencies the noise originates from the low-density areas, whereas in the case of high frequencies it stems from the high-density areas. Thus, the high-frequency noise is suppressed by the optimized test functions, because the current is sampled only in the low-density area (Fig. 8.2). As a result, the variance of the terminal current is much smaller for the optimized test functions. But this is of no consequence for the efficiency of the terminal current estimator, because due to the low-pass characteristics of the time average (8.1) the high-frequency components of the noise do not increase the variance of the terminal current averaged over a sufficiently long simulation time. Since in the low-frequency limit the spectral intensity does not depend on the test function, it does not matter where the terminal current is sampled. Particle counting at the highly doped contacts resulting in strong high-frequency fluctuations is therefore as efficient as sampling in low-density areas. This result can also be illustrated at the particle level. For the kth particle entering the device at the time t_{in} and leaving at t_{out} the contribution to the integral of the terminal

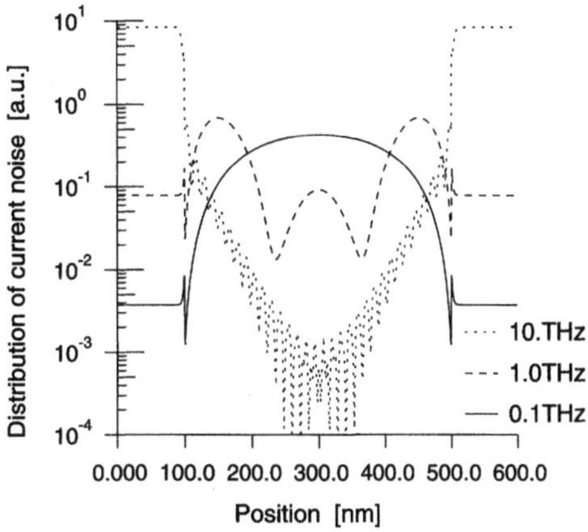

Figure 8.5: Spatial origin of the spectral intensity of the terminal current fluctuations for the 1D N^+NN^+ structure biased at 0.0V based on the Ramo-Shockley test functions

current over time is given by

$$\begin{aligned}\int_{t_{\text{in}}}^{t_{\text{out}}} I_l(t)dt &= \int_{t_{\text{in}}}^{t_{\text{out}}} Q_k \left(\nabla_{\mathbf{r}} h_l|_{\mathbf{r}=\mathbf{r}_k(t)}\right)^{\mathrm{T}} \mathbf{v}_k(t)dt \\ &= \int_{\mathbf{r}_k(t_{\text{in}})}^{\mathbf{r}_k(t_{\text{out}})} Q_k \left(\nabla_{\mathbf{r}} h_l|_{\mathbf{r}=\mathbf{r}_k(t)}\right)^{\mathrm{T}} d\mathbf{r} \\ &= Q_k \left(h_l(\mathbf{r}_k(t_{\text{out}})) - h_l(\mathbf{r}_k(t_{\text{in}}))\right) . \end{aligned} \tag{8.23}$$

Since the particles enter and leave the device at the contacts where the value of the test functions is given by the boundary conditions (6.73), which are the same for all types of test functions, this result is independent of the method used to determine the test functions. There are only three possible values for the integral: $-Q_k$, 0, and Q_k depending on where the particle enters and where it leaves the device. In the first case the particle enters the device at the lth contact $(h_l(\mathbf{r}_k(t_{\text{in}})) = 1)$ and leaves through a different contact $(h_l(\mathbf{r}_k(t_{\text{out}})) = 0)$. In the second case the particle either enters and leaves through contacts other than the lth contact $(h_l(\mathbf{r}_k(t_{\text{in}})) = h_l(\mathbf{r}_k(t_{\text{out}})) = 0)$ or enters and leaves through the lth contact $(h_l(\mathbf{r}_k(t_{\text{in}})) = h_l(\mathbf{r}_k(t_{\text{out}})) = 1)$. In the third case the particle enters through a contact other than the lth contact $(h_l(\mathbf{r}_k(t_{\text{in}})) = 0)$ and leaves through the lth contact $(h_l(\mathbf{r}_k(t_{\text{out}})) = 1)$.

It is worth noting that the above result only holds in the case of an exact integration in time. If instead of an exact path integral, for example, before-scattering statistics is used to sample the terminal current as in Ref. [8.4], for

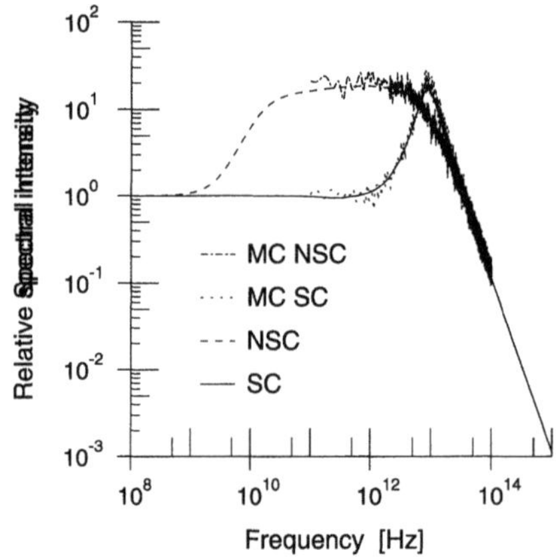

Figure 8.6: Relative spectral intensity of the terminal current for the one-dimensional resistor biased at 0.0V calculated with Eq. (8.18) based on the Ramo-Shockley test functions by self-consistent (SC) and nonself-consistent (NSC) device simulations and corresponding MC results (MC SC/MC NSC)

which the sampled particle current density under stationary bias conditions is no longer divergenceless, a different result is obtained. Since current continuity is violated by before-scattering statistics, the variance of the time average of the terminal current is artifically increased. This artifical increase can be reduced by using optimized test functions. Nevertheless, the variance of the terminal current estimated by before-scattering statistics will be larger than the one based on an exact integration in time. In addition, the exact integration in time can be performed by particle counting at the contacts, which is much more CPU efficient and simpler than before-scattering statistics based on test functions.

Another striking result shown in Fig. 8.3 is that the approximation (8.19) based on the so-called optimized test functions gives the correct noise at low frequencies (Fig. 8.3). This accidental result is due to the assumption of thermal equilibrium and uniformly weighted particles for which the Euler-Lagrange equation (8.20) for the optimized test functions equals Eq. (7.163) for the basis function of the quasifermi potential. Since the same boundary conditions are used, the optimized test functions and the basis functions are equal. In addition, these basis functions equal the corresponding Green's functions for the terminal current (cf. Eq. (7.174)), and the approximation (8.19) equals the exact expression (8.18).

An interesting consequence of Eqs. (8.6) and (8.8) is that SC and NSC device simulations yield the same current noise for zero frequency under equilibrium conditions, because Eqs. (8.6) and (8.8) are decoupled from the PE in this case. This behavior is reproduced by the simulations and in Fig. 8.6 the exact spectral intensities based on the Ramo-Shockley test functions for the SC and NSC case converge at low frequencies. This result is in contrast to Ref. [8.3], where a huge difference between the two cases is found. This might be due to the MC simulations used in Ref. [8.3] to obtain the spectral intensity of the current. Since it is impossible to evaluate the spectral intensity at zero frequency by MC simulations lasting only a finite time, the results of Ref. [8.3] must have been obtained at a nonzero frequency. In the case of the NSC simulation the maximum sampling frequency for which the correct low frequency behavior is obtained is more than two orders of magnitude smaller than the one for the SC simulations. Thus, a too high sampling frequency might explain the results in Ref. [8.3].

In addition, in Fig. 8.6 MC results are shown, and excellent agreement between the nonstochastic numerical and MC results is obtained. This is due to the fact that the only approximations in the derivation of the numerical model from the Langevin-type BTE are the assumption of linear response and constant transport coefficients.

As explained in Sec. 3.5 convergence estimation entails an unbiased estimator for the relative error of the quantity under consideration. The bias of the estimator (3.75) depends on the duration T_{sub} of the subhistories over which the quantity X is averaged in order to obtain uncorrelated samples. With the respective spectral intensity the estimated relative error (3.75) normalized to the unbiased value and sampled for a very long total simulation time T_{sim} reads

$$\frac{r\{\widehat{X}\}(T_{\mathrm{sub}})}{\frac{2\sigma\{\overline{X}\}}{\mathrm{E}\{\overline{X}\}}} = \sqrt{\frac{2}{\pi}\int_0^{\infty}\frac{S_{XX}\left(\frac{\nu}{T_{\mathrm{sub}}}\right)}{S_{XX}(0)}\frac{1-\cos(\nu)}{\nu^2}d\nu} \ . \tag{8.24}$$

In Fig. 8.7 the normalized relative error of the terminal current as a function of the subhistory length is shown. For subhistories shorter than 10ps both estimators, the one based on the Ramo-Shockley test functions and the one based on the so-called optimized test functions, are biased. While the estimator based on the Ramo-Shockley test functions tends to overestimate the relative error for subhistories longer than a few femtoseconds, the other estimator underestimates the relative error. Thus the use of subhistories of insufficient length might yield the incorrect result that the estimator based on the so-called optimized test functions is more efficient than the Ramo-Shockley estimator, because it apparently gives a smaller relative error.

The maximum length of the subhistories is limited due to the finite total simulation time. In addition, the number of subhistories must be sufficiently large in order to obtain a reliable estimate of the relative error (cf. Eq. (3.76)). In the case of the one-dimensional resistor it is still possible to simulate 100 subhistories of 10ps length, although the resultant error of the terminal current might be much smaller than necessary. In the case of a real two-dimensional device it is often not possible to simulate sufficiently long subhistories and the biased Ramo-Shockley estimator is used, because it gives an upper bound for the real error (cf. Sec 9.2.1). But the reader should keep in mind that in the case of different devices and bias

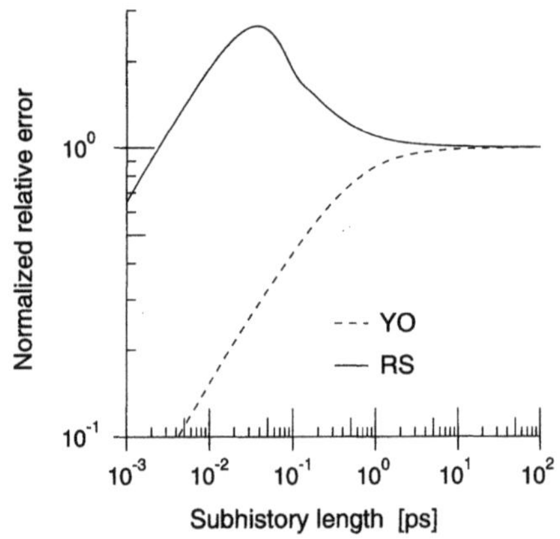

Figure 8.7: Normalized relative error of the terminal current of the one-dimensional resistor biased at 0.0V based on the Ramo-Shockley test functions (RS) and the optimized ones by Yoder et al. (YO) as a function of the subhistory length

conditions the error might be overestimated by orders of magnitude.

8.2 In-Advance CPU Time Estimation for Stationary MC Device Simulations

In the following again only MC algorithms will be investigated which are based on uniformly weighted particles (no statistical enhancement) and where the nonlinear impact of many-particle effects can be neglected. In this case the stochastic properties of the MC algorithm are described by the LBTE (cf. Sec. 2.4). Based on the LBTE the spectral intensities in (8.2) can be calculated. A disadvantage of this approach is that it is rather time consuming. Therefore, Langevin-type momentum-based models, the drift-diffusion (LDDM) and hydrodynamic model (LHDM) are used here, which are CPU-time efficient approximations of the LBTE (cf. Chap. 7). Of course, these models can only be used in those cases where stochastic variables are investigated which are covered by these models. These are quantities like the terminal currents or internal distributions like the particle density or velocity.

Since the momentum-based models were developed for the simulation of real electronic noise, the resultant spectral intensities have to be scaled in order to account for the difference in the charge of a simulation particle and a real electron

[8.2,8.5]

$$S_{XX}^{\mathrm{MC}}(\omega) = \frac{Q_{\mathrm{tot}}}{qN_{\mathrm{sim}}} S_{XX}^{\mathrm{phys}}(\omega) , \tag{8.25}$$

where the total particle charge $Q_{\mathrm{tot}} = \sum_{k=1}^{N_{\mathrm{sim}}} |Q_k^{\mathrm{sim}}|$ is the sum of the absolute values of the uniform particle charges (electrons and holes), N_{sim} the total particle number, q the physical electron charge, S_{XX}^{phys} the physical spectral intensity evaluated with the methods presented in Chap. 7, and S_{XX}^{MC} the spectral intensity of the corresponding MC quantity. The variance of an average $\overline{X}$ (3.51) is now given with Eq. (8.2) by

$$\sigma^2\{\overline{X}\} = \frac{Q_{\mathrm{tot}}}{2qN_{\mathrm{sim}}} \frac{S_{XX}^{\mathrm{phys}}(0)}{T_{\mathrm{sim}}} \tag{8.26}$$

for sufficiently large T_{sim}.

In the case that the CPU time for solving the PE is negligible compared to the CPU time of the MC simulation of the particles, the total CPU time T_{CPU} is proportional to the particle number N_{sim} and the time T_{sim} each particle is simulated

$$T_{\mathrm{CPU}} = \alpha_{\mathrm{cost}} N_{\mathrm{sim}} T_{\mathrm{sim}} , \tag{8.27}$$

where α_{cost} is the cost factor, which depends on the device structure, bias point, and CPU speed and can be determined by running an MC simulation for a short period of time. Solving (3.59) with (8.26), (8.27), and a given relative error r for the required CPU time yields

$$T_{\mathrm{CPU}} = \frac{2\alpha_{\mathrm{cost}}}{r^2} \frac{Q_{\mathrm{tot}}}{q} \frac{S_{XX}^{\mathrm{phys}}(0)}{\langle X \rangle^2} , \tag{8.28}$$

which depends on the relative error and the investigated quantity X.

The intrinsic simulation-time factor

$$P_X = \frac{r^2}{\alpha_{\mathrm{cost}}} T_{\mathrm{CPU}} = 2\frac{Q_{\mathrm{tot}}}{q} \frac{S_{XX}^{\mathrm{phys}}(0)}{\langle X \rangle^2} , \tag{8.29}$$

which is fully determinable by the momentum-based models, is discussed below.

With the simulation-time factor guidelines can be developed how to reduce the CPU time of an MC simulation beyond reducing the cost factor α_{cost} by improving the program code or using faster computers. One option is the reduction of the total particle charge Q_{tot}, which should be reduced as much as possible by limiting the maximum doping in the nonactive regions of the device and the size of the highly doped contacts. Whether these modifications of the device structure have a negative impact on the simulation results can be checked by momentum-based simulations with and without the modifications. The other option is the reduction of the power spectrum $S_{XX}^{\mathrm{phys}}(0)$. For example, in [8.6] an unbiased estimator for

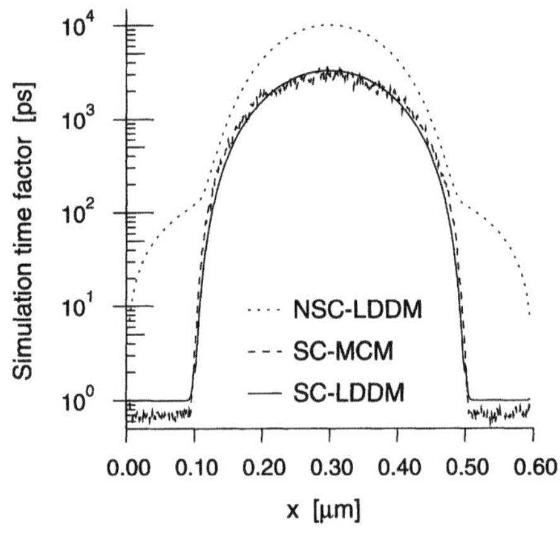

Figure 8.8: Simulation-time factors of the electron density for the N^+NN^+ structure of Sec. 9.1 biased at zero volts as obtained by self-consistent MC and DD simulations and nonself-consistent DD results

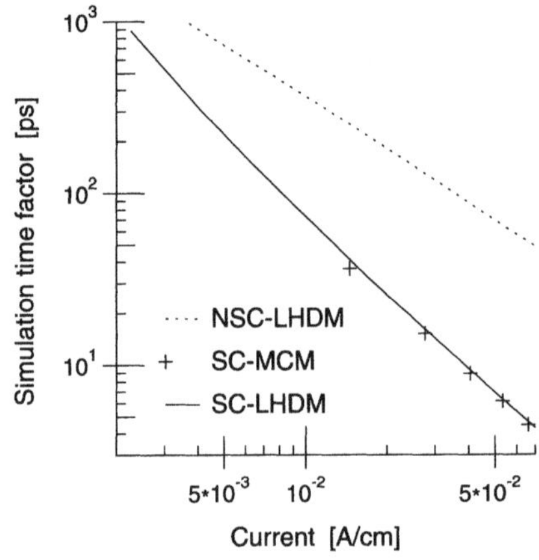

Figure 8.9: Simulation-time factors of the terminal current for the N^+NN^+ structure of Sec. 9.1 as obtained by self-consistent MC and HD simulations and nonself-consistent HD results

the substrate current of MOSFETs is given which reduces the spectral intensity compared to the physical estimator.

In Fig. 8.8 the simulation-time factor (8.29) of the electron density is shown for the N^+NN^+ structure of Sec. 9.1 at equilibrium and good agreement of the momentum-based simulator with the MC model is obtained for the SC results. Similar good agreement between the MC and momentum-based model is found for the simulation-time factor of the terminal current (Figs. 8.9 and 8.10) validating

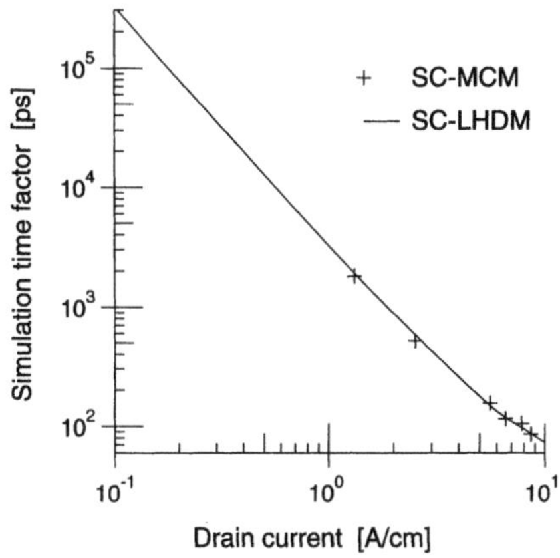

Figure 8.10: Simulation time factors for the drain current of the 80nm NMOSFET of Sec. 9.2 biased at $V_G = 1.5$V as obtained by MC and HD simulations

our new approach, where instead of the LDDM the LHDM was used, because it yields better results under nonequilibrium conditions. In addition, in Figs. 8.8 and 8.9 the simulation-time factors of NSC simulations are shown, where the electric field is kept constant during the MC simulation (frozen field). The NSC simulation-time factors are found to be much larger than the SC ones. Thus, any reduction obtained in the cost factor α_{cost} by the NSC approach might be canceled by the increase in the simulation-time factor and it is not clear whether the NSC approach is more efficient than the SC or not. At least in the case of the N^+NN^+ structure the NSC approach is the less efficient one.

Since the NSC approach is problematic in the case of deep sub-micron devices [8.7, 8.8], in [8.7] an iteration scheme of the NSC-MC method and a Poisson solver has been presented, which after a certain number of Gummel-loop-like iterations converges to the SC solution. This procedure increases the cost factor by the number of iterations necessary to achieve self-consistency compared to the NSC approach. In addition, the intrinsic simulation time factor is increased, because the NSC-MC simulations yield a much larger density noise than the SC simulations (Fig. 8.8). Since the NSC-MC density is the basis for the solution of the PE, its increased noise leads to a larger stochastic error of the electric field, which is fed back into the NSC-MC simulation[1]. Thus, the iteration scheme increases the noise of the NSC simulations further compared to the SC method. It is therefore doubtful, whether the iteration scheme is more efficient than the

[1]Since the electric field is kept constant during the NSC-MC simulation step of the iteration, its stochastic error is also constant. It leads therefore to a fluctuation of the expected values of the NSC-MC simulation, and this fluctuation can only be detected by investigating the results of the different iteration steps and not by investigating the stochastic error of an individual NSC-MC simulation.

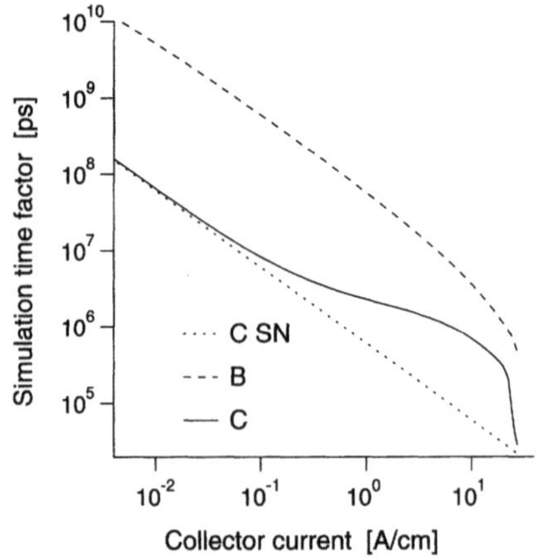

Figure 8.11: Simulation time factors for collector (C) and base (B) currents of the SiGe HBT of Sec. 9.3.3 biased at $V_{CE} = 2.0V$ and the shot noise limit for the collector current (C-SN)

direct SC approach as suggested in [8.7].

In Fig. 8.11 the simulation-time factors of the collector and base current of the npn SiGe HBT of Sec. 9.3.3 are shown for a constant collector/emitter bias, where the base current calculation requires much more CPU time than the collector current due to a difference in the magnitude of the currents. In the limit of small currents the collector current noise is shot-like

$$S^{\text{phys}}_{I_C I_C}(0) \approx 2qI_C \ , \tag{8.30}$$

and the simulation-time factor is inversely proportional to the current

$$P_{I_C} \approx \frac{4Q_{\text{tot}}}{I_C} \ . \tag{8.31}$$

Thus, the simulation of a ten times smaller current requires ten times more CPU time.

The deviation from the shot-noise formula of the collector-current simulation-time factor for large collector currents (Fig. 8.11) is due to hole scattering (cf. Sec 9.3.3). This increase in CPU time by the hole scattering can be avoided by solving a nonlinear PE together with a constant quasifermi potential for the holes instead of an MC simulation of the holes [8.9, 8.10]. In this case (8.31) also holds for large collector currents up to high injection and the simulation-time factor is considerably reduced.

In Fig. 8.12 results are shown for a constant base/emitter voltage. With increasing collector/emitter bias impact ionization becomes important and at very high voltages the collector simulation-time factor exceeds the one for the base, although the base current is still about ten times smaller than the collector current.

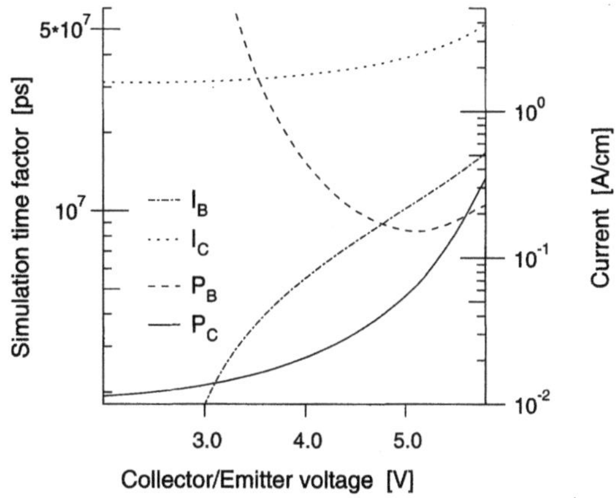

Figure 8.12: Simulation time factors (P) and currents (I) for collector (C) and base (B) of the SiGe HBT biased at $V_{\mathrm{BE}} = 0.85\mathrm{V}$

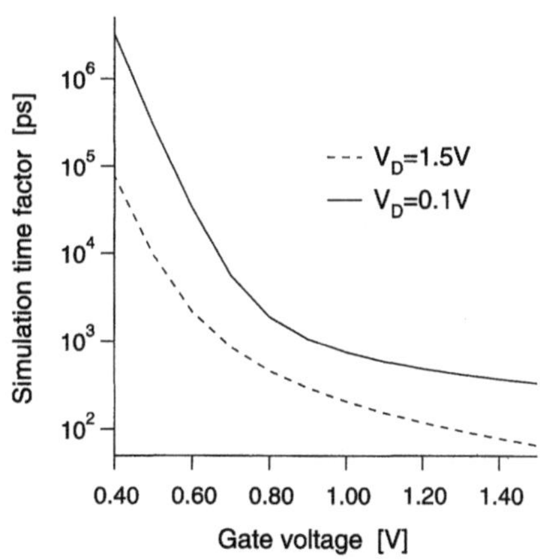

Figure 8.13: Simulation time factors for the drain current of the 80nm NMOSFET biased at $V_{\mathrm{D}} = 0.1\mathrm{V}$ and 1.5V

In Fig. 8.13 the simulation-time factor[2] is shown for the drain current of the NMOSFET for a constant drain bias as a function of the gate voltage. For example, with $\alpha_{\mathrm{cost}} = 2.5 \cdot 10^{10}$ (1GHz PC) and $V_{\mathrm{G}} = V_{\mathrm{D}} = 1.5\mathrm{V}$ a relative error of $r = 5\%$ is achieved within 860 CPU seconds. In the case of $V_{\mathrm{G}} = 0.5\mathrm{V}$ and $V_{\mathrm{D}} = 1.5\mathrm{V}$ the corresponding CPU time is 35 hours due to the small drain current. In the sub-threshold regime the noise is shot-like and the CPU time is inversely proportional to the drain current similar to (8.31). For larger gate voltages the

[2]Due to the lack of consistent noise sources for the HD model in the channel area, no surface mobility and noise models were used in these HD simulations.

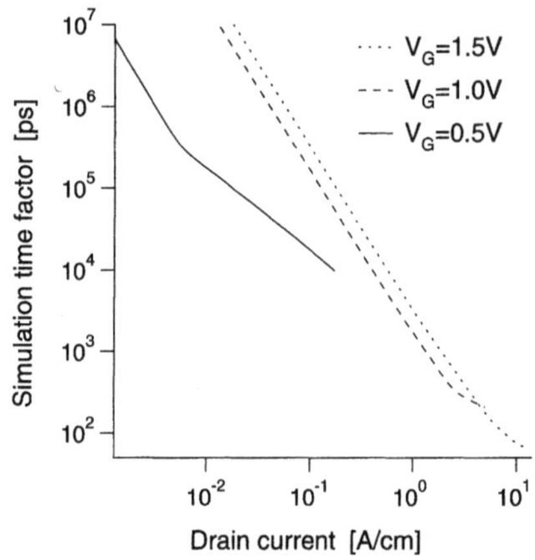

Figure 8.14: Simulation time factors for the drain current of the 80nm NMOSFET biased at $V_{\mathrm{G}} = 0.5$, 1.0, and 1.5V

noise becomes more like thermal noise

$$S^{\mathrm{phys}}_{I_{\mathrm{D}}I_{\mathrm{D}}}(0) = \gamma 4 k_{\mathrm{B}} T G_{\mathrm{D,D}} \ , \tag{8.32}$$

where γ is 2/3 for long channel MOSFETs in saturation [8.11] and slightly larger for short channel devices [8.12], $k_{\mathrm{B}}T$ is the thermal energy, and $G_{\mathrm{D,D}}$ the drain self-conductance at zero drain/source bias. In this case the CPU time is inversely proportional to the square of the drain current

$$P_{I_{\mathrm{D}}} = \frac{\gamma 8 k_{\mathrm{B}} T G_{\mathrm{D,D}} Q_{\mathrm{tot}}}{q I_{\mathrm{D}}^2} \ , \tag{8.33}$$

and a ten times smaller current requires a one hundred times larger CPU time to achieve the same relative error. The change from shot to thermal noise behavior is clearly discernible in Fig. 8.14, where the simulation-time factor is plotted versus the drain current for constant gate voltages. The simulation-time factor is first inversely proportional to the square of the drain current and then inversely proportional to the drain current.

The results in Figs. 8.10, 8.11, and 8.14 show that the simulation of small currents requires very large CPU times. In some cases the CPU times can be reduced by statistical enhancement [8.13, 8.14] (cf. Sec. 6.7), especially when the current is due to rare events. This is, for example, not the case for MOSFETs in strong inversion biased at a small drain voltage (Fig. 8.14), where the current is carried by the majority of the particles. In this case population control methods will not improve the efficiency of the MC simulation. Furthermore, statistical enhancement only reduces the increase of the simulation time for smaller currents, but it does not reverse this trend. Thus, simulations of small currents with statistical enhancement are still very CPU intensive compared to simulations of large currents. With respect to large currents the improvement in efficiency of the MC algorithm due to statistical enhancement is negligible according to our experience.

References

[8.1] P. D. Yoder, K. Gärtner, U. Krumbein, and W. Fichtner, "Optimized terminal current calculation for Monte Carlo device simulation", *IEEE Trans. Computer-Aided Des.*, vol. 16, pp. 1082–1087, 1997.

[8.2] C. Jungemann and B. Meinerzhagen, "Analysis of the stochastic error of stationary Monte Carlo device simulations", *IEEE Trans. Electron Devices*, vol. 48, no. 5, pp. 985–992, 2001.

[8.3] L. Varani, L. Reggiani, T. Kuhn, T. González, and D. Pardo, "Microscopic simulation of electronic noise in semiconductor materials and devices", *IEEE Trans. Electron Devices*, vol. 41, no. 11, pp. 1916–1925, 1994.

[8.4] F. M. Bufler, A. Schenk, and W. Fichtner, "Efficient Monte Carlo device modeling", *IEEE Trans. Electron Devices*, vol. 47, no. 10, pp. 1891–1897, 2000.

[8.5] C. Moglestue, "Monte-Carlo particle modelling of noise in semiconductors", in *International Conference on Noise in Physical Systems and 1/f Fluctuations*, 1983, pp. 23–25.

[8.6] J. M. Higman, K. Hess, C. G. Hwang, and R. W. Dutton, "Coupled Monte Carlo-drift diffusion analysis of hot-electron effects in MOSFET's", *IEEE Trans. Electron Devices*, vol. 36, no. 5, pp. 930–937, 1989.

[8.7] F. Venturi, R. K. Smith, E. C. Sangiorgi, M. R. Pinto, and B. Riccò, "A general purpose device simulation coupling Poisson and Monte Carlo transport with application to deep submicron MOSFET's", *IEEE Trans. Computer-Aided Des.*, vol. 8, no. 4, pp. 360–369, 1989.

[8.8] C. Jungemann and B. Meinerzhagen, "On the applicability of nonself-consistent Monte Carlo device simulations", *IEEE Trans. Electron Devices*, vol. 49, no. 6, pp. 1072–1074, 2002.

[8.9] M. V. Fischetti and S. E. Laux, "Monte Carlo analysis of electron transport in small semiconductor devices including band–structure and space–charge effects", *Phys. Rev. B*, vol. 38, pp. 9721–9745, 1988.

[8.10] C. Jungemann, B. Neinhüs, and B. Meinerzhagen, "Comparative study of electron transit times evaluated by DD, HD, and MC device simulation for a SiGe HBT", *IEEE Trans. Electron Devices*, vol. 48, no. 10, pp. 2216–2220, 2001.

[8.11] F. M. Klaassen and J. Prins, "Thermal noise of MOS transistors", *Philips Res. Repts*, pp. 505–514, 1967.

[8.12] A. J. Scholten, L. F. Tiemeijer, R. van Langevelde, R. J. Havens, V. C. Venezia, A. T. A Zegers van Duijnhoven, B. Neinhüs, C. Jungemann, and D. B. M. Klaassen, "Compact modeling of drain and gate current noise for RF CMOS", *IEDM Tech. Dig.*, pp. 129–132, 2002.

[8.13] C. Jacoboni, "Generalization of the Monte Carlo method for charge transport to biased simulations", in *Proceedings of the Sixth International Nasecode Conference*, 1989, pp. 226–231.

[8.14] M. G. Gray, T. E. Booth, T. J. T. Kwan, and C. M. Snell, "A multi-comb variance reduction scheme for Monte Carlo semiconductor simulators", *IEEE Trans. Electron Devices*, vol. 45, no. 4, pp. 918–924, 1998.

9 Results

In this chapter the transport models are applied to the simulation of three different types of devices to assess the accuracy of the transport models either by comparison to each other or experimental data. In addition, the transport in the different device types is analyzed with the appropriate simulation model. The first set of devices, unipolar 1D N^+NN^+ and P^+PP^+ structures, is mainly of academic interest and is frequently investigated because of its relative simplicity (e.g. [9.1–9.6]), which reduces the CPU time and makes it easier to compare the MC model to the momentum-based models. The second group of devices consists of CMOS devices, which are the standard devices of today's integrated circuits. With the introduction of the SiGe technology many new types of CMOS devices have appeared [9.7, 9.8]. The most important types are MOSFETs with strained Si channels [9.9–9.13]), which are investigated together with standard Si MOSFETs in the second part of this chapter. In addition, PMOSFETs with a strained SiGe channel, which have promising potential for low noise amplifiers, are discussed [9.14–9.18]. The third group of investigated devices consists of SiGe HBTs, which exhibit a superior RF and noise performance compared to standard Si BJTs [9.19–9.22].

9.1 N^+NN^+ and P^+PP^+ Structures

Due to their simplicity and unipolar nature N^+NN^+ and P^+PP^+ structures are often used for investigations of transport models [9.1–9.6]. In the first part of this section the MC model is used to analyze the charge transport in these structures at a microscopic level with an emphasis on momentum-based modeling. In the second part terminal current noise due to diffusion noise and II generation of secondary particles is investigated and the accuracy of the momentum-based models is assessed by comparison with the MC model.

The unipolar 1D structures consist of a 100nm wide region with a constant doping of $5 \cdot 10^{17}/\mathrm{cm}^3$ followed by 400nm of $2 \cdot 10^{15}/\mathrm{cm}^3$ and again 100nm of $5 \cdot 10^{17}/\mathrm{cm}^3$. The dopants are donors in the case of the N^+NN^+ structure and acceptors in the case of the P^+PP^+ structure. The real space is discretized on a

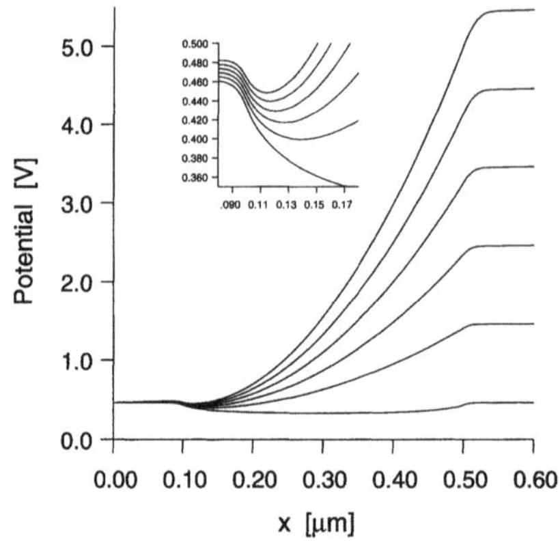

Figure 9.1: Electrostatic potential of the N^+NN^+ structure evaluated with the MC model for a bias of 0, 1, 2, 3, 4, and 5V

grid of 201 nodes with a constant spacing of 3nm. Since the MC model is based on a 2D grid in the real space, the width of the devices is set to 10nm resolved by two nodes. In order to ensure a 1D simulation, the particles cannot move into the direction of the width and are placed in the middle between those two grid nodes. The MC simulations have been performed, if not stated otherwise, with about 1000 uniformly weighted particles for more than 80ns, where the total particle number can fluctuate due to the ohmic contacts. The PE was solved every 10fs. The long CPU times and uniformly weighted particles are required by the noise simulations discussed further below. The simulations include phonon and impurity scattering, if not otherwise stated, and no apparent band gap narrowing for these artificial structures. The current flow is in ⟨100⟩ direction, and all results are for unstrained Si at room temperature in this section.

9.1.1 General Transport Properties

In Figs. 9.1 to 9.4 the potential, electron density, velocity, and dynamic temperature are shown for the N^+NN^+ structure biased with voltages between 0 and 5V. The barrier height of the N^+N junction is strongly reduced with increasing bias (Fig. 9.1). This leads to a strong punch through and the electron density rapidly increases with bias in the N region (Fig. 9.2). This behavior distinguishes the 1D N^+NN^+ structure from a well-designed MOSFET, in which punch through does not occur. In this sense the N^+NN^+ structure is not a 1D MOSFET. Due to the large driving fields in the case of the highest voltages a velocity overshoot occurs at the N^+N junction (Fig. 9.3). Since this junction is forward biased, the electron temperature is slightly below the lattice temperature, whereas at the NN^+ junction it exceeds the lattice temperature for nonzero bias due to the en-

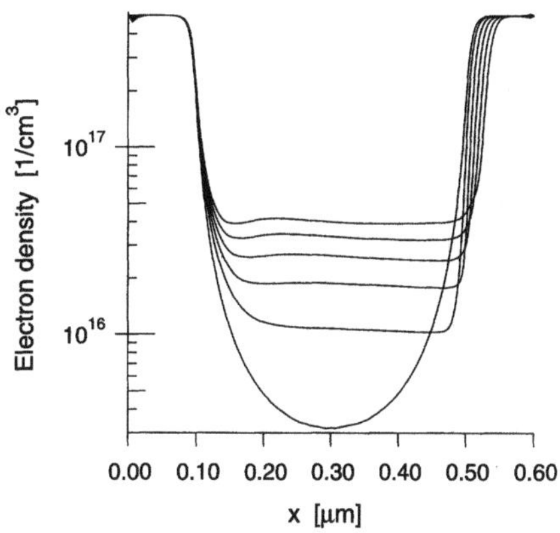

Figure 9.2: Electron density of the N^+NN^+ structure evaluated with the MC model for a bias of 0, 1, 2, 3, 4, and 5V

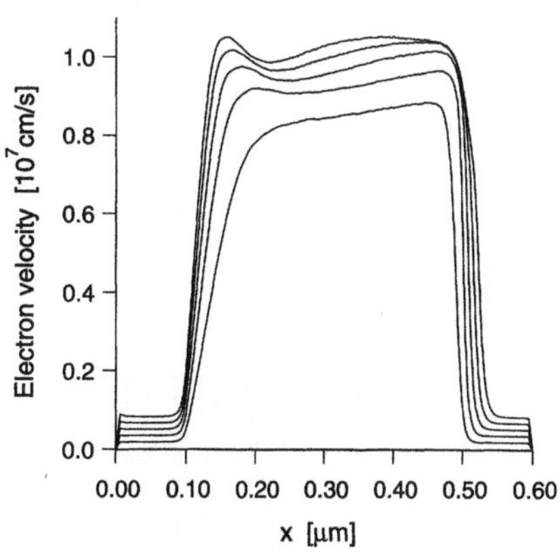

Figure 9.3: Electron velocity of the N^+NN^+ structure evaluated with the MC model for a bias of 0, 1, 2, 3, 4, and 5V

ergy gain in the electric field (Fig. 9.4). At zero bias the MC simulation shows a small deviation from the lattice temperature due to interaction with plasma oscillations enhanced by the time step length of 10fs (cf. Sec. 6.4). A shorter, more CPU-intensive time-step length reduces this deviation.

In Fig. 9.5 the corresponding electron energy distribution function is shown for a bias of 2V as an example of the application of the MR method. In this special case an ensemble of nonuniformly weighted particles is simulated. The MR method is applied to the PS by dividing the energy between 0 and 3.0eV into 60 bins of 50meV width for each cell of the RS grid. 10 particles were as-

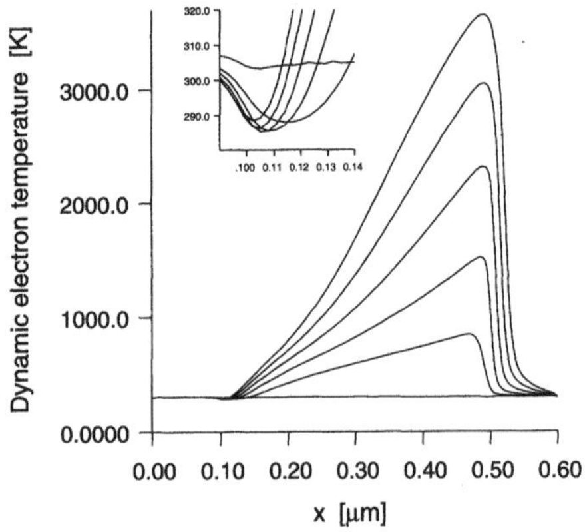

Figure 9.4: Electron dynamic temperature of the N^+NN^+ structure evaluated with the MC model for a bias of 0, 1, 2, 3, 4, and 5V

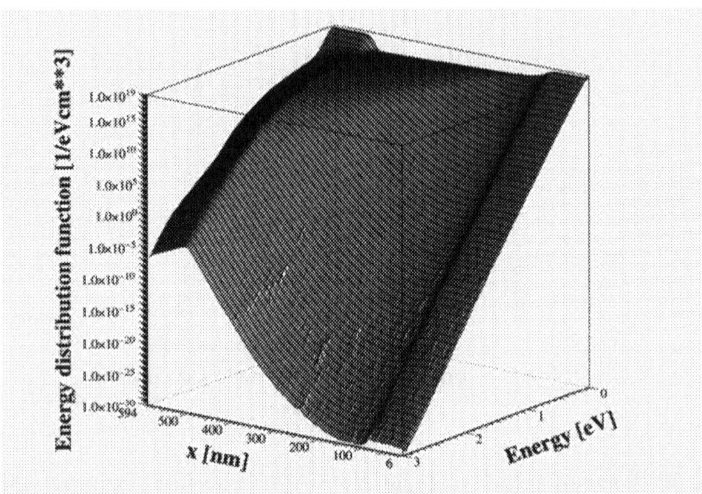

Figure 9.5: Electron energy distribution function evaluated with the MC mode for the 1D N^+NN^+ structure biased at 2.0V

signed to each bin. In the case of the energetically lowest bins at least 30 particles were simulated. The resultant energy distribution function is resolved over nearly 50 orders of magnitude with a quality otherwise only obtained by nonstochastic solvers for the BTE (e.g spherical harmonics solver [9.23]) but without the stringent approximations (spherical band structures, small number of harmonics, etc). In the region for $x < 100$nm the energy distribution function behaves more or less like an equilibrium Maxwell-Boltzmann distribution. In the N region the

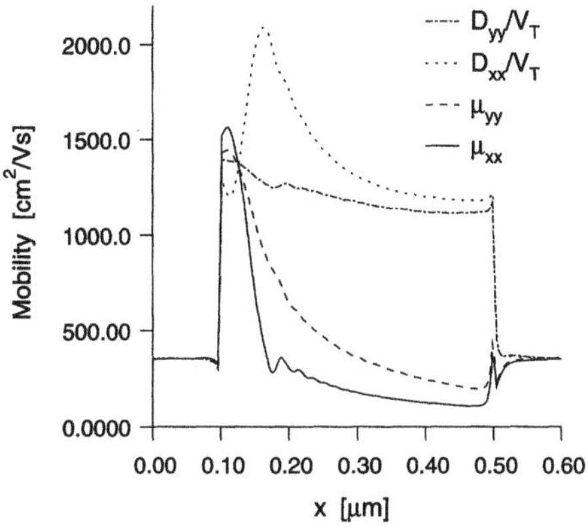

Figure 9.6: Electron mobility and diffusion constant divided by the thermal voltage (2.64) parallel (xx) and perpendicular (yy) to the direction of the current flow in the N^+NN^+ structure evaluated with the MC model for a bias of 2V

particles gain energy from the driving electric field and the distribution becomes nonmaxwellian and can no longer be characterized by a single temperature. This is also the case in the N^+ region for $x > 500$nm where the energy distribution function is a mixture of an equilibrium Maxwell-Boltzmann distribution and a hot-electron tail, which rapidly decays for increasing x.

In Fig. 9.6 the electron mobility and diffusion constant parallel and perpendicular to the current flow are shown evaluated with Eqs. (2.58) and (2.59) for a bias of 2V. In the N^+ region on the left-hand side the electron gas is more or less in equilibrium, and the mobility and diffusion constant are isotropic. Moreover, the Einstein relation (2.63) holds. The sharp changes in the mobilities and diffusion constants at the N^+N junction are due to the abrupt change of the doping concentration which determines the strength of impurity scattering. A similar effect is observed at the right junction. In the N region the mobility rapidly drops due to the heating of the electron gas, where the drop in the longitudinal mobility is larger than in the transverse. The diffusion constant shows the opposite behavior. While the relation between the mobility and diffusion constant is described for the transverse transport with a reasonable accuracy by the Einstein relation based on the local electron gas temperature (Fig. 9.7), this is not the case for the important longitudinal transport. Although the DD and HD models are based on Einstein relations (DD: Eq. (2.63), HD: Eq. (7.18)), which are shown by the MC model to fail, good agreement of the terminal currents of all three models is obtained (Fig. 9.8). In addition to the results for the N^+NN^+ structure, current data for the P^+PP^+ structure is shown in Fig. 9.8, and excellent agreement of all three models is found, because due to the stronger phonon coupling the heating of the hole gas is much smaller than in the electron case. (The maximum hole

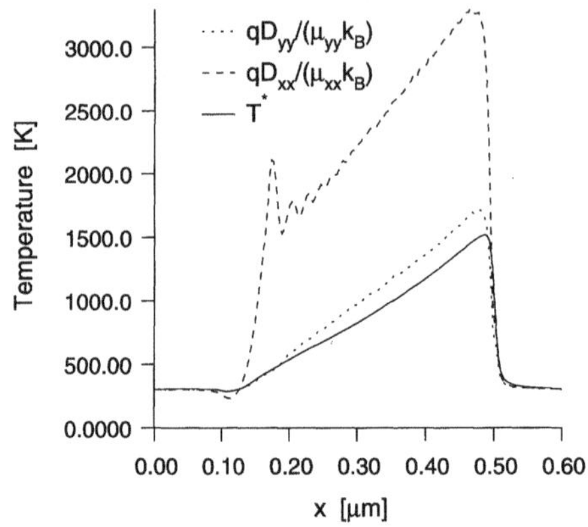

Figure 9.7: Dynamic electron temperature and the temperature resulting from a generalized Einstein relation parallel (xx) and perpendicular (yy) to the direction of the current flow in the N^+NN^+ structure evaluated with the MC model for a bias of 2V

temperature of 1586K for a bias of 5V is less than half of the 3651K for electrons.) This has another side effect. In the case of the holes the differential conductance increases with bias due to the punch through, whereas for electrons it decreases, because in the case of the electrons the increase by the punch through is canceled by the degradation of the mobility due to the strong electron gas heating. Thus, although the ohmic conductance of the N^+NN^+ structure at zero bias is more than twice as large as the one of the P^+PP^+ structure, the difference in current rapidly decreases with the applied bias.

In Fig. 9.9 the electron velocity is shown. Apart from the well-known spurious velocity overshoot on the right-hand side of the device, the HD model achieves the better overall agreement with the MC model compared to the DD model [9.2,9.3]. The spurious velocity overshoot is of minor importance for the terminal current, because the current is controlled by the injection on the left-hand side [9.24]. In Fig. 9.10 the dynamic electron temperature is shown and good overall agreement of the MC and HD models is obtained despite the stringent approximations used to derive the HD model. At the NN^+ junction the slope of the MC result is steeper than that of the HD. This is due to the fact that the HD model is based on a second order partial differential equation for the temperature, which imposes a certain limit on the second derivative of the temperature and therefore yields a less abrupt temperature profile.

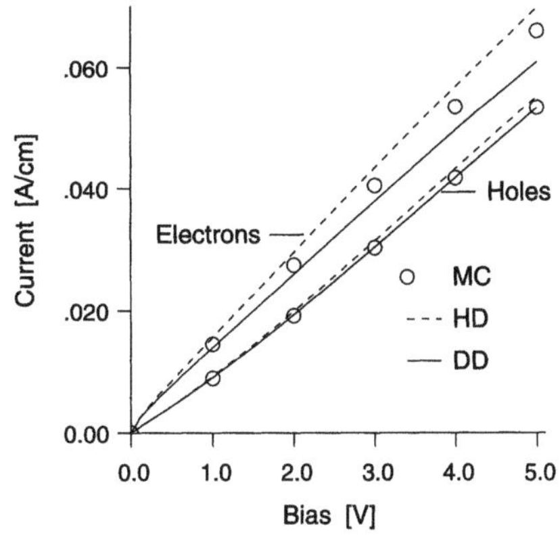

Figure 9.8: Terminal currents of the N^+NN^+ and P^+PP^+ structures evaluated with the DD, HD, and MC models

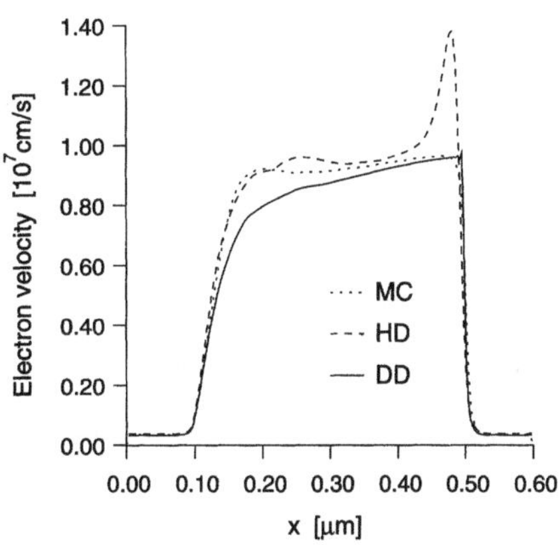

Figure 9.9: Velocity profile of the N^+NN^+ structure evaluated with the DD, HD, and MC models for a bias of 2V

9.1.2 Noise Simulations

In Fig. 9.11 noise results of the DD, HD, and MC models due to particle scattering (diffusion noise) are shown for the 1D structures, where the DD and HD models are based on the spectral intensities of the modified Langevin forces of Chap. 7. In addition, DD results based on the Einstein relation (7.155) are shown for electrons. The agreement of the DD and HD models based on the modified Langevin forces with the MC model is good for electrons and holes, whereas the standard approach based on the Einstein relation underestimates the MC results

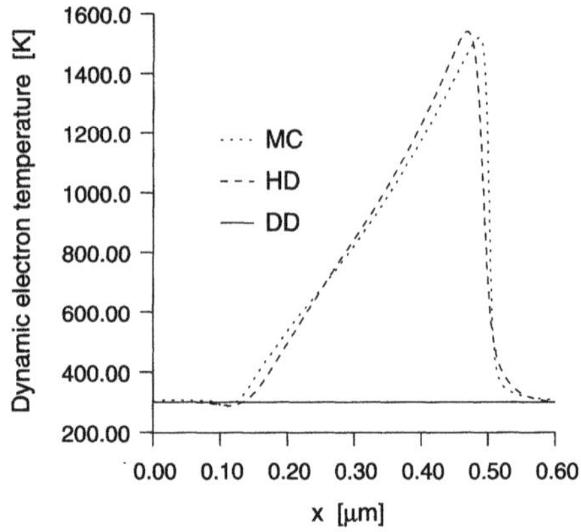

Figure 9.10: Temperature profile of the N^+NN^+ structure evaluated with the DD, HD, and MC models for a bias of 2V

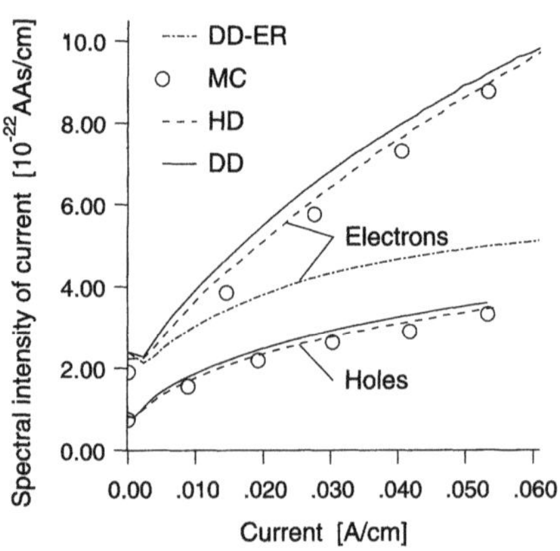

Figure 9.11: Spectral intensity of the terminal current fluctuations by DD, HD, and, MC simulations for electrons in the N^+NN^+ structure and for holes in the P^+PP^+ structure at 0Hz, and DD results based on the Einstein relation (DD-ER) for electrons

in the nonohmic regime. This is due to the Einstein relation, which fails to give the correct local noise sources far from equilibrium [9.25–9.27]. Therefore, approaches based on the Einstein relation (e.g. [9.28]) can be expected to yield accurate results only in those cases, where the particle gas is close to equilibrium in the regions, which mainly contribute to the terminal current noise. This is, for example, the case for MOSFETs with not too short channels [9.29] or BJTs with a large base length [9.30].

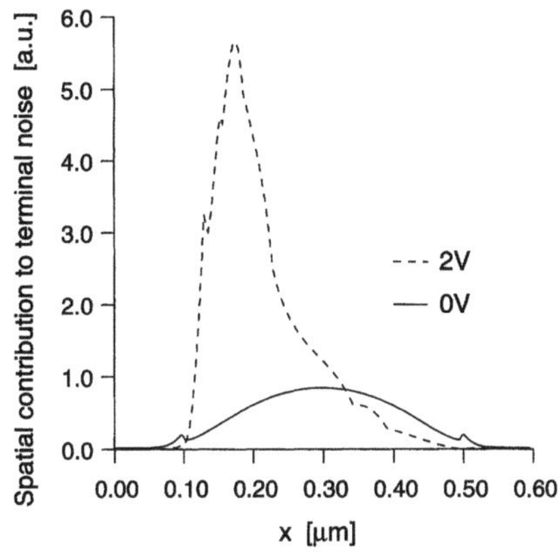

Figure 9.12: Spatial origin of the terminal current noise by HD simulations for the N^+NN^+ structure at 0Hz

Under equilibrium conditions (zero current and zero frequency) the Nyquist theorem (7.156) should hold, which relates the terminal current noise to the conductance. In the case of the Langevin-Boltzmann equation the proof is given in [9.26]. The proof for the DD model is given for multi-terminal, multi-dimensional devices in Sec. 7.4.1 [9.31, 9.32]. For the case of the HD model the proof of the Nyquist theorem has not been established. Nevertheless, the Nyquist relation is satisfied with high accuracy for the simulated devices, and in the case of the N^+NN^+ structure the deviation is less than 2%. The differences in the current noise found under equilibrium conditions for the three models (Fig. 9.11) are caused by the conductance, which is different for the different levels of approximation in the transport models [9.1]. The DD conductance is a few percent larger than the HD one, which in turn is larger than the MC result due to the interaction of the transport with the built-in fields. A similar behavior is found for the noise, because all three models more or less satisfy the Nyquist relation. Thus, at least under equilibrium conditions the small differences in the noise results are due to the inherent differences in the transport models and not due to the modeling of the noise sources.

In contrast to the MC model the momentum-based simulators allow to determine the spatial origin of the terminal current noise [9.30], and in Fig. 9.12 the integrand of Eq. (7.109) is shown for the N^+NN^+ structure evaluated by the HD model. For zero bias the spatial contribution to the terminal current noise is roughly inversely proportional to the conductivity $1/qn\mu$ and the maximum is found, where the electron density assumes its lowest value (Fig. 9.2). This is no longer the case, when a nonzero bias is applied. For a bias of 2V applied to the right terminal the distribution is strongly asymmetric and the maximum is located at the left-hand side of the lowly doped region, where the electron tem-

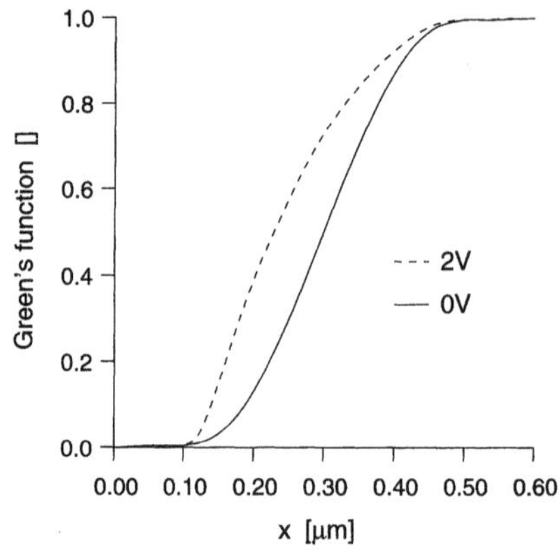

Figure 9.13: Terminal current Green's function for a particle generation source ($\underline{G}^{\dot{n}}_{\delta I_l}$) for the right-hand side terminal by HD simulations for the N^+NN^+ structure at 0Hz

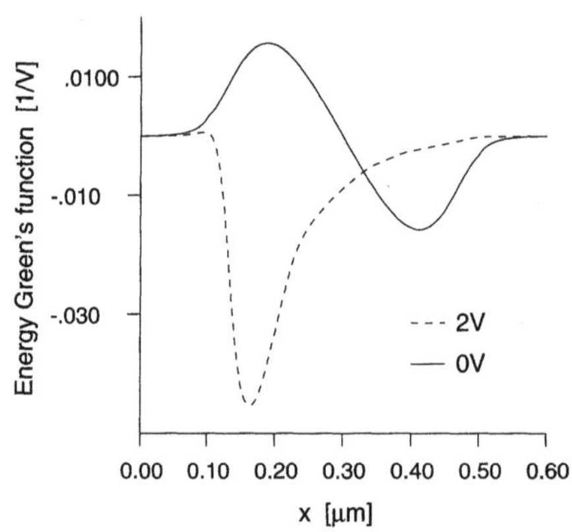

Figure 9.14: Terminal current Green's function for an energy generation source ($\underline{G}^{\dot{w}_n}_{\delta I_l}$) for the right-hand side terminal by HD simulations for the N^+NN^+ structure at 0Hz

perature is much lower than on the right-hand side (Fig. 9.10). This is due to the Green's functions, which are shown in Figs. 9.13 and 9.14. Because of the larger gradient of the Green's function $\underline{G}^{\dot{n}}_{\delta I_l}$ on the left-hand side of the lowly doped region and the larger value of the $\underline{G}^{\dot{w}_n}_{\delta I_l}$ there, the noise mainly originates from this region (cf. Eq. (7.103)). Compared to the equilibrium case the hot electrons on the right-hand side of the N region produce less noise although their density is higher.

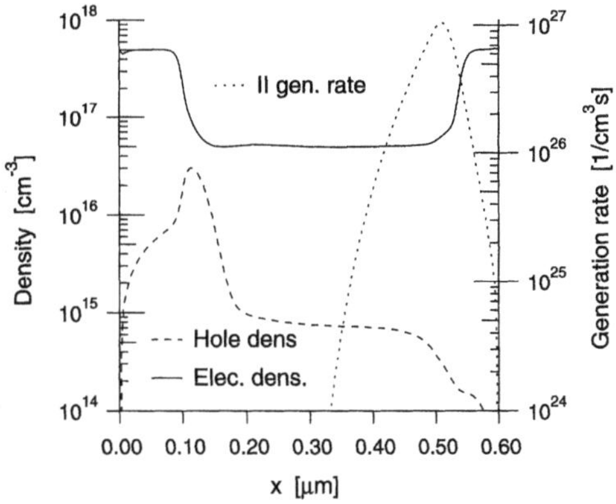

Figure 9.15: Electron, hole density, and II generation rate for the N^+NN^+ structure biased at 6V evaluated by MC simulation

In order to investigate II generation noise the N^+NN^+ structure was biased at 6V and generation of secondary particles by II was included in the simulation. The electron density, the density of secondary holes, and the II generation rate as calculated by MC simulation are shown in Fig. 9.15. The II generation rate of the secondary particles peaks at the right-hand side close to the position of the maximum electric field, whereas the secondary holes are swept to the left by the electric field and their density peaks on the left-hand side near the position where the electric field is zero. The autocorrelation function of the terminal current fluctuations is shown in Fig. 9.16 with and without generation of secondary particles. Up to nearly 0.5ps both correlation functions show the usual plasma oscillations [9.33]. In the case with II generation an additional, positive tail is found, which decays exponentially with a time constant of about 27ps over about 100ps. Although the absolute value of the tail is less than 2% of the maximum value of the autocorrelation function, it nevertheless dominates the low-frequency spectrum because of its slow decay (Fig. 9.17). The spectral intensity strongly depends on the frequency in the range from 1 to 100GHz, whereas the spectrum without II generation is white up to 100GHz. In addition, in Fig. 9.17 results of the DD and HD models are shown, where the parameters of the II models of the DD (local field model [9.34]) and HD (local temperature model [9.35, 9.36]) simulators were adjusted in such a way that the integral over the II generation rate of the MC model is reproduced. Comparison of the results of the three models reveals good agreement especially for the HD and MC models up to 100GHz showing that the omission of the derivatives with respect to time in Eqs. (7.17), (7.21), and (7.30) is justified below 100GHz.

In Fig. 9.18 the spatial distribution of the terminal current noise is shown for the three different noise sources. The electron diffusion noise still originates from

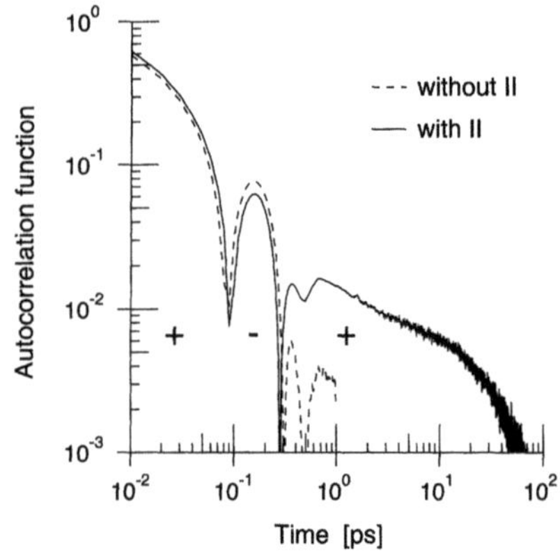

Figure 9.16: Absolute value of the autocorrelation function of the terminal current for the N^+NN^+-structure biased at 6V evaluated by MC simulations with and without II generation, where the sign of the correlation functions is indicated

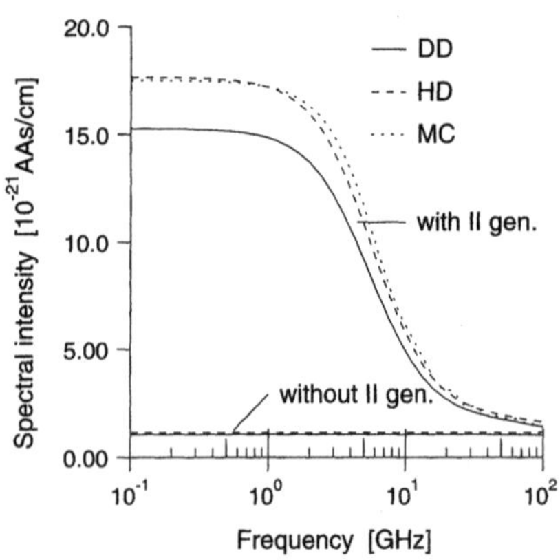

Figure 9.17: Spectral intensity of the terminal current for the N^+NN^+-structure biased at 6V evaluated by MC, HD, DD simulations with and without II generation

the region near the N^+N junction, although a small peak is observed close to the NN^+ junction which is not found in the case without generation of secondary particles. The contribution due to generation of secondary particles by II has its maximum at the peak of the II generation rate (Fig. 9.15) and the hole diffusion noise peaks at the maximum of the hole density.

The extremely long tail of the autocorrelation function (Fig. 9.16) necessitated the self-consistent MC simulation of about 1000 uniformly weighted particles for

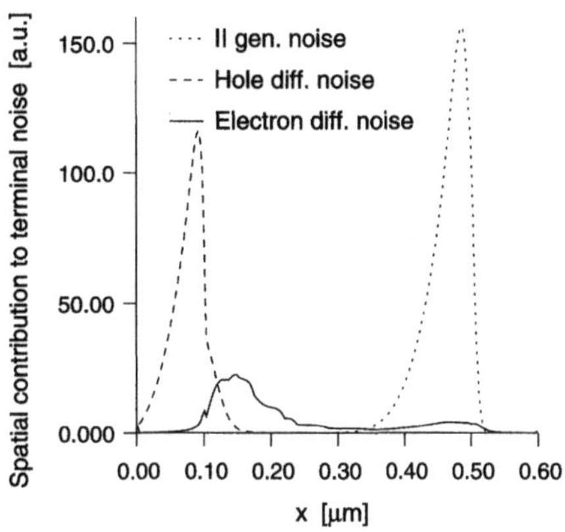

Figure 9.18: Spatial origin of the terminal current noise for the N^+NN^+-structure biased at 6V evaluated by HD simulations including generation of secondary particles by II at zero frequency

0.5μs resulting in a total CPU time of about 19 days on a 1GHz PC. Since 2D MC device simulations are in general more CPU intensive than 1D simulations, MC noise simulations of 2D devices are feasible only in very special cases, where the correlation functions decay on a sub-picosecond scale and the corresponding spectra are white up to at least 100GHz (see also [9.37]). It is worth noting that not only II generation produces these tails but also strong capacitive coupling and other device related effects as shown in Sec. 9.3.3. In the case of the momentum-based simulators the tail problem does not occur, because these work in the frequency-domain and even the frequency of 0Hz is accessible. The CPU time of the momentum-based models is less than a CPU minute for the results shown in Fig. 9.17, whereas the MC simulation required 19 CPU days. Thus, albeit the momentum-based models are not always as accurate as the MC model, they have a clear advantage over the time-domain-based MC model in the RF range, because their CPU times are orders of magnitude smaller and their numerical accuracy is almost independent of the frequency, in contrast to the MC method, where the tail of the correlation function, which determines the low-frequency behavior, can easily be missed in the stochastic noise yielding erroneous results.

9.2 MOSFETs

Transport in MOSFETs is dominated by the inversion layer, where the transport considerably differs from the bulk case [9.38], which is the basis of the hierarchical approach presented in Chap. 7. Since a similar hierarchical approach has not yet been developed for the so-called surface transport, the main tool for the investigation of the transport in MOSFETs is in this section the MC model. Only in the first part of this section an especially modified DD model is compared to the MC model to analyze the impact of the velocity overshoot on the device characteristics of Si NMOSFETs. In the second part the performance of N- and PMOSFETs with strained Si channels pseudomorphically grown on relaxed SiGe is estimated by comparison to conventional Si MOSFETs. A simpler device concept based on a strained SiGe channel is explored in the third part for SiGe PMOSFETs. In the fourth part the accuracy of NSC-MC approach for hot-electron simulations is assessed and in the next part it is used to investigate II in MOSFETs. In the last part of the section the NSC-MC approach is discussed for terminal current estimation.

9.2.1 Velocity Overshoot in Deep Sub-Micron NMOSFETs

The workhorse of industrial TCAD is the DD model [9.34,9.39–9.41]. As explained in Sec. 7.4 it does not contain any nonlocal transport effects of which the velocity overshoot is the most important one[1]. The impact of the velocity overshoot has been studied experimentally [9.43–9.47] and theoretically based on hydrodynamic models [9.48–9.54] and MC models [9.53–9.59].

Here, the quasiballistic transport phenomena are investigated by a comparison of consistent DD and MC models for a series of NMOSFETs with channel lengths ranging from 80 to 250nm [9.58]. The devices are designed in a 0.1μm-CMOS technology with a metal gate, ultra-shallow junctions, and a 3nm thick oxide (Fig. 9.19, Tab. 9.1) (e.g. [9.60]). They are scaled by adjusting the channel doping in such a way that a realistic I_{off} of 1μA/cm at $V_{\text{gate}} = 0$V and $V_{\text{drain}} = 1.5$V is achieved for the different gate lengths. A low I_{off} implies weak short channel effects, especially a small drain induced barrier lowering (DIBL) [9.61]. This is important, because in this case the drain current is mainly controlled by the source side injection [9.24, 9.50, 9.62]. A strong punch through on the other hand might artificially enhance the impact of the velocity overshoot leading to unrealistic results.

To ensure a meaningful comparison, the DD and the MC model must be as consistent as possible. The DD model used here is the one implemented in the simulator Galene III and reproduces under stationary homogeneous bulk conditions the results of the MC program [9.63]. But consistency under bulk conditions alone is not sufficient for a comparison in the case of MOSFETs, because the device characteristics are dominated by the transport in the channel which strongly differs from bulk transport [9.38]. In the case of the momentum-based models the problem of surface transport is not considered in Chap. 7, because an approach similar to the one for bulk transport, where all transport parameters are derived

[1] So-called augmented DD models containing a first-order correction of nonlocal effects via the derivative of the driving force are not considered here [9.42].

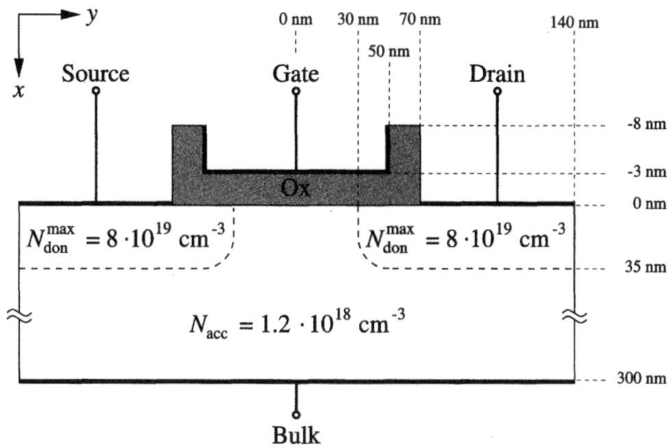

Figure 9.19: Device structure of the 100nm NMOSFET with a 3nm gate dielectric

Table 9.1: *Gate length, metallurgical channel length, channel doping, effective channel field, mobility and inversion layer density at* $V_{gate} = 1.5$V *under homogeneous channel conditions*

L_{gate} [nm]	L_{met} [nm]	N_{acc} [cm^{-3}]	E_{eff} [kV/cm]	μ_{eff} [cm^2/Vs]	N_{inv} [cm^{-2}]
80	40	$2.2 \cdot 10^{18}$	1170	200	$4.1 \cdot 10^{12}$
100	60	$1.2 \cdot 10^{18}$	1054	247	$5.6 \cdot 10^{12}$
130	90	$7.0 \cdot 10^{17}$	976	265	$6.5 \cdot 10^{12}$
180	140	$4.0 \cdot 10^{17}$	926	274	$7.2 \cdot 10^{12}$
250	210	$3.3 \cdot 10^{17}$	900	280	$7.5 \cdot 10^{12}$

from MC bulk simulations, has not yet been developed. On the other hand, surface transport must be included in a realistic simulation of MOSFETs and in the case of the momentum-based models the surface mobility model of Ref. [9.64] is used together with an analytical velocity-field relation [9.65]

$$\mu(|\nabla_{\mathbf{r}}\Phi_n|) = \frac{\mu_0}{\frac{1}{2} + \frac{1}{2}\left[1 + \left(\frac{2|\nabla_{\mathbf{r}}\Phi_n|\mu_0}{v_{\text{sat}}}\right)^{\beta}\right]^{\frac{1}{\beta}}}, \tag{9.1}$$

where μ_0 is calculated with the surface mobility model of Ref. [9.64], v_{sat} is the saturation drift velocity in the electron inversion layer, and is obtained together with β by matching MC results for homogeneous inversion layers.

The DD and MC models must not only be consistent under bulk conditions but also under homogeneous channel conditions. The surface scattering parame-

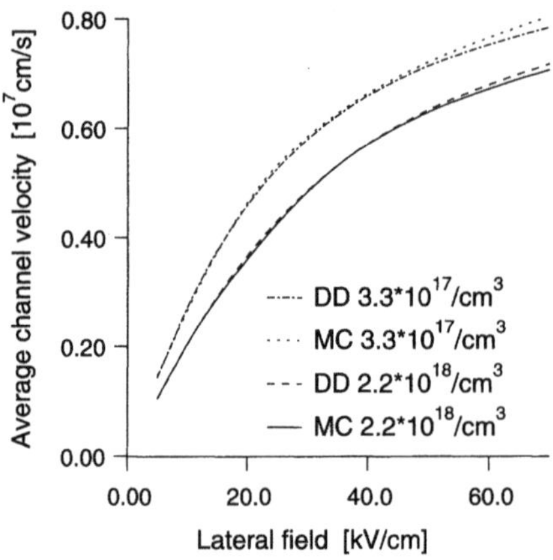

Figure 9.20: Average channel velocity versus the lateral electric field for a homogeneous channel ($V_{\text{gate}} = 1.5\text{V}$)

ters of both models are adjusted to exactly match the measured low-field mobility (Tab. 9.1) [9.66]. In the case of the MC model the probability for diffusive scattering at the oxide is reduced by a few percent compared to the value given in Sec. 4.5 to match the experimental mobilities at the high channel dopings for which the value of Sec. 4.5 is no longer universal. In the case of the DD model the surface roughness scattering parameter of the channel mobility model of Ref. [9.64] is changed accordingly. In addition, the velocity-field relation of the DD model (Eq. (9.1)) is tuned to reproduce the respective MC results. In Fig. 9.20 the average channel velocity is shown for devices with a constant doping and a homogeneous lateral electric field, and good agreement of the DD and MC models is achieved. The resultant saturation velocity equals the experimental value of 10^7cm/s for channels [9.67,9.68], and β is given by 2.0.

Unfortunately, these measures are not sufficient to guaranty the consistency of both models in the case of a 2D device simulation. In the channel area of a MOSFET near the source and drain junctions the electric field is considerably inhomogeneous and under these inhomogeneous conditions the two surface transport models still yield different results even under near-equilibrium conditions[2]. Although the differences in the drain current are at most a few percent, they can not be neglected for the NMOSFETs with the two shortest channel lengths, and the surface roughness scattering parameter of the DD model is further adjusted a little bit to match exactly the MC results for these two NMOSFETs. In Fig. 9.21 the average channel velocity is shown for the 80nm device biased with a drain voltage of 0.1V. The differences between both models are negligible, although the transport is already nonlinear. Moreover, the electron sheet density is shown and good agreement is found indicating the high degree of consistency of both models.

[2]In contrast to the N^+NN^+ structure the differences in the ohmic conductances evaluated

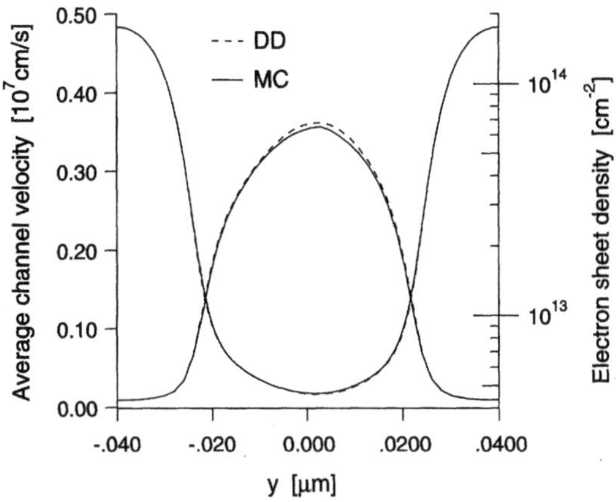

Figure 9.21: Electron density integrated perpendicular to the Si/SiO_2-interface and the average channel velocity for the 80nm device at $V_{\text{gate}} = 1.5\text{V}$ and $V_{\text{drain}} = 0.1\text{V}$

The drain current for the on-state at $V_{\text{gate}} = V_{\text{drain}} = 1.5\text{V}$ is shown in Fig. 9.22 versus the gate length. The drain current of the DD model has a maximum at about 100nm gate length, because at shorter gate lengths the decrease in the channel mobility and inversion layer density (Tab. 9.1) is larger than the increase in the average channel velocity caused by the stronger lateral electric field. In the case of the MC simulations the velocity overshoot removes this maximum, but nevertheless the gain in I_{on} is strongly reduced at short gate lengths. The ratio of the drain currents of the MC and DD model monotonically decreases from 1.3 for the 80nm device to 1.05 for the 250nm device. The differences between both models decrease with increasing channel length due to the constant drain voltage resulting in a reduction of the lateral electric field and quasiballistic transport effects.

In Fig. 9.23 the drain currents are shown as a function of the drain voltage for both models and they agree well in the linear regime. As expected, the discrepancy increases with the drain voltage due to the velocity overshoot. For a gate length of 80nm both models already strongly differ for drain voltages larger than 0.3V. In the light of these findings the results of Ref. [9.57] appear somewhat doubtful, because that comparison was based on an MC model of which the surface roughness scattering parameter was calibrated at a drain voltage of 0.5V to match the DD result. But for this drain bias the impact of the velocity overshoot is not negligible and the calibration might have masked a considerable part of the impact of the velocity overshoot. This might be one of the reasons, why the impact of the velocity overshoot was found to be small in that investigation.

---

with the DD and MC models due to the built-in fields is negligible in MOSFETs.

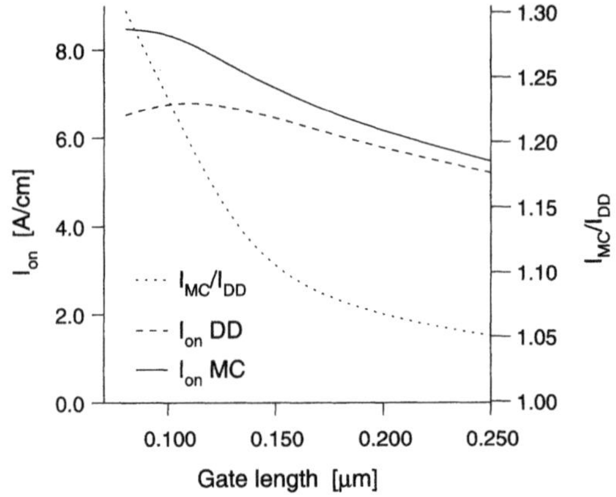

Figure 9.22: Drain current versus gate length for $V_{\mathrm{gate}} = V_{\mathrm{drain}} = 1.5\mathrm{V}$

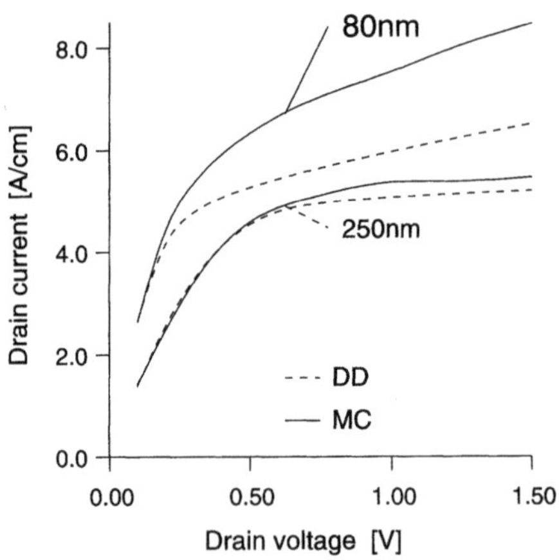

Figure 9.23: Drain current versus drain voltage for the 80 and 250nm device at $V_{\mathrm{gate}} = 1.5\mathrm{V}$

In the case of the 250nm device the increase of I_{on} is only 5%. But this does not mean that the velocity overshoot is small in this device. A maximum overshoot of 1.5 times the bulk saturation velocity is found near the drain side of the channel (Fig. 9.24). But this increase of the effective velocity is counterbalanced by a decrease of the electron sheet density, because the total current in the channel must be constant due to current continuity. A similar situation is found in the case of the 80nm device, where the maximum overshoot is 2 times the bulk saturation velocity (Fig. 9.25), and the enhancement of I_{on} is only 30%. The increase of the drain current is mainly due to the enhanced velocity at the source side which depends on the strength of the lateral electric field at the beginning of the channel

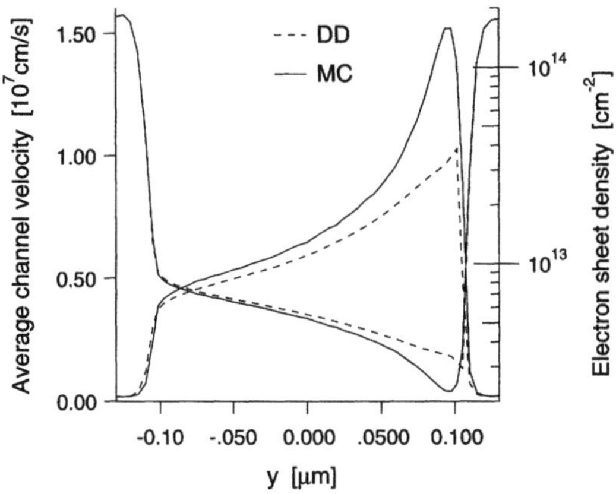

Figure 9.24: Average channel velocity and electron sheet density of the 250nm NMOSFET for $V_{\mathrm{gate}} = V_{\mathrm{drain}} = 1.5\mathrm{V}$

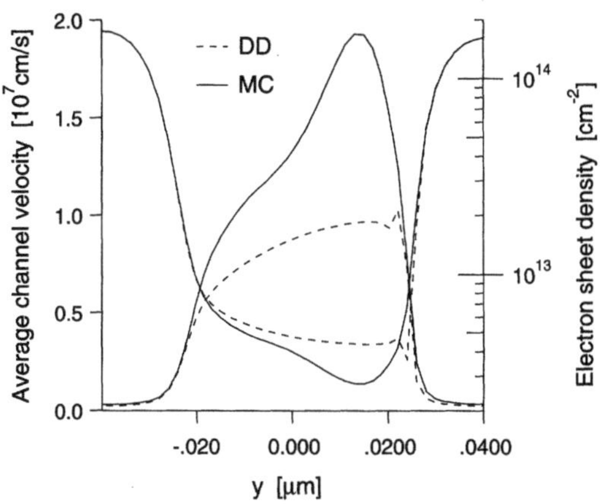

Figure 9.25: Average channel velocity and electron sheet density of the 80nm NMOSFET for $V_{\mathrm{gate}} = V_{\mathrm{drain}} = 1.5\mathrm{V}$

[9.24,9.50,9.62]. This field depends on the space charge distribution near the drain side via the PE which in turn depends on the transport model (cf. Sec. 9.2.6). On the other hand, the impact of the drain side on the source-side injection field is limited in a well-designed MOSFET in order to avoid a too high I_{off} and DIBL limiting the impact of the velocity overshoot on I_{on} in the case of devices scaled with a low I_{off}.

In Fig. 9.26 the two-dimensional electron density distribution is shown evaluated by the DD and MC models. Good agreement is found between both models for very high densities in the source and drain, whereas huge differences are found

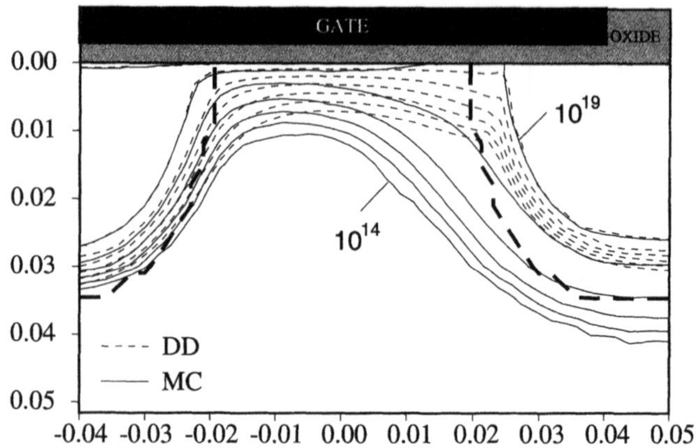

Figure 9.26: Electron density [cm^{-3}] of the 80nm NMOSFET for $V_{\text{gate}} = V_{\text{drain}} = 1.5$V

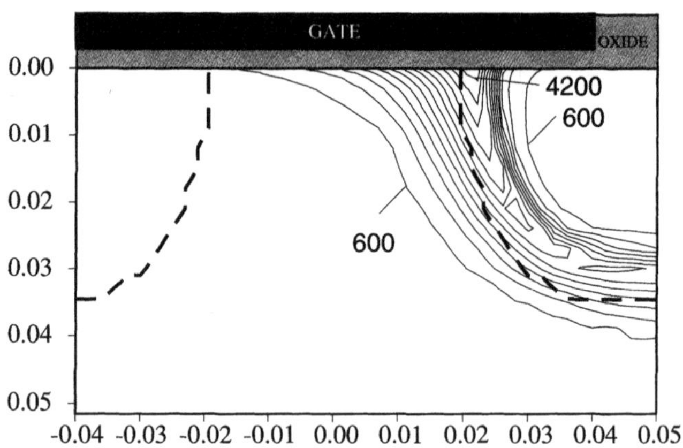

Figure 9.27: Electron temperature [K] of the 80nm NMOSFET for $V_{\text{gate}} = V_{\text{drain}} = 1.5$V simulated by the MC method (contour lines for every 300K)

in the channel region. In the MC case the electrons near the drain junction spread deep into the bulk due to their high temperature (Fig. 9.27). A maximum temperature of more than 4200K is found near the drain junction, while in the case of the DD model the electron gas temperature is given by the lattice temperature and the diffusion into the bulk is strongly suppressed.

In modern integrated circuits not only the maximum drive current I_{on} is important but also the high-frequency performance of the devices. The cutoff frequency f_T is defined in analogy to the definition for bipolar transistors given in Ref. [9.69] for MOSFETs as the frequency for which the current gain calculated in the quasi

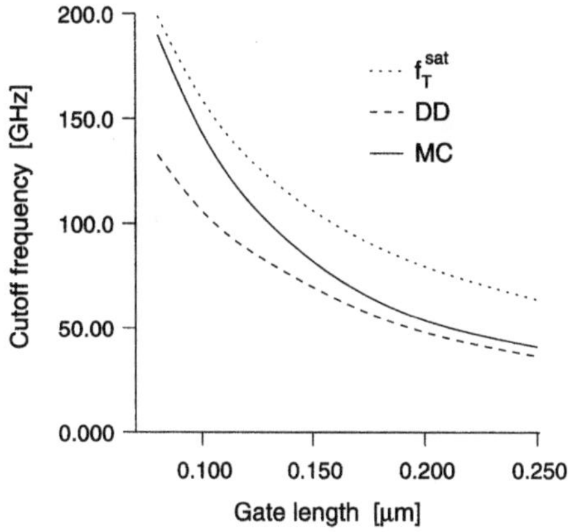

Figure 9.28: Cutoff frequency versus gate length for the on-state evaluated with the MC and DD model, and the saturated velocity cutoff frequency

stationary approximation becomes equal to unity [9.70]

$$f_{\mathrm{T}} = \frac{1}{2\pi}\frac{dI_{\mathrm{drain}}}{dQ_{\mathrm{gate}}} \approx \frac{1}{2\pi}\frac{I_{\mathrm{drain}}(V_{\mathrm{gate}}+\Delta V) - I_{\mathrm{drain}}(V_{\mathrm{gate}}-\Delta V)}{Q_{\mathrm{gate}}(V_{\mathrm{gate}}+\Delta V) - Q_{\mathrm{gate}}(V_{\mathrm{gate}}-\Delta V)}\,, \tag{9.2}$$

where only the gate voltage is varied and all other terminal voltages are kept constant. Since the dependence of the drain current I_{drain} and gate charge Q_{gate} on the gate voltage is almost linear in strong inversion, ΔV can be rather large compared to the thermal voltage and a value of 0.3V is used. In Fig. 9.28 the cutoff frequency is shown for the on-state. The MC result exceeds the DD result by up to 43%, because the drain current is larger and the inversion layer density reduced compared to the DD model (Figs. 9.22, 9.25, and 9.26). In addition, in Fig. 9.28 the saturated velocity cutoff frequency is shown which is defined as [9.70]

$$f_{\mathrm{T}}^{\mathrm{sat}} = \frac{v_{\mathrm{sat}}}{L_{\mathrm{gate}}}\,. \tag{9.3}$$

In the case of the shortest NMOSFET the MC cutoff frequency is nearly as large as this approximate result, which ignores the impact of the source and drain transit times, clearly indicating the strong influence of the velocity overshoot.

In the case of the MC model 98.000 particles were simulated for 10ps to calculate the drain current. The time step length of the iteration loop for the SC solution of the MC model and the PE was 0.5fs (cf. Sec. 6.4). Statistics were enhanced by the MR method applied to each cell of the RS grid without resolving the regions in energy. 20 simulation particles were assigned to each cell and 30000 particles were distributed according to the electron density calculated with the DD model in order to increase the number of particles per cell in the regions with a very high electron concentration to achieve a collisionless plasma

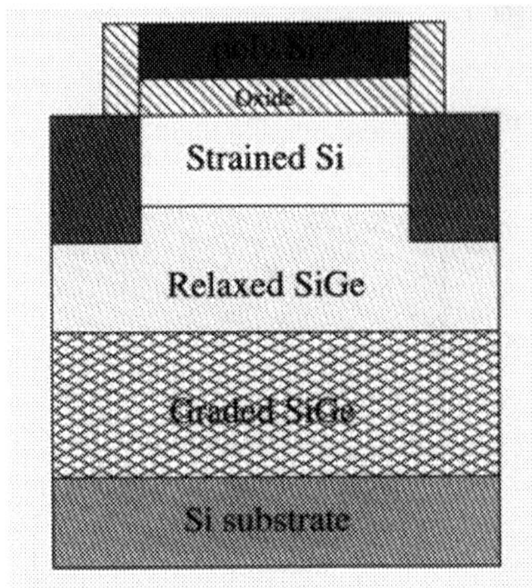

Figure 9.29: Structure of a MOSFET with a strained Si channel

(cf. Sec. 6.7). Only in the case of Fig. 9.27 the results were obtained by dividing the energy up to 1.0eV into ten bins with five particles per bin. In the case of the bin with the lowest energy the same number of particles was simulated as in the other simulations. The partitioning in energy was necessary to calculate the correct electron temperature in the regions with a very low electron density ($< 10^{10}/\text{cm}^3$) (cf. [9.71]), although these regions do not influence any of the other results shown here .

9.2.2 MOSFETs with Strained Si Channels

In order to improve the performance of CMOS devices the use of strained Si channels has been proposed for N- and PMOSFETs [9.9–9.13]. The biaxial tensile strain is induced by growing a thin Si layer pseudomorphically on relaxed SiGe, which is grown on a SiGe layer in which the Ge content decreases towards the Si substrate (Fig.9.29). This layer is so thick that the strain relaxes within this layer. The strain reduces the symmetry of the Si crystal and the six minima of the first conduction band are no longer equivalent. The four valleys on the principal axes within the plane parallel to the Si/SiGe-interface (in-plane) are lifted upwards in energy, whereas the two valleys on the perpendicular principal axis of the KS are lowered [9.8]. Due to the lower energy most electrons populate those two valleys under equilibrium conditions. In these valleys the scattering rate is reduced due to the reduced number of final states for intervalley scattering. In addition, the conductivity mass decreases for particle motion within the plane. Therefore, the in-plane mobility increases with the strain, which in turn increases with the Ge content of the relaxed SiGe layer [9.72, 9.73]. This increase of the in-plane electron bulk mobility is shown in Fig. 9.30 and taking into account the uncertainty of the experimental results good agreement of the simulations and

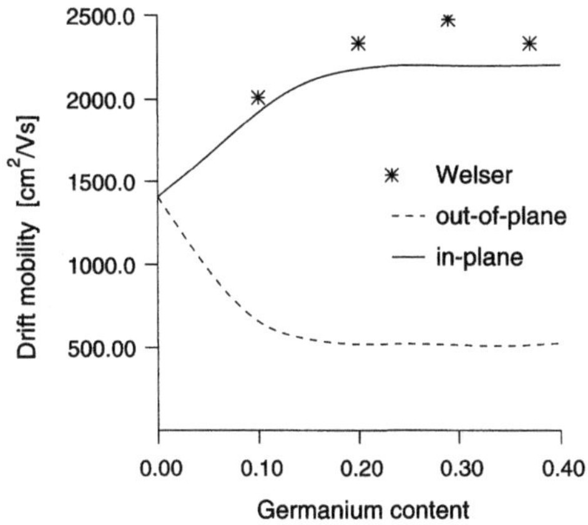

Figure 9.30: Electron bulk mobility for undoped strained Si grown on relaxed SiGe and experimental results by J. Welser et al. [9.9]

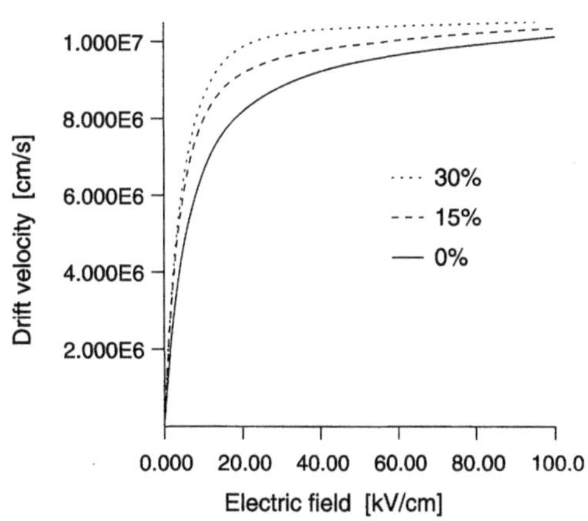

Figure 9.31: Electron bulk drift velocity for undoped strained Si grown on relaxed SiGe for an electric field in ⟨011⟩ direction (in-plane)

experiments is found. The out-of-plane mobility on the other hand decreases, because the conductivity mass increases for this direction. The in-plane drift velocity is shown in Fig. 9.31 and although the low-field mobility increases with the strain, the saturation velocity remains the same.

Based on these bulk results it is difficult to estimate the potential performance of strained Si NMOSFETs. Especially, since bulk and inversion transport are quite different in their microscopic nature [9.38]. In Fig. 9.32 the inversion layer mobility is shown calculated under the assumption that the probability for a diffusive scattering event at the Si/SiO_2 interface does not change with the strain

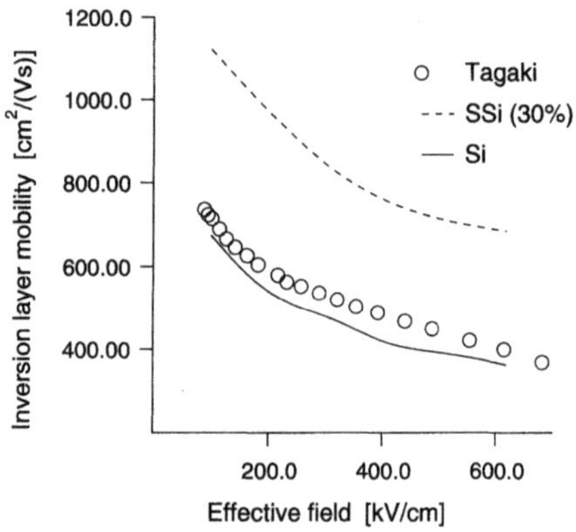

Figure 9.32: Electron inversion layer mobility for strained Si (30% Ge in the relaxed SiGe layer), unstrained Si, and experimental results by S. Takagi et al. [9.66]

level (Sec. 4.5) [9.74]. This transport model, which neglects the quasi 2D nature of the electron gas in the inversion layer, yields in agreement with the experimental results a strong enhancement of the mobility also for high effective fields in the channel. This is in contrast to the more physics-based quasi 2D electron gas models, which by now for a given surface roughness fail to reproduce the experimental results at high effective fields [9.75].

Due to the lack of understanding of the microscopic transport in the inversion layer, here quantization effects in the channel are neglected and the empirical surface scattering model of Sec. 4.5 is used for device simulation because it shows good agreement with the experimental results. The investigated NMOSFETs are the same ones as in Sec. 9.2.1 besides the strain in the Si bulk (Fig. 9.19, Tab. 9.1). Instead of simulating the full structure shown in Fig. 9.29 it is assumed that the whole bulk consists of strained Si [9.74]. Although this configuration is impossible in reality due to strain relaxation, simulations of this artifical structure yield the same results as simulations with a 10nm thick strained Si layer and a relaxed SiGe substrate, and the simulations are performed for the simpler structure. The results of the strained Si NMOSFETs are compared to the data of the NMOSFETs of Sec. 9.2.1. Since the strain changes the threshold voltage, the workfunction difference between the gate and Si has been adjusted in such a way that the same inversion layer density is obtained for low drain voltages with and without strain (Fig. 9.33). The on-current of the strained Si NMOSFETs has been simulated for channel lengths from 80 to 250nm and is shown in Fig. 9.34 together with the results for the unstrained case. In the long-channel case, where the current is mainly determined by the near-equilibrium transport at the source side of the channel [9.24], the largest increase in current with strain is found. This increase in the on-current degrades for shorter channel devices. This is due to the strain-

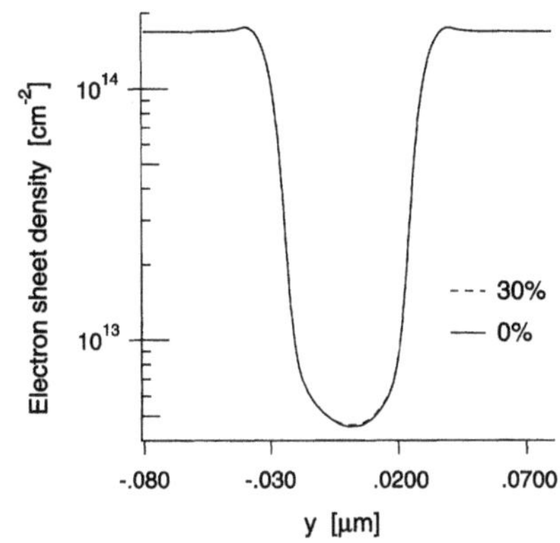

Figure 9.33: Electron inversion layer density for the 80nm NMOSFET with a strained Si (30% Ge in the relaxed SiGe layer) channel and unstrained Si ($V_{\text{gate}} = 1.5\text{V}, V_{\text{drain}} = 0.1\text{V}$)

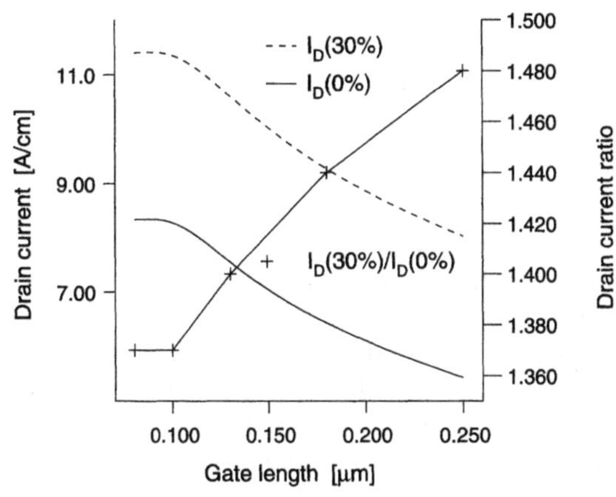

Figure 9.34: On-current for strained Si (30% Ge in the relaxed SiGe layer) and unstrained Si NMOSFETs ($V_{\text{gate}} = V_{\text{drain}} = 1.5\text{V}$)

independent saturation velocity, which results in a decrease of the improvement with the driving field (Fig. 9.31) [9.74]. Since the driving field increases with decreasing channel length, the performance gain degrades. This degradation levels off in the case of the shortest channel device, and the question remains, whether shorter channel devices will be affected in the same way or not.

The impact of strain on the performance of NMOSFETs with ultra-short channels is investigated for the 25nm device proposed by Y. Taur et al. [9.76]. The oxide thickness is 1.5nm and the device is designed for a supply voltage of 1V.

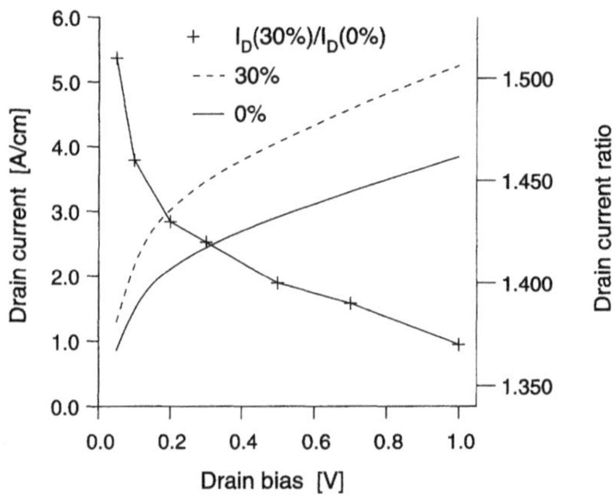

Figure 9.35: On-current for a strained Si (30% Ge in the relaxed SiGe layer) and unstrained Si 25nm NMOSFETs ($V_{\text{gate}} = 1.0\text{V}$)

It is again simulated with a strained and unstrained Si bulk. The Ge content of the relaxed SiGe layer is 30% for the strained case. In Fig.9.35 the drain current is shown as a function of the drain bias. Due to the increasing driving fields the performance gain decreases with increasing drain bias. The improvement for the on-current at 1V drain bias is over 35% and similar to the gain in the case of the 80nm device (Fig. 9.34). Thus, also in the case of ultra-short channels strain can improve the device performance.

The performance gain for the 25nm device was unexpected because of the strain-independent saturation drift velocity (Fig. 9.31). A possible explanation is nonstationary transport in the channel and in Fig. 9.36 the channel velocity and inversion layer density are shown for the strained and unstrained cases. A strong velocity overshoot is found in the strained case on the source side. This higher velocity on the source side leads to the larger current, which in turn increases the inversion layer density on the drain side of the channel.

In Fig. 9.37 the drain current versus drain voltage is shown for a 25nm PMOSFET, designed similar to the above 25nm NMOSFET. Again, a performance gain is found in the strained case of about 40% for the on-current. Since similar to the electron case the saturation drift velocity does not improve with the strain [9.77], the gain is also partly due to nonstationary transport.

9.2.3 PMOSFETs with a Strained SiGe Channel

For low noise applications PMOSFETs with a strained SiGe channel have been proposed because of an improved hole mobility and better noise characteristics due to the higher quality of the Si/SiGe interface compared to the Si/SiO_2 interface [9.8, 9.14–9.18]. These devices have the advantage over heterodevices fabri-

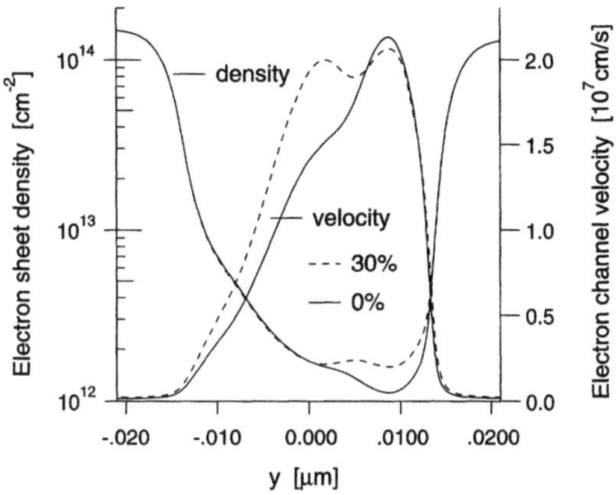

Figure 9.36: Inversion layer drift velocity and density for the strained Si (30% Ge in the relaxed SiGe layer) and unstrained Si 25nm NMOSFETs ($V_{\text{drain}} = V_{\text{gate}} = 1.0\text{V}$)

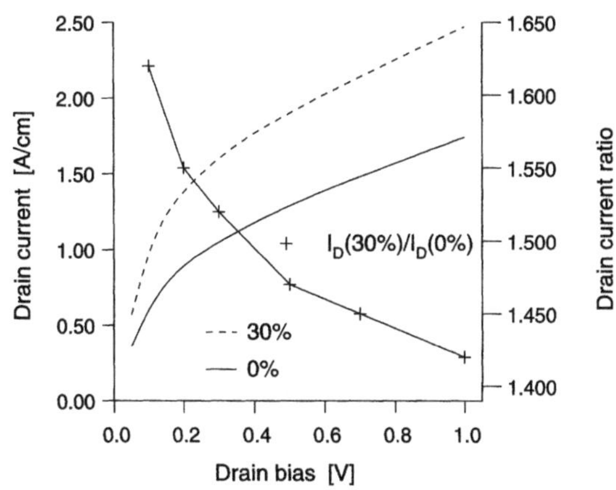

Figure 9.37: On-current for a strained Si (30% Ge in the relaxed SiGe layer) and unstrained Si 25nm PMOSFETs as a function of the negative drain bias ($V_{\text{gate}} = -1.0\text{V}$)

cated with group III/V semiconductors that they are largely compatible with the standard CMOS fabrication process. The structure of the PMOSFET is outlined in Fig. 9.38. Due to the lower valence band edge in the SiGe layer inversion occurs first in the SiGe layer and only for very high gate voltages at the Si surface. The current is therefore mostly confined to the SiGe layer. A similar NMOSFET concept is not possible, because the conduction band shift in the strained SiGe is too small for an effective confinement of electrons in the SiGe layer.

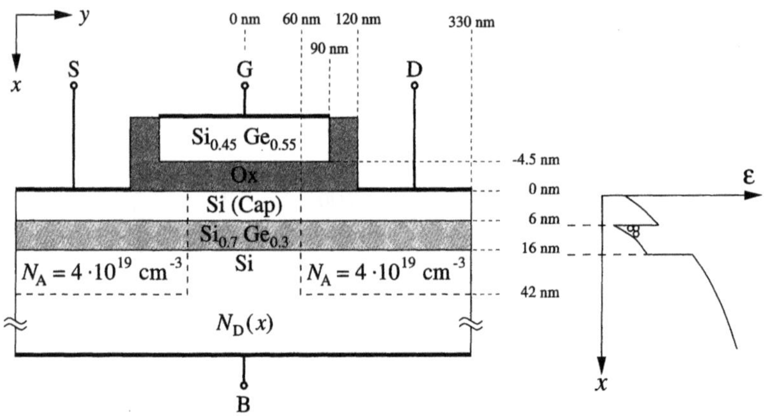

Figure 9.38: 0.12μm-Si/SiGe-PMOSFET

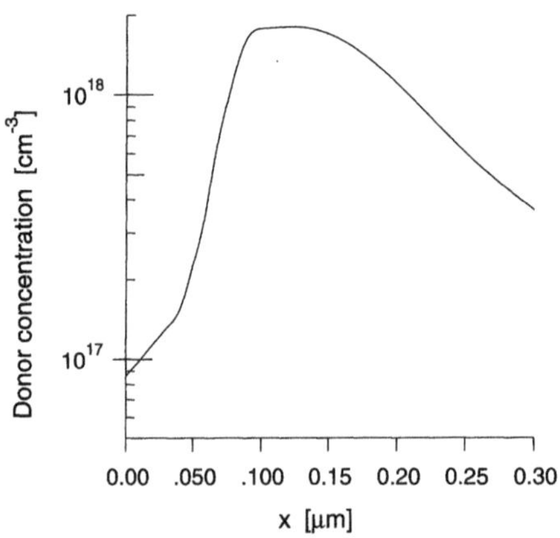

Figure 9.39: Bulk doping profile [9.15]

On the one hand, the effective low-field hole mobility of PMOSFETs with a strained SiGe-layer has been shown to improve with the Ge-content [9.8, 9.15–9.18]. On the other hand, theoretical results show that the saturation drift velocity in strained SiGe is unchanged or even reduced similar to the case of strained Si (Sec. 9.2.2) [9.78, 9.79]. The goal of this section is to investigate by device simulation how these contrary effects influence the performance of PMOSFETs with strained SiGe channels. In order to account for the warping of the valence bands and other details of the band structure the FB-MC model is employed [9.80].

In Fig. 9.38 the device structure is shown, which is simulated with and without a strained 10nm thick $Si_{0.7}Ge_{0.3}$-layer in the channel area to estimate the expected performance improvement. The difference in threshold voltage with and

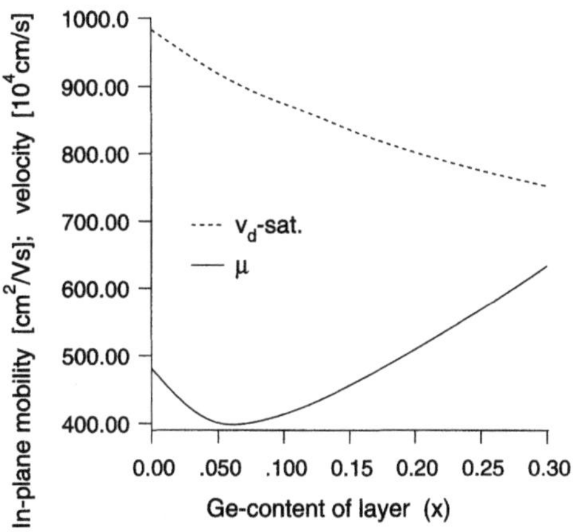

Figure 9.40: Hole in-plane mobility and saturation velocity for an undoped strained $Si_{1-x}Ge_x$-layer on unstrained Si

without a SiGe-channel is negligible. The doping profile is modeled similar to Refs. [9.15, 9.16] (Fig. 9.39). In order to estimate the performance improvement under ideal conditions (undoped channel) impurity scattering has been neglected in the channel. The drawn gate length is 0.18μm and the metallurgical channel length is 0.12μm. The oxide thickness is 4.5nm. The valence band offset at the Γ-point between the strained $Si_{0.7}Ge_{0.3}$ and the unstrained Si is 228meV [9.8]. surface roughness scattering is neglected at the Si/SiGe-interface in accordance with experimental findings [9.8].

In Fig. 9.40 the in-plane mobility and saturation velocity of holes are shown for undoped strained SiGe on unstrained Si [9.79]. For a Ge content of 30% the low-field mobility is enhanced and the saturation velocity reduced. This is shown in detail in Fig. 9.41. The drift velocity in Si exceeds the one in strained $Si_{0.7}Ge_{0.3}$ for electric fields above 5kV/cm. However, from these simulations under homogeneous conditions it cannot be concluded to which extent the lower saturation velocity will degrade the device performance.

Another important factor for the mobility improvement is the confinement of the holes in the SiGe-channel by which the scattering at the rough Si/SiO_2-interface is avoided. In Fig. 9.42 the effective channel mobilities are shown with and without a SiGe channel for a homogeneous doping of $10^{16}/cm^3$. The respective experimental results for unstrained Si agree well with the simulated data [9.66]. In the case of the SiGe channel the effective mobility is much higher, though it degrades with increasing effective field due to a redistribution of the particles towards the Si/SiO_2-interface. A similar behavior of the effective mobility is found in the case of a long channel version of the PMOSFET as shown in Fig. 9.43. The maximum effective field at -1.5V gate-voltage in Fig. 9.43 is 440kV/cm and similar to the maximum one in Fig. 9.42.

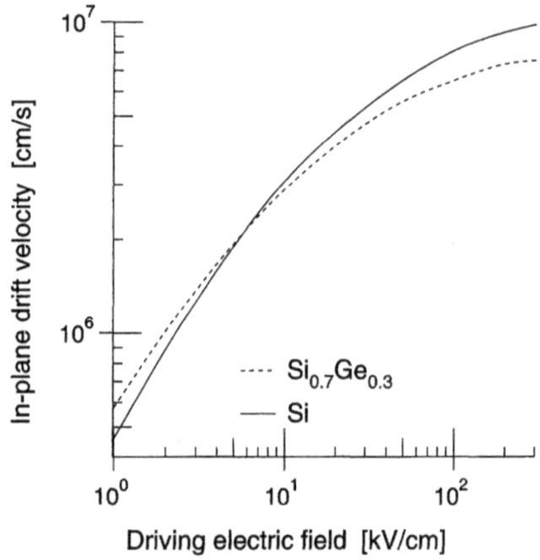

Figure 9.41: Hole in-plane drift velocity for undoped $Si_{0.7}Ge_{0.3}$ on unstrained Si and for relaxed Si

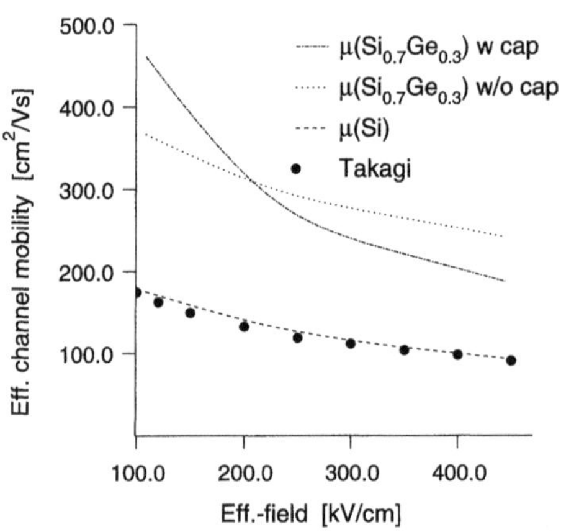

Figure 9.42: Effective channel mobility for holes as measured by S. Takagi et al. [9.66] and simulation results for relaxed Si and for a $Si_{0.7}Ge_{0.3}$-layer on unstrained Si with and without a 5nm thick cap layer

In Fig. 9.42 the effective mobility is also shown for the artificial case of a strained SiGe-channel without a cap layer. In this case the holes are scattered at the rough oxide, where the same diffusive scattering probability as for Si is used. Nonetheless the effective mobility is much higher than in the case of unstrained Si. These unexpected high differences in the effective channel mobilities can be explained by the warped band structure of the holes, where the warping depends on the strain (Fig. 9.44). In Fig. 9.45 a section of the first valence band is shown for relaxed Si and strained SiGe. The strain reduces the warping near the Γ-point

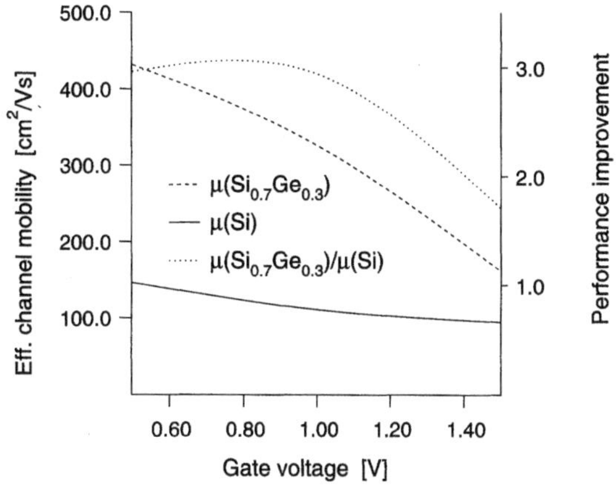

Figure 9.43: Effective channel mobility for a long channel version of the PMOSFET of Fig. 9.38 as a function of the negative gate voltage

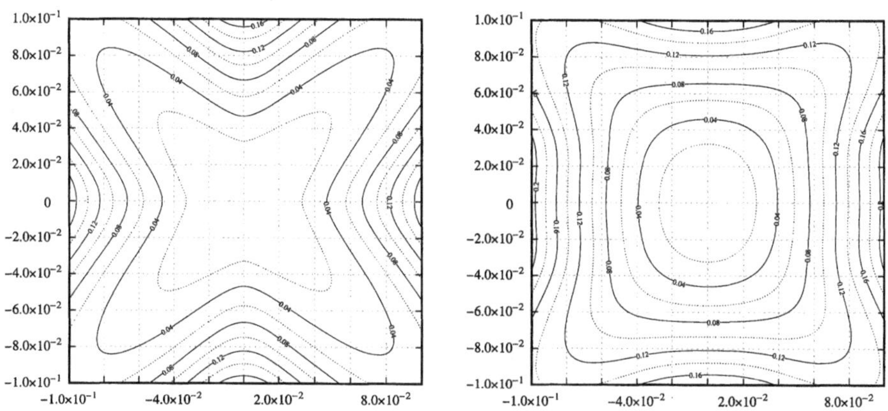

Figure 9.44: Equienergy contours [eV] of the heavy hole bands in the k_y, k_x-plane for unstrained Si (left) and strained $Si_{0.7}Ge_{0.3}$ (right) (wave vector in units of $2\pi/a_0$)

and this changes the behavior of the holes under confinement. When a particle in a parabolic band structure is reflected at a potential barrier, the component of its wave vector perpendicular to the barrier is reversed (specular reflection). Due to the symmetry of the band structure the velocity parallel to the barrier does not change. This is similar to the case of holes in strained SiGe as indicated in Fig. 9.45. In the case of holes in a warped band structure the situation is more complex. In Fig. 9.45 the reflection of a hole in unstrained Si at a barrier in the y, z-plane is shown, which is calculated under the condition of energy and parallel

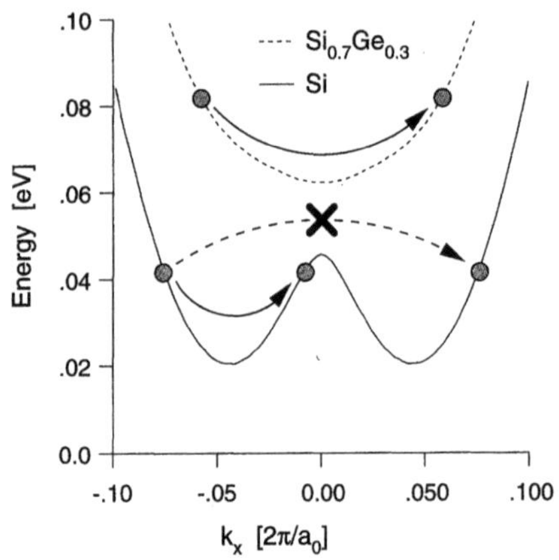

Figure 9.45: A section of the first valence band in a strained $Si_{0.7}Ge_{0.3}$ (100)-layer on unstrained Si and in relaxed Si in $\langle 100 \rangle$-direction (k_x) with $k_y = 0.05 \cdot 2\pi/a_0, k_z = 0$ (see Fig. 9.46)

momentum conservation. In this case the perpendicular component of the wave vector is not reversed. In Fig. 9.46 the respective change of the perpendicular component of the wave vector and of the group velocity in the channel direction is shown. Due to the warped band structure the particle ends up in a state, for which the velocity in the channel direction is reduced and its sign reversed. This effect reduces the effective channel mobility of holes in unstrained Si under confinement compared to the bulk mobility. On the other hand, the mobility reduction for holes confined in strained SiGe is smaller because of the less warped band structure. This effect does not only occur in the case of a reflection at a step-like potential barrier, but also when a particle is reflected by a strong confining field.

A comparison of the channel mobilities in Fig. 9.42 for the lowest effective field reveals that the effective channel mobility is strongly enhanced in the case of the strained SiGe layer by the reduced warping of the band structure in addition to the improvement due to the reduced scattering. However, for a short channel PMOSFET with $L_{\text{eff}} = 0.12\mu\text{m}$ the performance improvement due to the strained SiGe-layer (Fig. 9.47) is much smaller than it is expected from the long channel case (Fig. 9.43) and it rapidly degrades with increasing drain voltage. This is due to the fact that the saturation velocity in the strained SiGe layer is not enhanced but reduced (Figs. 9.40, 9.41). The average drift velocity in the channel is shown in Fig. 9.48. This figure demonstrates that for a drain bias of -1.5V the peak velocity is highest for the device without the strained SiGe layer. On the other hand, in the case of low driving fields in the channel ($V_D = -0.1$V) the expected result is found, that the average drift velocity in the entire channel is higher for the device with a strained SiGe layer.

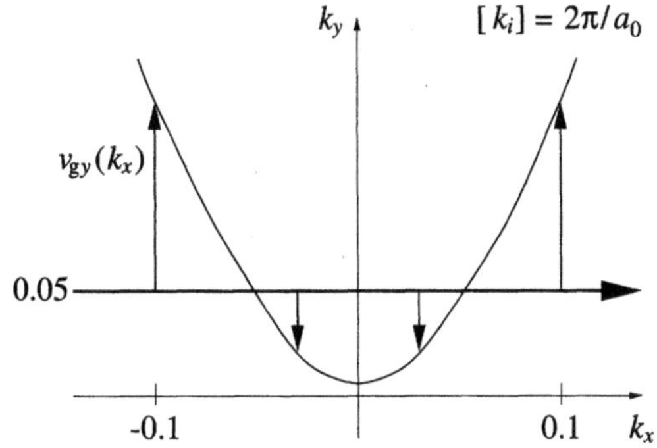

Figure 9.46: Schematic display of the group velocity in channel direction (v_{gy}) along the section in **k**-space of Fig. 9.45 with $k_z = 0$ for unstrained Si

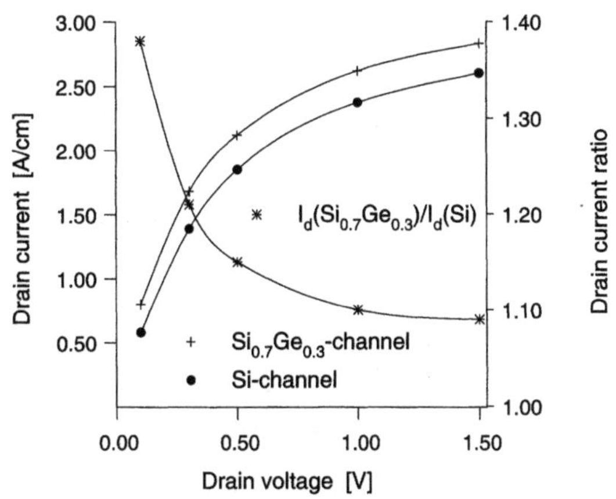

Figure 9.47: Negative drain current for the 0.12μm-PMOSFET with and without a strained $Si_{0.7}Ge_{0.3}$ (100)-layer and the respective drain current ratio for a gate voltage of -1.5V as a function of the negative drain bias

In Fig. 9.49 the drain current is shown as a function of the gate length. The performance improvement degrades when the channel length is reduced, because the driving channel field under saturation conditions increases for shorter channel lengths. Thus, the promising results found in the case of longer channel devices [9.15] can not be expected in the case of devices with short channels. Similar results were found in an experimental study [9.81].

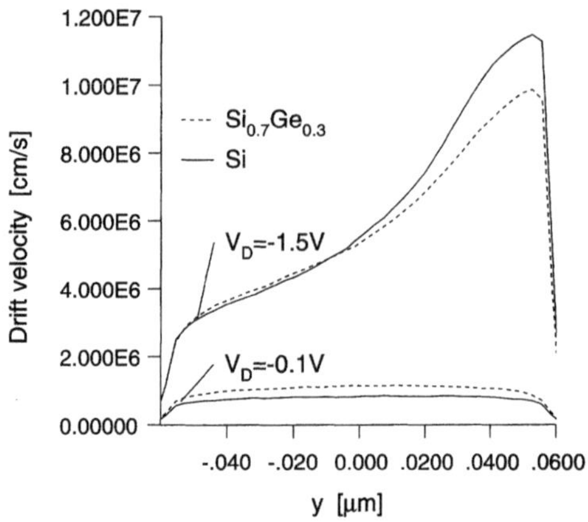

Figure 9.48: Average drift velocity in the channel of the 0.12μm-PMOSFET with and without a strained $Si_{0.7}Ge_{0.3}$ (100)-layer for a drain voltage of -0.1V and -1.5V

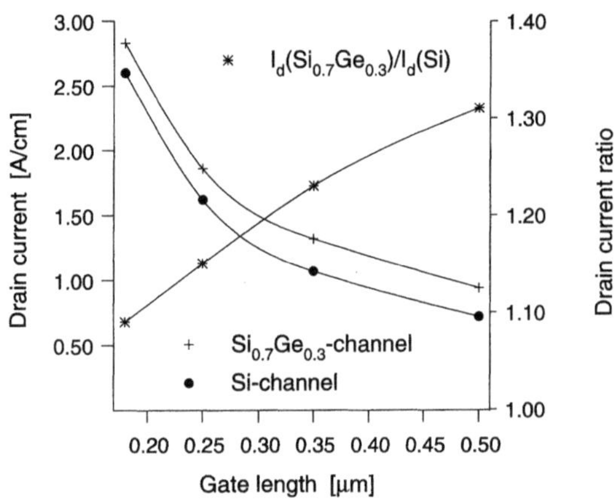

Figure 9.49: Negative drain current as a function of the gate length with and without a strained $Si_{0.7}Ge_{0.3}$ (100)-layer and the respective drain current ratio for a gate and drain voltage of -1.5V

9.2.4 Efficient MC Simulation of Hot-Electron Effects

Device degradation caused by hot-electrons is of severe concern for all manufacturers of sub-micron MOS devices [9.82]. The substrate current due to II is a key quantity for monitoring device degradation in NMOSFETs [9.83,9.84]. Therefore, efficient simulation of hot-electron effects and especially of the substrate current is most important. In Ref. [9.85] a NSC-MC method is presented which avoids

Table 9.2: *Substrate current and CPU time (SUN SPARC 20) for self-consistent and nonself-consistent simulations*

		Nonself-consistent		Self-consistent	
V_{gate}	V_{drain}	CPU time	I_{sub}	CPU time	I_{sub}
[V]	[V]	[s]	[A/cm]	[s]	[A/cm]
1.0	1.5	1592	$1.0 \cdot 10^{-5}$	416949	$1.3 \cdot 10^{-5}$
1.0	3.5	117	0.102	297568	0.103
3.0	3.5	182	0.171	135628	0.170

the CPU intensive SC simulation of the MC model and PE and allows to restrict the MC simulation domain to a small window of the active device region neglecting most parts of the high particle density areas of the source and drain regions (cf. Sec. 6.6). But in Ref. [9.85] no proof is given that the NSC-MC method yields the correct hot-electron distribution and in Ref. [9.86] this is denied.

In order to investigate the validity of the NSC approach SC and NSC simulations of a real NMOSFET were performed [9.87]. The device was fabricated in a quarter micron CMOS technology of the 1GBit DRAM generation with shallow highly doped source/drain extensions [9.88]. The polygate length is 220nm, the metallurgical channel length 160nm, and the oxide thickness 6.3nm. The device geometry and doping profile were obtained by careful inverse modeling under consideration of extensive C-V measurements (gate-, gate/drain-capacitance), subthreshold and linear I-V characteristics, and data like the real polygate length and implant doses [9.89–9.91]. The I-V characteristics simulated with the HD model based on this device description agree well with experimental results in subthreshold, linear, and saturation region for polygate lengths from 0.22μm to 20μm [9.90, 9.91].

The SC simulations were performed with a sufficient number of particles in such a way that the particle gas is approximately collisionless [9.92]. The statistics were enhanced with the MR method (Sec. 6.7) in the case of the SC simulations but not in the case of the NSC. The simulations were performed until the substrate current was converged within ±10% with a probability of 95.5%. In Tab. 9.2 the resultant substrate currents and accordant CPU times are shown. Both methods practically yield the same results for the substrate current and other hot-electron quantities not shown here, whereas the CPU times differ by about three orders of magnitude. Thus, at least in the case of a so-called collisionless plasma NSC simulations are a reliable and efficient alternative.

In Fig. 9.50 the measured and nonself-consistently calculated substrate current is shown for the NMOSFET and the simulation results are in excellent agreement with the experimental data. These results were obtained without fitting any parameters of the transport models on device level in contrast to Ref. [9.93],

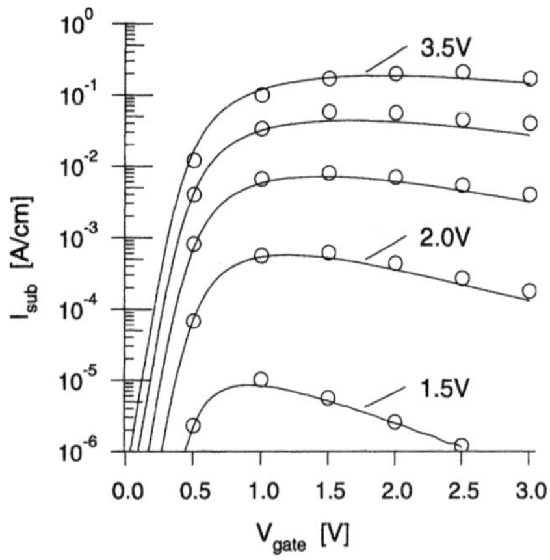

Figure 9.50: Experimental (—) and simulated (∘) substrate current of the shallow source/drain extension NMOSFET (V_{drain} = 1.5, 2.0, 2.5, 3.0, 3.5V)

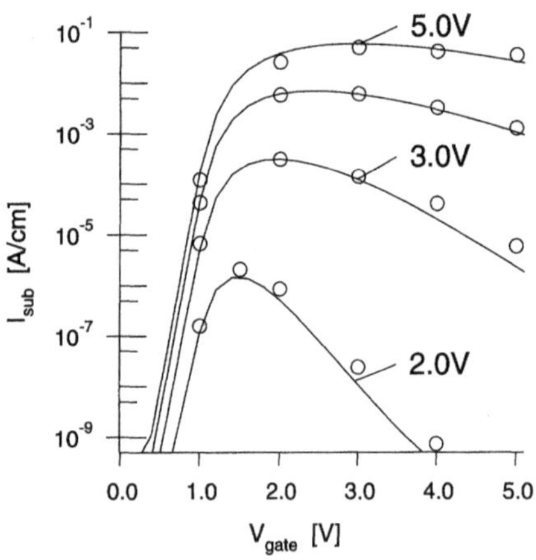

Figure 9.51: Experimental (—) and simulated (∘) substrate current of the single source/drain NMOSFET with a metallurgical channel length of 0.87μm and an oxide thickness of 20.4nm (V_{drain} = 2.0, 3.0, 4.0, 5.0V)

where similar good results were obtained after fitting one parameter of the II model to match the substrate current results. Similar excellent agreement as in Fig. 9.50 was found for many other devices fabricated with various technologies [9.90, 9.91, 9.94–9.96]. Two additional examples are shown here. In Fig. 9.51 substrate current data are shown for a device with a metallurgical channel length of 0.87μm and an oxide thickness of 20.4nm fabricated with a single source/drain technology [9.95]. The simulations reproduce the experimental results over eight orders of magnitude. In Fig. 9.52 substrate current data are shown for a device of

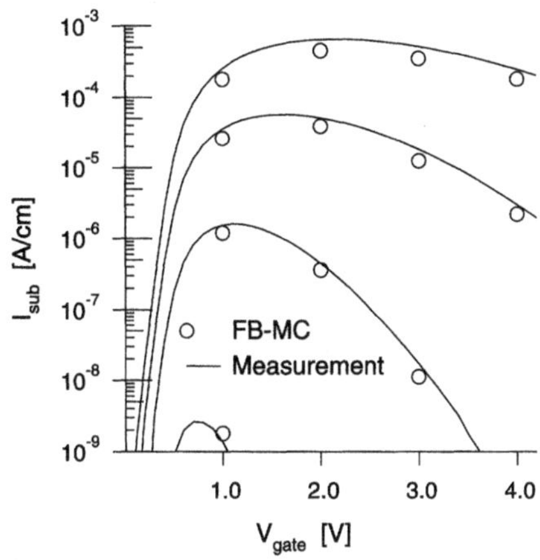

Figure 9.52: Substrate current versus gate voltage for the single source/drain NMOSFET with a metallurgical channel length of 1.8μm and an oxide thickness of 42nm (V_{drain} = 2.0, 3.0, 4.0, 5.0V)

the 256KBit DRAM generation. The metallurgical channel length is 1.8μm and the oxide thickness 42nm. In this case the doping profile was generated by process simulation [9.97]. Again, simulation and experiment are in good agreement. These excellent results for devices fabricated with three distinctly different technologies over a period of 15 years on the one hand and the good agreement with the fundamental physical experiments shown Sec. 4.3 on the other hand clearly demonstrate that the NSC MC method is not only a reliable and efficient tool for hot-electron simulations, but is also predictive, because adjustment of model parameters on device level is not necessary.

9.2.5 Surface Impact Ionization

The excellent results for MOSFETs found in Sec. 9.2.4 and the good agreement with the basic physical experiments demonstrated in Ref. [9.93] and Sec. 4.3 are in strong contrast to nearly every other work based on simpler models, like local field (LF) models [9.98, 9.99], local temperature models [9.35, 9.100–9.103], or nonlocal field models [9.36, 9.104–9.107]. While many of these simpler models work well for a certain range of device technologies, they all do not reproduce the II coefficient found under homogeneous bulk conditions which can be expressed with Chynoweth's law [9.108]

$$\alpha^{\mathrm{II}} = \alpha^{\mathrm{II}}_{\infty} \exp\left(-\frac{E_{\mathrm{C}}}{E}\right) , \tag{9.4}$$

where $\alpha^{\mathrm{II}}_{\infty}$ and the critical field E_{C} are fitting parameters. At room temperature and under bulk conditions the critical field is given by about 1.2MV/cm [9.99,

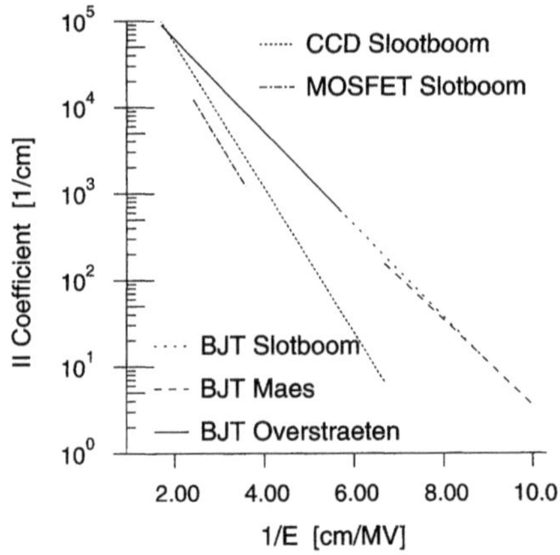

Figure 9.53: II coefficient for electrons at room temperature as determined from measurements on BJTs, MOSFETs, and CCDs (Slotboom et al. [9.99], Maes et al. [9.110], and van Overstraeten et al. [9.109])

9.109, 9.110]. In the case of the simpler models this value is increased to match measured MOSFET results, and values of the critical field up to 2.3MV/cm have been reported. This increase of the critical field results in a strong suppression of II in MOS devices. This effect is called surface II, because it seems to appear in the vicinity of the surface of the silicon bulk of a MOS device [9.99]. In Fig. 9.53 II coefficients for CCDs and MOSFETs are shown which were obtained by reverse engineering based on the LF-DD model [9.99]. Comparison with the bulk case reveals a considerable reduction of the II coefficient close to the interface. Different reasons were given for this effect and will be discussed in the following.

In Refs. [9.99, 9.104] the difference between surface and bulk II is attributed to the reduced mean free path of the carriers in the vicinity of the interface. Within the Lucky Electron concept [9.111] the critical field in Chynoweth's law (9.4) is inversely proportional to the mean free path λ

$$E_{\mathrm{C}} = \frac{\varepsilon_{\mathrm{TH}}}{q\lambda} \, , \tag{9.5}$$

where $\varepsilon_{\mathrm{TH}}$ is the threshold energy. A reduction in the mean free path, due to the enhanced scattering at the surface, results in a larger critical field and a reduced II coefficient. In the MC model the mean free path at the MOS interface is smaller due to surface roughness scattering. But it has been doubted, that this effect is responsible for the strong suppression of II in MOS devices [9.86]. In Fig. 9.54 substrate current modeling results are shown, which were calculated with and without surface roughness scattering, and the observed differences are far too small to explain the effect of surface II seen in Fig. 9.53.

To continue exploring the reasons for the different findings in Ref. [9.99] and

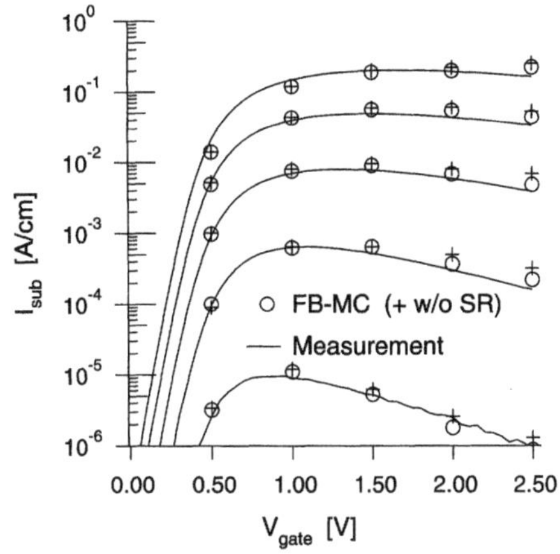

Figure 9.54: Substrate current versus gate voltage for a shallow source/drain extension NMOSFET with a metallurgical channel length of 160nm and an oxide thickness of 5.3nm ($V_{\rm drain}$ = 1.5, 2.0, 2.5, 3.0, 3.5V), MC simulations with and without surface roughness scattering

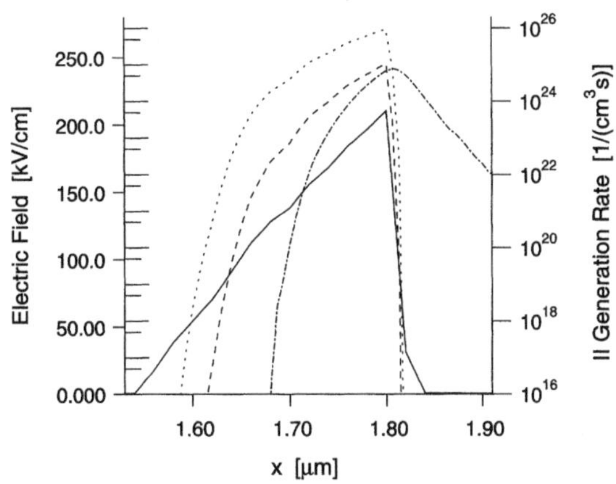

Figure 9.55: Electric field (—) and II generation rates for the case B of the CCD experiments of Ref. [9.99] calculated with the LF model based on bulk parameters (···), based on surface parameters (– – –) and MC results (– · – · –)

this work, the CCD experiment of Ref. [9.99] is reinvestigated. The lateral electric field given in Ref. [9.99] for case B of the CCD experiments and the corresponding spatial II generation rates are shown in Fig. 9.55. The generation rates were calculated with the LF model with bulk parameters, the LF model with the surface II parameters of Ref. [9.99] and the MC model. In Ref. [9.99] it is stated that the LF model can be applied to this case, because the electric field does not

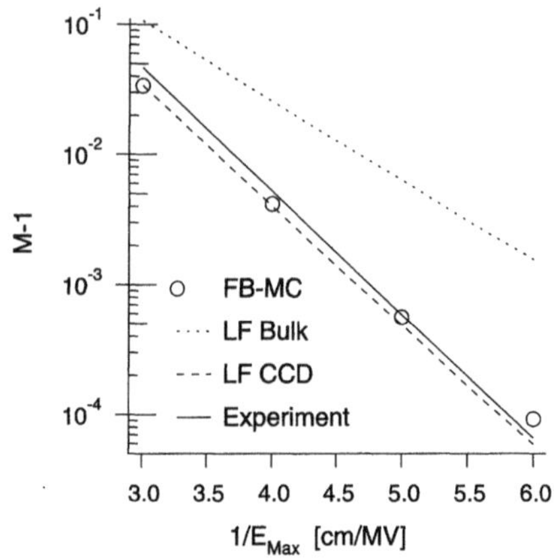

Figure 9.56: Multiplication factor for the case B of the CCD experiments [9.99]

vary much over distances comparable to one mean free path length which is in the order of 10nm. The more physics-based MC results show that this statement is not correct. If it were correct, the results of the LF model with bulk parameters and MC should be the same for the field distribution in the CCD experiment, because MC reproduces the bulk results under homogeneous conditions (see Fig. 4.6). But as is shown in Fig. 9.55 the maximum MC II generation rate is strongly reduced compared to the LF model with bulk parameters showing that the LF approximation fails for the rapidly varying electric field present in the CCD. The LF approximation is only valid, if the electric field is nearly constant over the distance, which the electrons have to cover to gain the energy necessary for II. The threshold energy of II is at least as big as the gap energy (1.1eV). To gain this energy from the electric field given in Fig. 9.55 the electrons have to travel along a distance of more than 70nm, along which the electric field strongly varies (roughly a factor of 2). Thus, the condition of a nearly constant electric field is violated and the LF approximation fails.

By reducing the II coefficient (switching from bulk to surface II) the LF model can reproduce the reduction of the II generation rate[3], but the microscopic details are still poorly represented (see Fig. 9.55). The LF model can neither reproduce the threshold characteristics of II nor the slow decay of the II generation rate in the field free space on the right-hand side.

The experimentally accessible multiplication factor can be calculated by integrating the generation rate over the device area, and in Fig. 9.56 the results are shown for the three different II models. For comparison also the experimental results of Ref. [9.99] are shown[4]. The results of the LF model relying on surface

[3] This effect has been discussed in detail for exponentially growing fields in Refs. [9.112,9.113].

[4] The assumed electric fields may not exactly reproduce the original ones of Ref. [9.99]. Nonetheless, the electric fields are similar to the ones of Ref. [9.99] and it is clearly shown

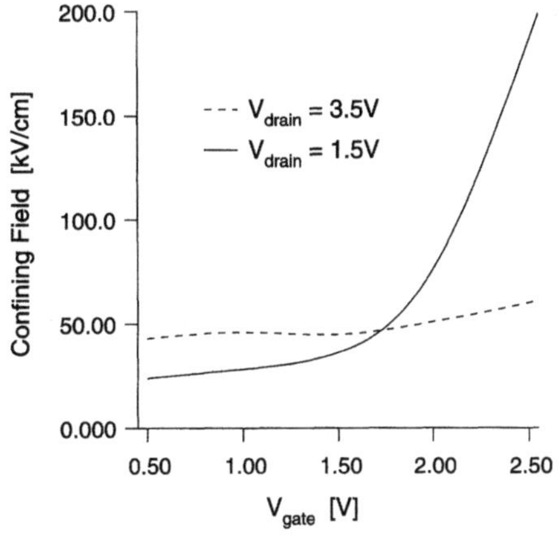

Figure 9.57: Confining electric field averaged with the II generation rate for the 160nm NMOSFET of Fig. 9.54

II parameters agree well with the ones of the MC model while the LF model with bulk parameters consistently yields higher values. This means that the suppression of the II coefficient in the case of the LF model is necessary to counterbalance the overestimation of the II generation rate caused by the failure of the LF model under inhomogeneous field conditions. The effect of surface II found in Ref. [9.99] is therefore due to the inadequate LF model and not due to a difference of physics in bulk silicon and at the surface of MOS devices.

In Ref. [9.86] a theoretical model for II in NMOSFETs is proposed which considers the quantization of the inversion layer. It is found that the II coefficient in the inversion layer decreases with an increasing electric field perpendicular to the interface. For very high electric fields perpendicular to the interface and under otherwise homogeneous conditions this II coefficient is similar to the results found in Ref. [9.99] (see Fig. 11 of Ref. [9.86]). But the situation in a MOSFET is different, because II mainly occurs behind the pinch-off point, where the electric fields are too weak for a considerable confinement of the carriers. In Fig. 9.57 the average value of the confining field, at the position where II occurs, is shown for the shallow source/drain extension NMOSFET. Only confining (positive) electric fields perpendicular to the interface ($E_\perp$) are taken into account and the confining field is defined as

$$E_{\text{conf}} = \frac{\int E_\perp(x,y)\Theta(E_\perp(x,y))G^{\text{II}}(x,y)dxy}{\int G^{\text{II}}(x,y)dxy}, \tag{9.6}$$

where G^{II} is the II generation rate. Only in the case where the gate voltage is larger than the drain voltage a substantial confining field arises. But these fields

that the LF model fails due to the nonlocal nature of the transport caused by these strongly inhomogeneous electric fields.

are still small compared to the fields necessary to reproduce surface II with the model proposed in Ref. [9.86]. Moreover, the strong dependence on the confining field implies that the effect should be negligible for small gate voltages. This is in contrast to the effect of surface II which reduces substrate current even for gate voltages near the threshold voltage. This means that the genuinely new effect proposed in Ref. [9.86] does not explain surface II as described in Ref. [9.99] and is only of minor importance regarding the problem of surface II.

Another explanation given for surface II is mechanical stress at the interface due to the formation of the oxide. The magnitude of the stress should vary with the oxide thickness and accordingly the impact on II. Since the reported MC simulations are in close agreement with the experiment for the full range of oxide thickness (5.3nm to 42.0nm) without accounting for mechanical stress, it has to be concluded that the impact of this stress on II seems to be a minor one. The same can be said for other unknown effects which might depend on the specific process technology. The investigated devices belong to distinctly different technology generations (256KBit – 1GBit) and good results are obtained irrespectively of the technology and without any fitting of model parameters on device level. It is therefore concluded that the so-called surface II is due to nonlocal transport and not to any differences in the transport physics compared to the bulk case [9.96].

9.2.6 Accuracy of the NSC Approach for Terminal Currents

In Secs. 9.2.4 and 9.2.5 substrate currents due to II are calculated based on the NSC approach, which are in good agreement with experimental results and data obtained by SC simulations. These NSC simulations were based on the electrostatic potential evaluated by HD simulations, because potentials calculated by the DD model failed to yield good results, at least in the case of strong inversion. In Refs. [9.114,9.115] NSC simulations based on electric fields evaluated by DD simulations are proposed for the calculation of drain currents of NMOSFETs without investigating the accuracy of the NSC approach for this kind of calculations. In addition, in Ref. [9.116] NSC simulations are applied to determine the drain current noise of NMOSFETs, again without checking the accuracy of the NSC approach for this type of calculation. Already in Ref. [9.117] it was shown that the NSC approach fails to capture important aspects of the nonlinear transport in MOSFETs, which have a strong impact on the drain current. The impact of the NSC approximation on the drain current and the corresponding noise is investigated in more detail in the following [9.118]. In Sec. 8.2 the CPU efficiency of the NSC approach is investigated for these types of calculations.

In order to investigate the NSC approximation for terminal currents, SC and NSC simulations of the MC and HD models were performed, where the NSC simulations are based on electric fields evaluated by SC-DD simulations as in Ref. [9.115]. The transport parameters of the DD and HD models are obtained as far as possible by MC simulations to ensure a high degree of consistency in

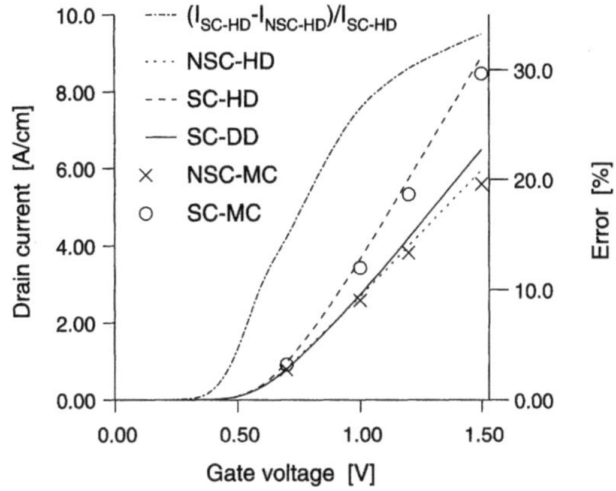

Figure 9.58: Drain current of an NMOSFET with a gate length of 80nm calculated by SC-DD, SC-HD, SC-MC, NSC-HD, and NSC-MC simulations for a drain bias of 1.5V

the simulation hierarchy (cf. Chap. 7)[5] [9.119]. The HD model is investigated, because it has the advantage of being more CPU efficient than the MC model allowing the investigation of a wider range of bias conditions and devices in less time. In addition, the HD model has a higher degree of consistency with the DD model, because it is based on the same mobility models (also in the channel region) and boundary conditions as the DD model[6]. The latter point is very important, because the boundary conditions have a much stronger impact on the results in the case of an NSC simulation than in the SC case. Therefore, comparison of SC-HD and NSC-HD simulations allows to identify the intrinsic differences of the SC and NSC methods excluding additional effects due to differences in the momentum and particle-based simulators.

In Fig. 9.58 the drain current of the NMOSFET with 80nm gate length of Sec. 9.2.1 is shown as a function of the gate voltage. In addition to the results of the momentum-based simulators, MC results are shown and good agreement is found between HD and MC results in both cases (NSC and SC), respectively. The NSC-HD (NSC-MC) current is smaller than the SC-HD (SC-MC) one and the error increases with the gate voltage. In the case of the on-state current ($V_{\text{drain}} = V_{\text{gate}} = 1.5\text{V}$) the error is larger than 30%. On the other hand, the NSC-HD and NSC-MC currents more or less reproduce the SC-DD current. This is expected, because the drain current of a MOSFET is determined by the source-side

[5] In the case of the inversion layer transport the same surface mobility and velocity models as in Sec. 9.2.1 are used here for the momentum-based models.

[6] The contact boundary conditions of the MC model are the ones of Ref. [9.120] (Sec. 6.1) similar to Ref. [9.115]. The momentum and particle based models yield the same drain current in the linear regime for MOSFETs (Sec. 9.2.1) [9.58].

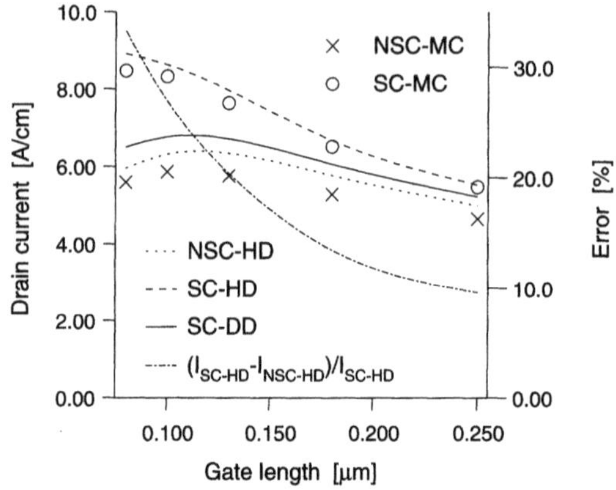

Figure 9.59: On-state drain current of NMOSFETs calculated by SC-DD, SC-HD, and NSC-HD simulations as a function of the gate length for a gate and drain bias of 1.5V

injection into the channel [9.24]. The injection condition given by the electrostatic field pattern at the source side depends on the particle density profile on the drain side via the PE. Hot-carrier effects, only captured by the MC and HD models, lead to a larger spreading of the particle density near the drain junction because of increased diffusion (Fig. 9.26) and to a reduction in the channel charge due to the velocity-overshoot compared to the DD case (Fig. 9.25) [9.58,9.106,9.121]. Although the particle density is also modified in the NSC approach, the feedback via the PE is missing and the injection condition is not modified. Therefore, the NSC drain current results deviate only slightly from the respective results of the SC simulations from which the electric field has been derived. The magnitude of the feedback effect depends on the channel charge. In the case of weak inversion the channel charge is small and a redistribution of it by hot-carrier effects has a negligible impact on the electrostatic potential and the error of the NSC method vanishes (Fig. 9.58). The feedback effect also decreases with increasing gate length. But still an error of about 10% (HD) and 15% (MC) is found for the on-state current of an NMOSFET with a gate length of 250nm (Fig. 9.59). In the case of collector currents of SiGe HBTs a similar behavior of SC and NSC simulations is found (not shown). Thus, for drain and collector current simulations the accuracy of the NSC approach is typically not better than the accuracy of the method upon which it is based. It is worth noting that the CPU time for an SC-DD simulation is less than one CPU minute, whereas the NSC-MC simulation roughly requires one CPU hour without improving the modeling accuracy. This is in contrast to NSC-MC simulations of parasitic currents, which are dominated by hot carrier effects like gate or substrate currents in MOSFETs (Sec. 9.2.4), where the NSC-MC model yields much better results than DD simula-

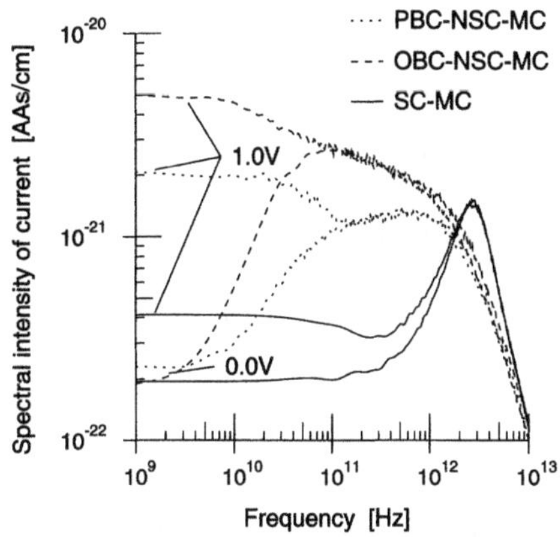

Figure 9.60: Spectral intensity of the terminal current of an N^+NN^+-resistor for a bias of 0.0 and 1.0V based on SC-MC and NSC-MC simulations with open and periodic boundary conditions (OBC/PBC)

tions [9.87, 9.93, 9.122]. The efficiency of the NSC-MC calculation of the parasitic currents can be improved, if the drain current is not evaluated, because in this case the simulation domain can be confined to the hot-carrier region of the device, whereas for the drain current the whole device domain has to be simulated.

In Refs. [9.116, 9.123] terminal current noise simulations with the NSC-MC method are reported. In Fig. 9.60 the spectral intensity of the terminal current of the N^+NN^+-resistor of Sec. 9.1 is shown calculated by SC-MC and NSC-MC simulations under equilibrium and nonequilibrium conditions, where the NSC-MC simulations are based on the SC-MC potential in order to ensure that the SC and NSC simulations yield the same average terminal current. Since in Refs. [9.116, 9.123] periodic boundary conditions for the particles were used, simulations were performed not only with the open boundary conditions (OBC) described in Sec. 6.1 but also with the periodic ones (PBC) of Ref. [9.116]. As already pointed out in Sec. 8.1 for equilibrium simulations, SC and NSC simulations show very different frequency dependences of the spectral intensity. In the case of the equilibrium simulation (zero bias) at least the important low-frequency noise is the same for both approaches. But in the case of the nonequilibrium simulation the NSC method overestimates the low-frequency noise by a factor of 5 (PBC) or 12 (OBC) for a bias of 1V.

Thus, NSC-MC simulations can be expected to yield results significantly different from SC-MC simulations for terminal currents and their noise in many cases. In addition, it does not matter, whether the BTE is solved by the MC method or a numerical method.

9.3 SiGe HBTs

In this section the properties of the momentum and particle-based transport models are investigated for SiGe HBTs, which are pseudomorphically grown on unstrained Si. The base region contains a position-dependent Ge content resulting in compressive biaxial strain which modifies the band edges. The position-dependant band edges induce an additional driving force for the electrons entering the base and block the holes from spilling into the emitter making a higher base doping possible, which improves the RF performance of the HBTs [9.19–9.22]. In the first part the stationary transport in HBTs and in the next part the electron transit times are discussed with respect to the three transport models. In the last part of this section terminal current noise is investigated for a realistic SiGe HBT.

In bipolar transistors apparent bandgap narrowing due to heavy doping cannot be neglected and is included for the HBT simulations in all three transport models (DD, HD, and MC) as described in Sec. 6.2. In order to reduce the CPU time of the MC simulations, a 1D device approximation is used in Secs. 9.3.1 and 9.3.2. In addition, the statistics are enhanced with the Multiple Refresh method in regions with low electron densities (Sec. 6.7). The missing base contact is simulated by setting the quasifermi potential of the holes in the center of the base to the value of the base voltage [9.124,9.125]. Generation/recombination processes are neglected, because they are difficult to include into an MC simulation. In the case of the MC model, no holes are simulated and their contribution to the space charge is calculated based on the DD hole quasifermi potential in conjunction with the nonlinear PE (Sec. 6.5).

In the case of noise simulations the 1D approximation is not possible, because the holes produce a significant part of the total terminal current noise, and in Sec. 9.3.3 a 2D SiGe HBT is investigated with the DD and HD models including generation/recombination processes. MC simulations are not performed for this 2D structure, because they are too CPU intensive.

9.3.1 Stationary Transport

In Fig. 9.61 the doping profile and the graded Ge profile of the investigated 1D SiGe HBT are shown [9.126]. The corresponding band edges are shown in Fig. 9.62, where the Ge content mainly modifies the valence band edge. The maximum donor concentration is limited to $4 \cdot 10^{19}/\mathrm{cm}^3$ to improve the time stability of the self-consistent MC device simulations (Sec. 6.4) [9.127]. It was checked by HD simulations that the reduction of the maximum donor concentration does not impair the results of the investigations described below. In the vicinity of the base a lowly doped emitter with a doping of $6 \cdot 10^{18}/\mathrm{cm}^3$ is included to avoid junction leakage due to tunneling and to keep the emitter/base capacitance low. The metallurgical base width is 24nm and the maximum acceptor concentration $5 \cdot 10^{19}/\mathrm{cm}^3$. Minimal collector doping is $4 \cdot 10^{17}/\mathrm{cm}^3$. The maximum Ge concentration is 20% and extends 12nm into the lowly doped collector. The real space

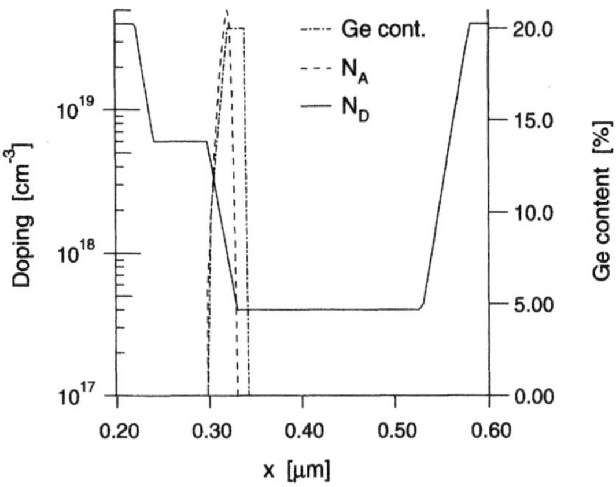

Figure 9.61: Doping profile and Ge content of the HBT

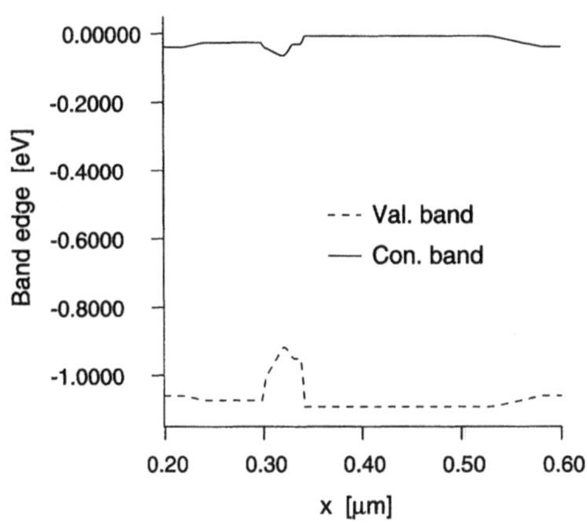

Figure 9.62: Conduction and valence band edges of the HBT

is resolved with a grid of 216 nodes where the grid spacing varies between 1 and 5nm. In the case of the MC simulation at least 50 particles are simulated in each grid cell and an additional set of 10000 particles is distributed according to the electron density resulting in a total of about 19500 particles. The PE is solved every 1fs.

The electron density and drift velocity are shown for the three simulation models in Figs. 9.63 and 9.64, respectively. Quasiballistic transport occurs in the base manifested through a strong velocity overshoot in the MC result which is well reproduced by the HD model. On the other hand, at the beginning of the lowly doped collector a spurious velocity overshoot occurs in the case of the HD model. This spurious velocity overshoot could be reduced by reducing the heat flux in

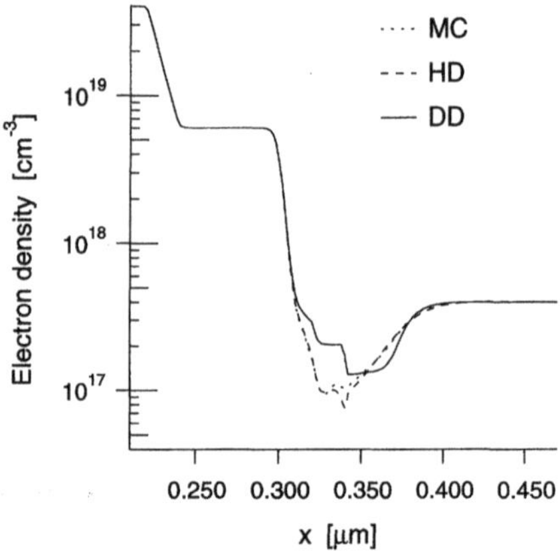

Figure 9.63: Electron density for $I_C = 2.02\text{mA}/\mu\text{m}^2$ and $V_{CE} = 0.8\text{V}$

the HD model [9.2, 9.128, 9.129], but these modifications of Fourier's law have a detrimental side effect on the noise simulations and are therefore not used in this investigation. In addition, the overall effect of this spurious velocity overshoot on other quantities shown below is rather small. The DD model on the other hand fails in the whole base region and in the space charge region of the lowly doped collector. At the end of the base the MC velocity is about twice as large as the DD one, which results due to current continuity in a much lower electron density in the case of the MC and HD models.

In Fig. 9.65 Gummel plots for the three different device models are shown. The overall agreement of the DD and HD model with the MC model is very good. This is astonishing in the case of the DD model, because the velocity of this model strongly deviates from the other two models within the base (Fig. 9.64).

9.3.2 Transit Times

In the case of a 1D device approximation, the electron transit time for a region extending from x_b to x_e reads [9.130]

$$\tau(x_b, x_e) = \int_{x_b}^{x_e} q \left. \frac{dn}{dI_C} \right|_{V_{CE}=\text{const}} dx \ , \tag{9.7}$$

where q is the electron charge, n the electron density, I_C the collector current per device area, and V_{CE} the collector/emitter voltage. The derivative is calculated by varying the base/emitter bias. The total transit time is inversely proportional to the common emitter cutoff frequency: $f_T = 1/2\pi\tau(0, L)$, where L is the device length. The transit time and cutoff frequency are calculated in the quasistationary approximation in accordance with the definition given in Ref. [9.69].

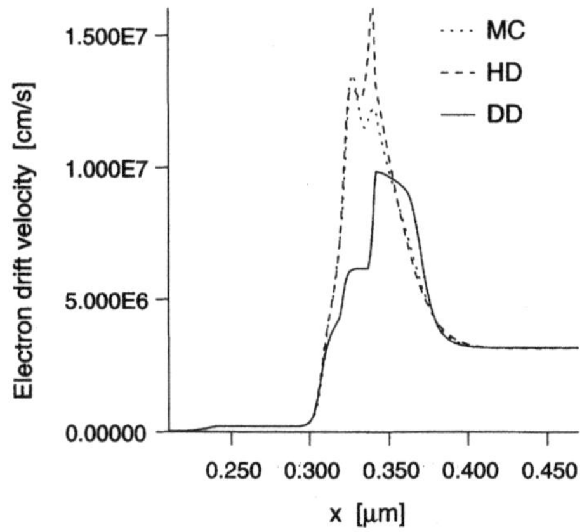

Figure 9.64: Drift velocity for $I_C = 2.02\text{mA}/\mu\text{m}^2$ and $V_{CE} = 0.8\text{V}$

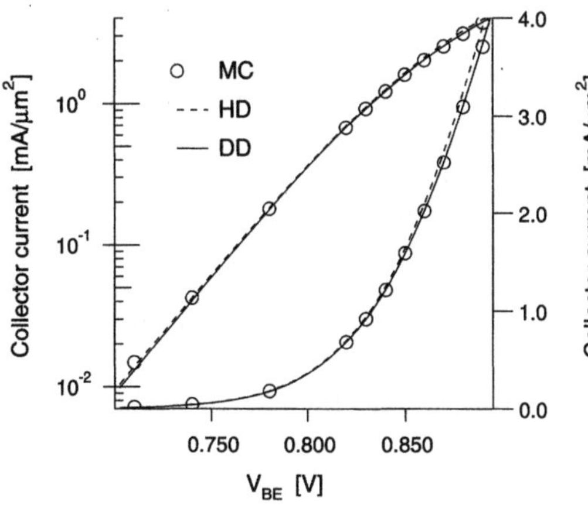

Figure 9.65: Collector current for $V_{CE} = 0.8\text{V}$

In order to avoid the difficulties of an MC small-signal analysis the derivative in Eq. (9.7) is evaluated with the finite-difference technique and in the case of the MC model a change in the base bias of 10mV is used. Due to the small change in the base/emitter voltage necessary to avoid nonlinear effects the accordant change in the output signals of the MC simulator is also small and the relative stochastic error of the output is strongly amplified by the finite differences. In order to obtain finite-difference results with an acceptable relative error, the noise level of the MC large signal quantities must be very low. To this end, MC simulations are performed until the relative error of the collector current is below 0.7% with a probability of 95.5%.

The standard procedure for optimizing the RF performance of a bipolar tran-

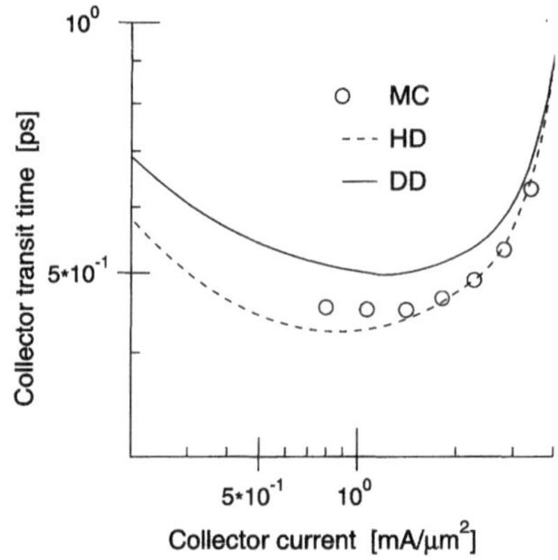

Figure 9.66: Collector transit times for $V_{\mathrm{CE}} = 0.8\mathrm{V}$

sistor starts with the evaluation of the transit times by Eq. (9.7) in the emitter, base and collector regions followed by subsequent design modifications aimed at a reduction of the dominating transit time. The transit times resulting from DD, HD, and MC are evaluated for the emitter ($x < 0.304$nm), base ($0.304\text{nm} \leq x \leq 0.328\text{nm}$), and collector ($x > 0.328$nm) regions, and are shown together with the total transit time in Figs. 9.66-9.69 as a function of the collector current. In the case of the collector transit time (Fig. 9.66) good agreement of the HD and MC model is found. The DD collector transit time is larger than the MC result because of the Ge profile overlap with the collector (Fig. 9.61) resulting in a strongly reduced saturation velocity within the overlap region compared to pure silicon. The lower velocity (Fig. 9.64) results in a larger contribution of this region to the transit time (Fig. 9.70), whereas in the Ge free region of the collector relatively good agreement of the DD model with the other two models is found.

Again, good agreement is obtained between the HD and MC model for the base transit time (Fig. 9.67). Since the DD model yields a velocity which is considerably smaller than the one of the other two models (Fig. 9.64), the base transit time of the DD model is the largest one. Consequently, the DD model indicates that the base transit time is dominant for the collector current levels where the HBT shows its peak performance, whereas this is not the case for the HD and MC models. This shows that the application of the DD model results in misleading design guidelines, at least for the optimization of this HBT.

In the case of the emitter transit time (Fig. 9.68) the DD and MC model agree well, whereas the HD model yields a transit time being lower than the DD result. Analytic modeling of the emitter transit time shows that the transit time is strongly related to the Gummel plot (Fig. 9.65) [9.70]. The collector current calculated by the HD model is larger than the corresponding DD value. Consequently, for a given collector current and a fixed emitter/collector voltage, which are the constraints in Fig. 9.68, the base/emitter voltages are different for

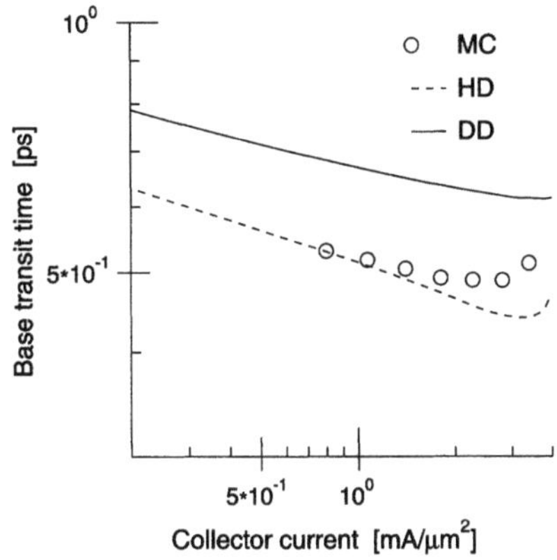

Figure 9.67: Base transit times for $V_{\mathrm{CE}} = 0.8\mathrm{V}$

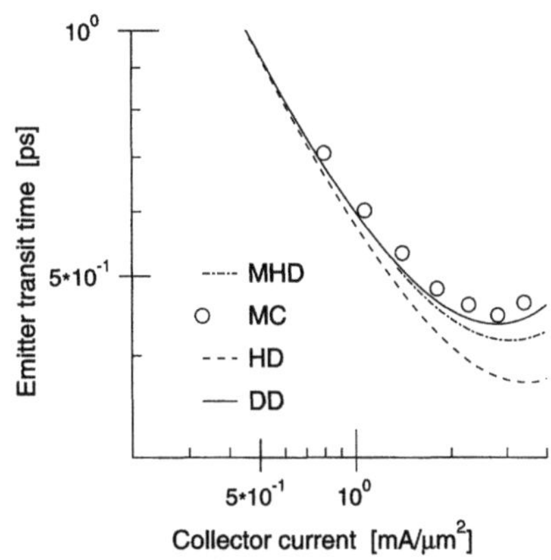

Figure 9.68: Total and emitter transit times for $V_{\mathrm{CE}} = 0.8\mathrm{V}$

the two models. Although the difference in the collector current is less than 5% it can have a strong impact on the emitter transit time. Therefore, in order to further clarify the behavior of the emitter transit time, simulations with the HD model were performed, where the collector/emitter voltage was adjusted in such a way that the collector current of the DD model was matched for the same base/emitter voltage. In Fig. 9.69 the resultant emitter transit time of these modified HD (MHD) simulations is shown and it appears that the different transit time of the HD model is predominantly due to the rather small differences in the Gummel plot (Fig. 9.65).

In contrast to the base and collector regions for which the qualitative behavior of the transit times of the different simulation models is generally valid, no generally valid conclusions can be drawn for the emitter region. In Ref. [9.129]

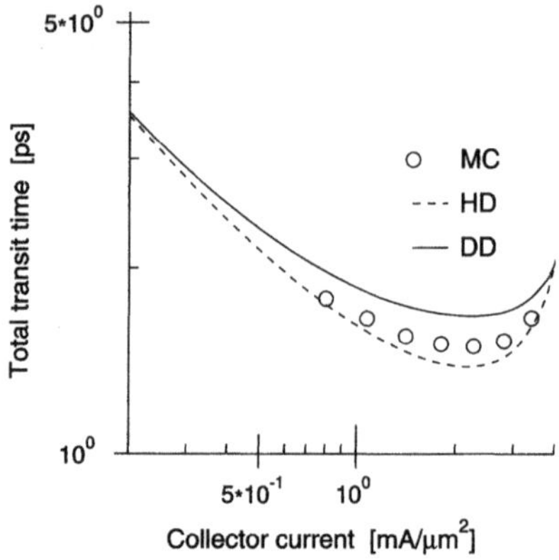

Figure 9.69: Total and emitter transit times for $V_{\mathrm{CE}} = 0.8\mathrm{V}$

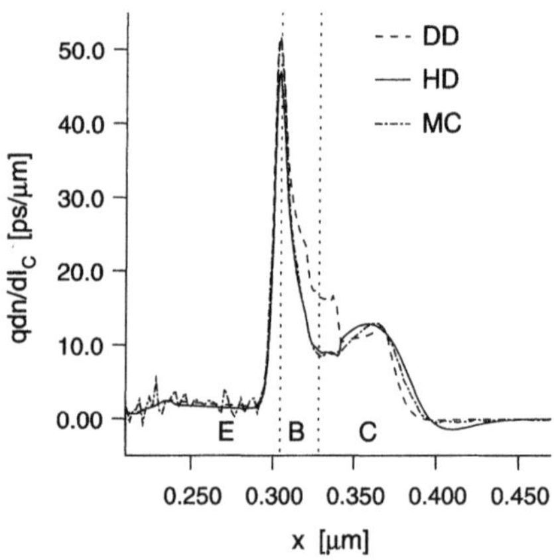

Figure 9.70: $d\tau/dx = qdn/dI_{\mathrm{C}}$ for $I_{\mathrm{C}} = 2.27\mathrm{mA}/\mu\mathrm{m}^2$ and $V_{\mathrm{CE}} = 0.8\mathrm{V}$

results are shown for a different device, where the MC model yields a much larger emitter transit time than the DD and HD model which in turn agree quite well.

The cutoff frequency is shown in Fig. 9.71 and all models peak at about the same collector current of $2.0\mathrm{mA}/\mu\mathrm{m}^2$. The MC model peaks at 106GHz and the HD model at 114GHz. The larger cutoff frequency of the HD model is mainly due to its lower emitter transit time (Fig. 9.68). As expected, the DD model yields the smallest cutoff frequency of about 95GHz due to the higher transit times in the base and in the Ge overlap region of the collector. Since the other transit times are of a similar magnitude, the impact of this error on the cutoff frequency is only about 10%.

In order to ensure that the remaining differences in the MC and HD results are not caused by the different treatment of the holes, further simulations were

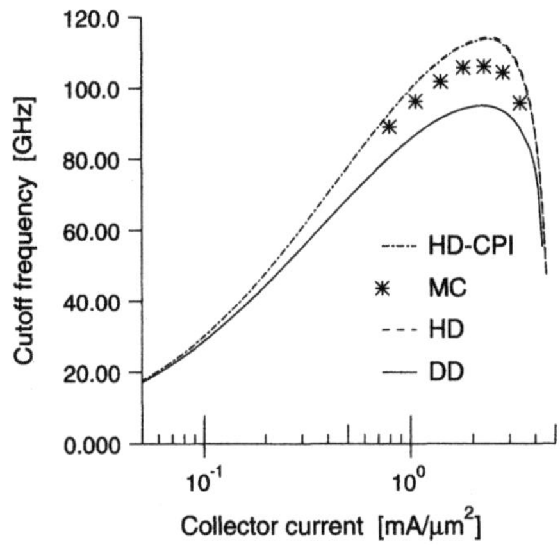

Figure 9.71: Cutoff frequency for $V_{\mathrm{CE}} = 0.8\mathrm{V}$

performed. Modified HD (HD-CPI) simulations based on the DD hole quasifermi potential (similar to the MC simulations) were performed with negligible changes in the results compared to the self-consistent HD simulations for electrons and holes validating the utilized approach (cf. Fig. 9.71).

9.3.3 Noise

In the case of noise the hole motion in the base perpendicular to the electron current significantly contributes to the total terminal current noise [9.131]. Therefore, the 1D approximation is not feasible for noise simulations and 2D simulations have to be performed. In Fig. 9.72 the 2D structure of a realistic SiGe NPN HBT is shown [9.132], which is discretized with a grid of 194×25 nodes with a nonuniform spacing in the range from 1 to 50nm. Since the structure is symmetric with respect to $y = 0$, only the part shown in Fig. 9.72 is simulated. The emitter width of this structure is 0.5μm. The doping and germanium profiles are shown in Fig. 9.73 and the device has a base length of about 30nm with a maximum germanium content of 20% and doping of $8 \cdot 10^{19}/\mathrm{cm}^3$. The Gummel plots evaluated by the DD and HD transport models including Shockley-Read-Hall recombination [9.133] are shown in Fig. 9.74, where the reduction of the current gain at small base/emitter voltages due to Shockley-Read-Hall recombination is clearly visible [9.22, 9.70]. The peak cutoff frequency f_{T} of this device for a collector/emitter bias of 2V is 45GHz at a collector current of 3.9A/cm and similar to the measured value [9.132, 9.134] (Fig. 9.75).

In Fig. 9.76 the low-frequency spectral intensities of the base/base ($S_{I_{\mathrm{B}}I_{\mathrm{B}}}$), base/collector ($S_{I_{\mathrm{B}}I_{\mathrm{C}}}$), and collector/collector ($S_{I_{\mathrm{C}}I_{\mathrm{C}}}$) current fluctuations are shown for the SiGe HBT calculated with the DD and HD noise models of Chap. 7 including noise due to Shockley-Read-Hall recombination [9.135]. The CPU time

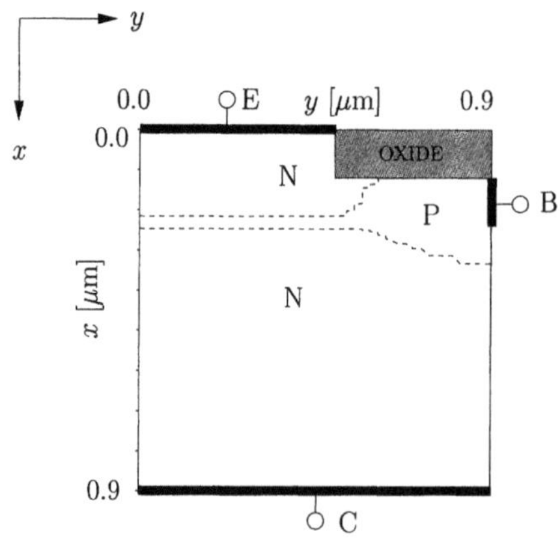

Figure 9.72: SiGe NPN HBT structure [9.132]

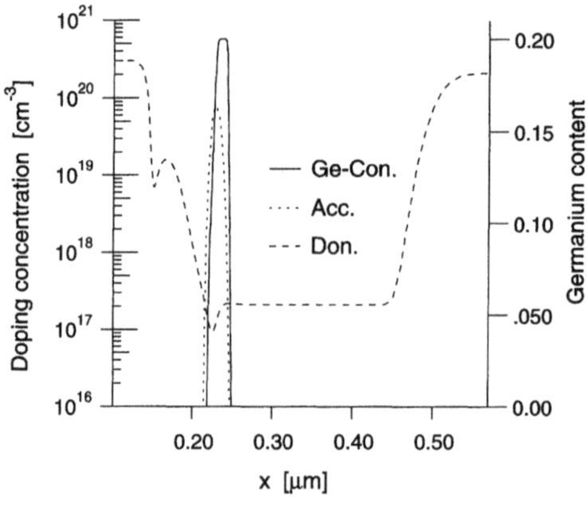

Figure 9.73: Doping and germanium profile of the SiGe HBT

per bias point is 18 CPU seconds for the DD model and 80 for the HD model on a 1GHz PC.

For the base current fluctuations both models yield similar results. In the case of the collector current fluctuations below high-injection the DD model yields more noise than the HD model, which is also reflected in the base/collector correlation. For small collector currents the collector current noise evaluated by the HD model is shot-like ($\approx 2qI_C$), whereas the DD model overestimates the noise by about a factor of four (Fig. 9.76). In order to determine the reason for the failure of the DD noise model, additional simulations have been performed. Simulations with coarser and finer grids have shown that this effect is not related to the grid spacing. Since in Ref. [9.30] results for a BJT with 1μm base length are

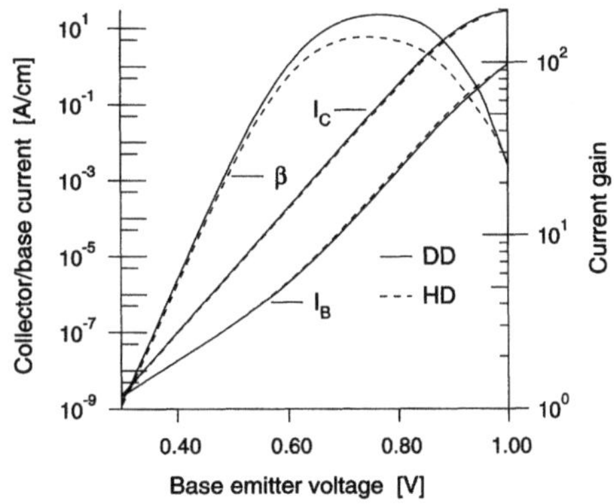

Figure 9.74: Collector (I_C) and base (I_B) currents by the DD and HD models for the SiGe HBT at room temperature and $V_\mathrm{CE} = 2\mathrm{V}$ and the corresponding current gain (β)

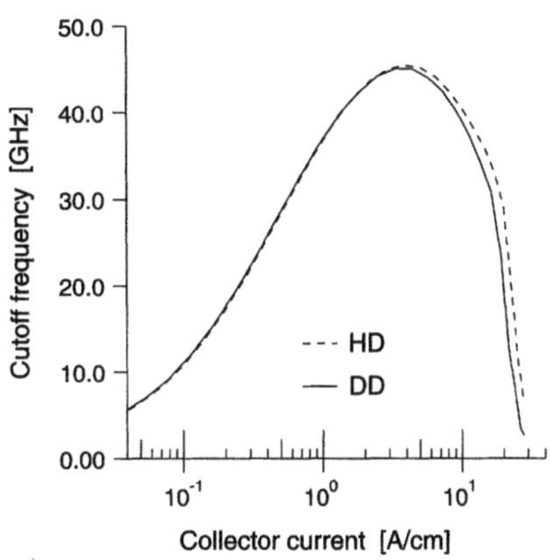

Figure 9.75: Cutoff frequency by the DD and HD models for the SiGe HBT at room temperature and $V_\mathrm{CE} = 2\mathrm{V}$

shown, where the noise calculated with the DD model is shot-like, simulations with the DD model of Chap. 7 have performed for a similar BJT. In this case the DD model reproduces the shot noise results for low collector currents. The deviation from the shot noise relation in the case of the small base length is due to the large driving forces within the base, which lead to velocity saturation in the case of the DD model. In order to check this assumption, simulations were performed for the BJT with 1μm base length, where the saturation velocity of the electrons was reduced without changing the low-field mobility. For saturation

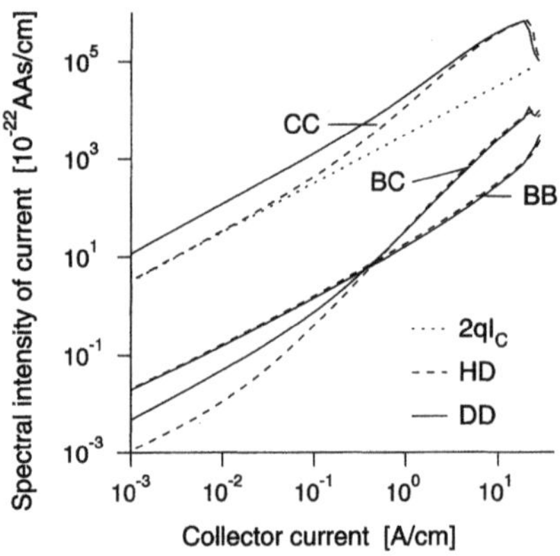

Figure 9.76: Spectral intensity of the fluctuations of the base/base (BB), base/collector (BC), and collector/collector (CC) currents by DD and HD simulations for the SiGe HBT at room temperature, $V_{\mathrm{CE}} = 2\mathrm{V}$, and 0Hz

velocities reduced to such a level that significant velocity saturation occurs at the beginning of the base, these simulations also overestimated the noise compared to the shot noise formula. Thus, this artefact of the DD model seems to be due to the failure of the DD approximation under quasiballistic transport conditions, which are encountered in modern SiGe HBTs (cf. Sec. 9.3.1 and 9.3.2). Therefore, only the HD model is used in the remaining part of this section.

For small collector currents ($< 0.1\mathrm{A/cm}$) the low-frequency collector current noise of the HD model is shot-like, because for this bias range the noise is mainly due to electron scattering (Fig. 9.77). Even beyond 0.1A/cm the electron scattering contribution to the collector current noise stays shot-like until the electron density becomes comparable to the hole density in the base and the shot noise is suppressed by the interaction with the mobile space charge. For collector currents larger than 0.1A/cm the shot noise formula underestimates the total collector current noise due to the increasing contribution of hole scattering to the noise (Fig. 9.77). This noise originates from the base region, where it has a nontrivial 2D distribution (Fig. 9.78). It can be described by thermal noise due to an equivalent base noise resistance. The corresponding analytical expression for the total collector current noise reads

$$S_{I_\mathrm{C} I_\mathrm{C}} = 2q|I_\mathrm{C}| + 4k_\mathrm{B}T_0 R_{\mathrm{B},n}|Y_{\mathrm{C,B}}(\omega)|^2, \tag{9.8}$$

where the equivalent noise resistance of the base $R_{\mathrm{B},n}$ is bias and frequency independent and not necessarily equal to the base resistance as in Ref. [9.136]. Collector current noise caused by Shockley-Read-Hall recombination is found to be negligible for the given bias range (Fig. 9.77).

In Fig. 9.79 the contributions of the different noise sources to the base current

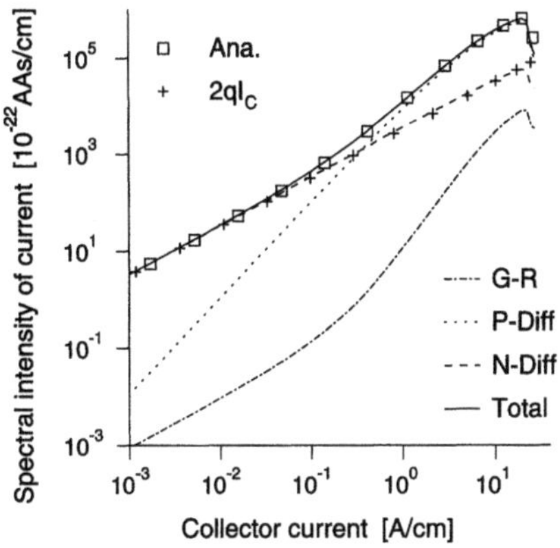

Figure 9.77: Spectral intensity of the collector current fluctuations due to electron scattering (N-Diff), hole scattering (P-Diff), generation/recombination noise (G-R), the total noise (Total), and shot noise ($2qI_{\mathrm{C}}$) and the analytical result (9.8) (Ana.) for the SiGe HBT at room temperature, $V_{\mathrm{CE}} = 2\mathrm{V}$, and 0Hz evaluated by HD simulations

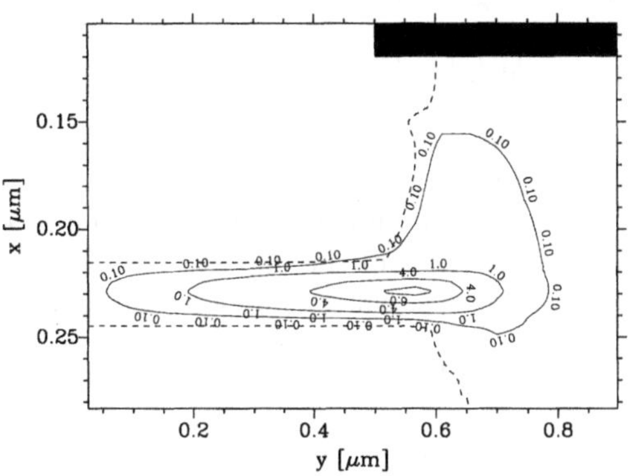

Figure 9.78: Origin of the contribution of hole scattering to the spectral intensity of the collector current fluctuations for the SiGe HBT at room temperature, $V_{\mathrm{BE}} = 0.7\mathrm{V}$, $V_{\mathrm{CE}} = 2\mathrm{V}$, and 0Hz in arbitrary units

noise are shown. The most important noise sources are due to hole scattering and Shockley-Read-Hall recombination, where the impact of the latter noise source vanishes for large collector currents. The total low-frequency base noise is reproduced by the shot noise formula with reasonable accuracy. On the other hand, in the GHz range the base noise is dominated by thermal noise [9.137] and the total

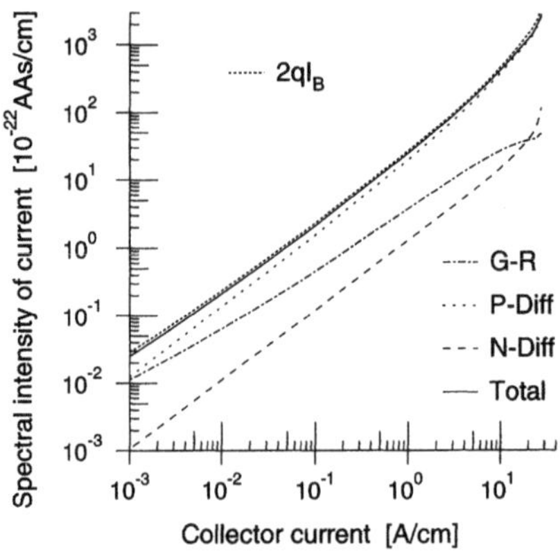

Figure 9.79: Spectral intensity of the base current fluctuations due to electron scattering (N-Diff), hole scattering (P-Diff), generation/recombination noise (G-R), shot noise ($2qI_{\mathrm{B}}$), and the total noise (Total) for the SiGe HBT at room temperature, $V_{\mathrm{CE}} = 2\mathrm{V}$, and 0Hz evaluated by HD simulations

base noise is approximately given by

$$S_{I_{\mathrm{B}}I_{\mathrm{B}}} = 2q|I_{\mathrm{B}}| + 4k_{\mathrm{B}}T_0\Re\left\{Y_{\mathrm{B,B}}(\omega) - Y_{\mathrm{B,B}}(0)\right\} , \tag{9.9}$$

where the correlation between the shot and thermal noise is accounted for by subtracting from the base self-admittance its zero-frequency value [9.30, 9.138]. Good agreement of the analytical and HD results is found in Fig. 9.80.

The thermal noise of the holes in the base leads to a correlation of the base and collector current fluctuations, which is approximately given by

$$\underline{S}_{I_{\mathrm{C}}I_{\mathrm{B}}} = 4k_{\mathrm{B}}T_0R_{\mathrm{B},n}Y_{\mathrm{C,B}}(\omega)Y^*_{\mathrm{B,B}}(\omega) , \tag{9.10}$$

where the much smaller correlation due to electrons has been neglected. In Fig. 9.81 the results of the analytical and HD model are shown and reasonable agreement is obtained.

The impact of II on the terminal current noise is demonstrated in Fig. 9.82 for the HBT. The noise is evaluated with the HD model for a constant base/emitter bias of 0.85V as a function of the collector/emitter voltage, where the local temperature II model reproduces the II coefficients for BJTs given in Refs. [9.99, 9.110]. Due to II the base current reverses its sign at a collector/emitter voltage of 2.7V. The terminal current noise strongly increases with the collector/emitter bias, and neither the collector nor the base current noise are anymore shot-like at zero frequency.

The noise results for the SiGe HBT obtained with the momentum-based simulators (Fig. 9.80) are in strong contrast to the MC results presented in

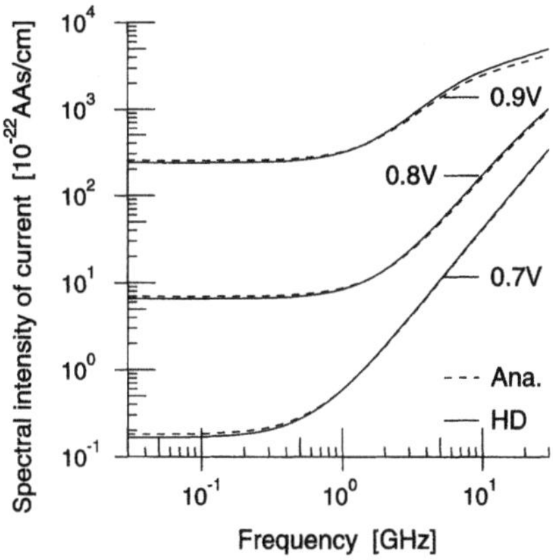

Figure 9.80: Spectral intensity of the fluctuations for the base by HD simulations and the analytical expression (9.9) (Ana.) for the SiGe HBT at room temperature and $V_{\mathrm{CE}} = 2\mathrm{V}$ for three base voltages $V_{\mathrm{BE}} = 0.7, 0.8, 0.9\mathrm{V}$

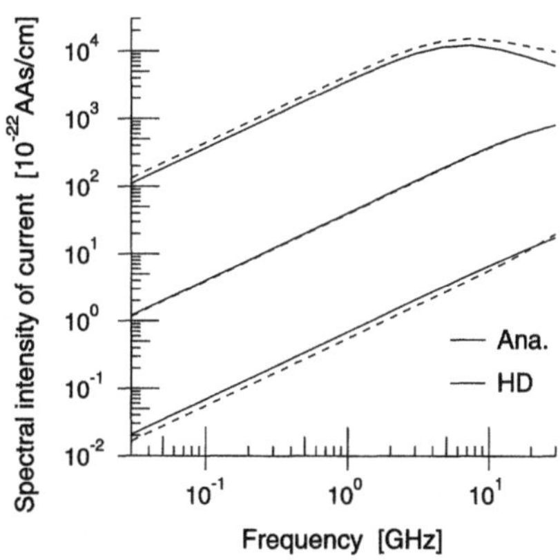

Figure 9.81: Negative imaginary part of the cross power of the fluctuations for the collector and base by HD simulations and the analytical expression (9.10) (Ana.) for the SiGe HBT at room temperature and $V_{\mathrm{CE}} = 2\mathrm{V}$ for three base voltages $V_{\mathrm{BE}} = 0.7, 0.8, 0.9\mathrm{V}$

Refs. [9.139–9.141], where it was found that the current noise of an HBT is white below 150GHz. In order to understand this discrepancy the HBT of Ref. [9.140] was simulated with the HD model, and in Fig. 9.83 the terminal current noise for the collector and base is shown for a frequency of 0 and 150GHz. In the case of the base current noise the results for 0 and 150GHz differ up to three orders of magnitude. Thus, according to the HD model the base current noise is not white

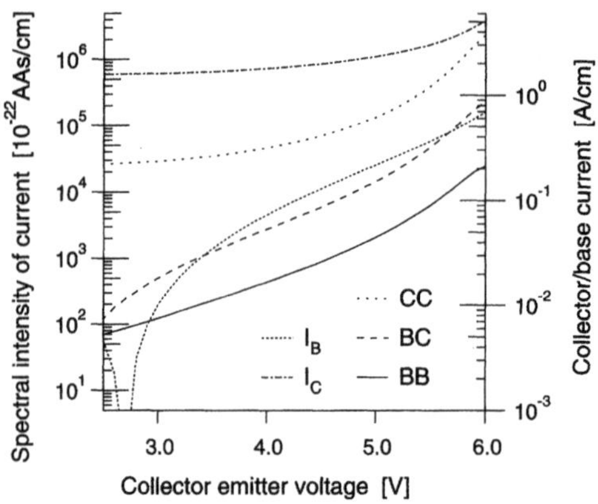

Figure 9.82: Spectral intensity of the fluctuations of the base/base (BB), and collector/collector (CC) currents by HD simulations for the SiGe HBT at room temperature, $V_{BE} = 0.85V$, and 0Hz, and the absolute value of the base/collector current correlation (BC), the collector current (I_C), and the absolute value of the base current (I_B)

below 150GHz. This can be explained based on Eq. (9.9). In the high frequency range the small-signal equivalent circuit for $Y_{B,B}$ is approximately given by the base resistor in series with the base/emitter junction capacitor [9.142]. At medium frequencies ($\approx$ 1GHz) $Y_{B,B}$ is small, because the admittance of the capacitor being proportional to the frequency is small, whereas at high frequencies ($\approx$ 100GHz) the admittance is very high and the total admittance is dominated by the base resistance. Therefore, $Y_{B,B}$ strongly depends on the frequency[7]. With respect to noise this means that although the local noise sources are white, the base current noise decreases with decreasing frequency, because the local fluctuations induce smaller and smaller base current fluctuations due to the increasing impedance of the base/emitter junction.

The strong dependence of the noise on the frequency in the GHz range (Figs. 9.83 and 9.80) indicates that the correlation function of the base current fluctuations needs to be accurately evaluated for time differences of up to 100ps depending on the base/emitter bias. This is similar to the case shown in Fig. 9.16 with the difference that the tail of the base fluctuations is negative. In addition to the HD results, the low-frequency MC results of [9.140] are shown in Fig. 9.83 and they are similar to the HD results for 150GHz and not to the ones for 0Hz.

[7] The Y-parameter and the spectral intensity of the corresponding terminal current fluctuations can be calculated with the same Green's functions [9.26]. Therefore, the Y-parameter and the spectral intensity can be expected to exhibit a somewhat similar dependence on the frequency, and already the small-signal parameters of a simple pn-junction can show a rather complex frequency behavior [9.143].

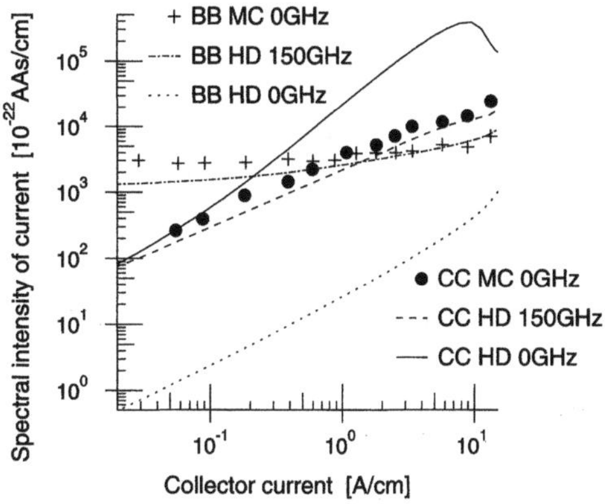

Figure 9.83: Spectral intensity of the fluctuations of the base (BB) and collector (CC) currents by the HD model for 0GHz and 150GHz for the SiGe HBT of Ref. [9.140] at room temperature and $V_{\mathrm{CB}} = 1\mathrm{V}$ and low-frequency MC results of Ref. [9.140] (MC)

A probable reason for this discrepancy might be that in Refs. [9.139–9.141] the tails of the correlation functions beyond about 1ps were missed resulting in white noise below 150GHz. Moreover, in Refs. [9.139–9.141] the low-frequency base current noise exceeds the collector current noise at low injection. This is not only in contrast to the results shown here but also in contrast to results by other groups [9.28, 9.30, 9.137, 9.144], where the compact models were shown to yield good agreement with experimental data [9.137, 9.144].

The problems with the tails of the correlation functions discussed above and in Sec. 9.1 are the reason why no attempt was made to verify the momentum-based models for 2D HBTs by MC device simulations. Simulation of the terminal current correlation functions including the tails requires a prohibitive amount of CPU time not only because of the large decay times but also because of the very small absolute values of the tails, since an improvement by a factor of 10 in the accuracy of an MC simulation requires a 100 times larger simulation time (cf. Sec. 3.5).

References

[9.1] M. Nekovee, B. J. Geurts, H. M. J. Boots, and M. F. H. Schuurmans, "Failure of extended-moment-equation approaches to describe ballistic transport in submicrometer structures", *Phys. Rev. B*, vol. 45, no. 12, pp. 6643–6651, 1992.

[9.2] B. Meinerzhagen, R. Thoma, H. J. Peifer, and W. L. Engl, "On the consistency of the hydrodynamic and the Monte Carlo models", in *Proc. IWCE*, Illinois, 1992, Urbana–Champaign, pp. 7–12.

[9.3] C. L. Gardner, "Hydrodynamic and Monte Carlo simulation of an electron shock wave in a 1-μm n^+-n-n^+ diode", *IEEE Trans. Electron Devices*, vol. 40, no. 2, pp. 455–457, 1993.

[9.4] L. Reggiani, P. Shiktorov, V. Gružinskis, E. Starikov, L. Varani, T. González, M. J. Martín, and D. Pardo, "Hydrodynamic modeling of transport and noise spectra in n^+n^{++} semiconductor structures", in *Proc. ESSDERC*, Bologna, 1996, vol. 26, pp. 295–298.

[9.5] B. Fischer, "A full-band Monte Carlo charge transport model for nanoscale silicon devices including strain", Doctor thesis, University Hannover, Hannover, 1999.

[9.6] T. W. Tang and H. Gan, "Two formulations of semiconductor transport equations based on spherical harmonic expansion of the Boltzmann transport equation", *IEEE Trans. Electron Devices*, vol. 47, no. 9, pp. 1726–1732, 2000.

[9.7] U. König, "Electronic Si/SiGe devices: Basics, technology, performance", in *Advances in Solid State Physics (Festkörperprobleme)*, vol. 32, pp. 199–220. Vieweg, Braunschweig, 1992.

[9.8] F. Schäffler, "High–mobility Si and Ge structures", *Semicond. Sci. Technol.*, vol. 12, pp. 1515–1549, 1997.

[9.9] J. Welser, J. L. Hoyt, S. Takagi, and J. F. Gibbons, "Strain dependence of the Performance Enhancement in Strained-Si n-MOSFETs", in *IEDM Tech. Dig.*, 1994, pp. 373–376.

[9.10] K. Rim, J. L. Hoyt, and J. F. Gibbons, "Fabrication and analysis of deep submicron strained-Si N-MOSFET's", *IEEE Trans. Electron Devices*, vol. 47, no. 7, pp. 1406–1415, 2000.

[9.11] K. Rim, S. Koester, M. Hargrove, J. Chu, P. M. Mooney, J. Ott, T. Kanarsky, P. Ronsheim, M. Ieong, A. Grill, and H.-S. P. Wong, "Strained Si NMOSFETs for high performance CMOS technology", *Symposium on VLSI Technology Digest of Technical Papers*, pp. 59–60, 2001.

[9.12] J. L. Hoyt, H. M. Nayfeh, S. Eguchi, I. Aberg, G. Xia, T. Drake, E. A. Fitzgerald, and D. A. Antoniadis, "Strained silicon MOSFET technology", *IEDM Tech. Dig.*, pp. 23–26, 2002.

[9.13] S. Thompson, N. Anand, M. Armstrong, C. Auth, B. Arcot, M. Alavi, P. Bai, J. Bielefeld, R. Bigwood, J. Brandenburg, M. Buehler, S. Cea, V. Chikarmane, C. Choi, R. Frankovic, T. Ghani, G. Glass, W. Han, T. Hoffmann, M. Hussein, P. Jacob A. Jain, C. Jan, S. Joshi, C. Kenyon, J. Klaus, S. Klopic, J. Luce, Z. Ma, B. Mcintyre, K. Mistry, A. Murthy, P. Nguyen, H. Pearson, T. Sandford, R. Schweinfurth, R. Shaheed, S. Sivakumar, M. Taylor, B. Tufts, C. Wallace, P. Wang, C. Weber, and M. Bohr, "A 90nm logic technology featuring 50nm strained silicon channel transistors, 7 layers of Cu interconnects, low k ILD, and 1um^2 SRAM cell", *IEDM Tech. Dig.*, pp. 61–64, 2002.

[9.14] S. Verdonckt-Vandebroek, E. F. Crabbé, B. S. Meyerson, D. L. Harame, P. J. Restle, J. M. C. Stork, and J. B. Johnson, "SiGe-channel heterojunction p-MOSFET's", *IEEE Trans. Electron Devices*, vol. 41, no. 1, pp. 90–101, 1994.

[9.15] L. Risch, H. Fischer, F. Hofmann, F. Schäfer, M. Eller, and T. Aeugle, "Fabrication and electrical characterization of Si/SiGe p–channel MOSFETs with a delta doped boron layer", in *Proc. ESSDERC*, Bologna, 1996, vol. 26, pp. 465–468.

[9.16] P. Bouillon, T. Skotnicki, S. Bodnar, C. Morin, J.-L. Regolini, P. Gouagout, and P. Dollfus, "Experiments with 0.18 μm SiGe channel pMOSFETs and p^+ poly–SiGe gate", in *Proc. ESSDERC*, Bologna, 1996, vol. 26, pp. 473–476.

[9.17] S. Okhonin, M. A. Py, B. Georgescu, H. Fischer, and L. Risch, "Dc and low-frequency noise characteristics of SiGe p-channel FET's designed for 0.13-μm technology", *IEEE Trans. Electron Devices*, vol. 46, no. 7, pp. 1514–1517, 1999.

[9.18] A. Asai, J. Sato-Iwanaga, A. Inoue, Y. Hara, Y. Kanzawa, H. Sorada, T. Kawashima, T. Ohnishi, T. Takagi, and M. Kubo, "Low-frequency noise characteristics in sige channel heterostructure dynamic threshold pMOSFET (HDTMOS)", *IEDM Tech. Dig.*, pp. 35–38, 2002.

[9.19] S. S. Iyer, G. L. Patton, J. M. C. Stork, B. S. Meyerson, and D. L. Harame, "Heterojunction bipolar transistors using Si-Ge alloys", *IEEE Trans. Electron Devices*, vol. 36, no. 10, pp. 2043–2064, 1989.

[9.20] J. D. Cressler, "Re-engineering silicon: Si-Ge heterojunction bipolar technology", *IEEE Spectrum*, vol. 3, pp. 49–55, 1995.

[9.21] A. Schüppen, "SiGe–HBTs for mobile communication", *Solid–State Electron.*, vol. 43, pp. 1373–1381, 1999.

[9.22] C. K. Maiti and G. A. Armstrong, *Applications of Silicon–Germanium Heterostructure Devices*, Series in Optics and Optoelectronics. Institute of Physics Publishing, Bristol, Philadelphia, 2001.

[9.23] K. A. Hennacy and Neil Goldsman, "A generalized Legendre polynimial/sparse matrix approach for determining the distribution function in non-polar semiconductors", *Solid–State Electron.*, vol. 36, pp. 869–877, 1993.

[9.24] M. Lundstrom, "Elementary scattering theory of the Si MOSFET", *IEEE Electron Device Lett.*, vol. 18, no. 7, pp. 361–363, 1997.

[9.25] J.-P. Nougier, "Fluctuations and noise of hot carriers in semiconducor materials and devices", *IEEE Trans. Electron Devices*, vol. ED-41, no. 11, pp. 2034–2049, 1994.

[9.26] Sh. Kogan, *Electronic Noise and Fluctuations in Solids*, Cambridge University Press, Cambridge, New York, Melbourne, 1996.

[9.27] C. Jungemann, B. Neinhüs, and B. Meinerzhagen, "Hierarchical 2–D DD and HD noise simulations of Si and SiGe devices: Part I—Theory", *IEEE Trans. Electron Devices*, vol. 49, no. 7, pp. 1250–1257, 2002.

[9.28] F. Bonani, G. Ghione, M. R. Pinto, and R. K. Smith, "An efficient approach to noise analysis through multidimentional physics-based models", *IEEE Trans. Electron Devices*, vol. 45, no. 1, pp. 261–269, 1998.

[9.29] J.-S. Goo, C. Choi, F. Danneville, E. Morifuji, H. Sasaki Momose, Z. Yu, H. Iwai, T. H. Lee, and R. W. Dutton, "An accurate and efficient high frequency noise simulation technique for deep submicron MOSFETs", *IEEE Trans. Electron Devices*, vol. 47, no. 12, pp. 2410–2419, 2000.

[9.30] F. Bonani and G. Ghione, *Noise in Semiconductor Devices, Modeling and Simulation*, Advanced Microelectronics. Springer, Berlin, Heidelberg, New York, 2001.

[9.31] W. Shockley, John A. Copeland, and R. P. James, "The impedance field method of noise calculation in active semiconductor devices", in *Quantum theory of atoms, molecules and solid state*, P. O. Lowdin, Ed., pp. 537–563. Academic Press, 1966.

[9.32] E. V. Sukhorukov and D. Loss, "Noise in multiterminal diffuse conductors:

Universality, nonlocality, and exchange effects", *Phys. Rev. B*, vol. 59, pp. 13054–13066, 1999.

[9.33] L. Varani, T. Kuhn, L. Reggiani, and Y. Perles, "Current and number fluctuations in submicron n^+nn^+ structures", *Solid-State Electron.*, vol. 36, pp. 251–261, 1993.

[9.34] S. Selberherr, *Analysis and Simulation of Semiconductor Devices*, Springer, Wien, 1984.

[9.35] R. K. Mains, G. I. Haddad, and P. A. Blakey, "Simulation of GaAs IMPATT diodes including energy and velocity transport equations", *IEEE Trans. Electron Devices*, vol. 30, no. 10, pp. 1327–1337, 1983.

[9.36] H. J. Peifer, B. Meinerzhagen, R. Thoma, and W. L. Engl, "Evaluation of impact ionization modeling in the framework of hydrodynamic equations", in *IEDM Tech. Dig.*, 1991, pp. 131–134.

[9.37] A. Cappy, F. Danneville, and G Dambrine, "Noise modelling in linear and nonlinear devices", *IEICE Trans. on Electronics*, vol. E82–C, no. 6, pp. 900–907, 1999.

[9.38] T. Ando, A. B. Fowler, and F. Stern, "Electronic properties of two-dimensional systems", *Rev. Mod. Phys.*, vol. 54, pp. 437–672, 1982.

[9.39] W. V. Roosbroeck, "Theory of the flow of electrons and holes in germanium and other semiconductors", *Bell System Technical Journal*, pp. 561–607, 1950.

[9.40] W. Shockley, *Electrons and Holes in Semiconductors*, van Nostrand, Princeton, New Jersey, 1950.

[9.41] W. L. Engl, H. K. Dirks, and B. Meinerzhagen, "Device modeling", *Proc. IEEE*, vol. 71, pp. 10–33, 1983.

[9.42] P. A. Blakey and K. Joardar, "An analytic theory of the impact of velocity overshoot on the drain characteristics of field-effect transistors", *IEEE Trans. Electron Devices*, vol. 39, no. 3, pp. 740–742, 1992.

[9.43] S. Y. Chou, D. A. Antoniadis, and H. I. Smith, "Observation of electron velocity overshoot in sub-100-nm-channel MOSFET's in silicon", *IEEE Electron Device Lett.*, vol. 6, no. 12, pp. 665–667, 1985.

[9.44] G. A. S. Halasz, M. R. Wordeman, D. P. Kern, S. Rishton, and E. Ganin, "High transconductance and velocity overshoot in NMOS devices at the 0.1-μm gate-length level", *IEEE Electron Device Lett.*, vol. 9, no. 9, pp. 464–466, 1988.

[9.45] F. Assaderaghi, P. K. Ko, and C. Hu, "Observation of Velocity Overshoot in Silicon Inversion Layers", *IEEE Electron Device Lett.*, vol. 14, no. 10, pp. 484–486, 1993.

[9.46] F. Assaderaghi, D. Sinitsky, J. Bokor, P. K. Ko, H. Gaw, and C. Hu, "High-field transport of inversion-layer electrons and holes including velocity overshoot", *IEEE Trans. Electron Devices*, vol. 44, no. 4, pp. 664–671, 1997.

[9.47] B. Cheng, V. R. Rao, and J. C. S. Woo, "Exploration of velocity overshoot in a high–performance deep sub–0.1–μm SOI MOSFET with asymmmetric channel profile", *IEEE Electron Device Lett.*, vol. 20, no. 10, pp. 538–540, 1999.

[9.48] R. K. Cook and J. Frey, "An efficient technique for two-dimensional simulation of velocity overshoot effects in Si and GaAs devices", *COMPEL Int. J. Comput. Math. Electr. Electron. Eng.*, vol. 1, pp. 65–87, 1982.

[9.49] G. Baccarani and M. R. Wordeman, "An investigation of steady-state velocity overshoot in silicon", *Solid-State Electron.*, vol. 28, pp. 407–416, 1985.

[9.50] T. Kobayashi and K. Saito, "Two-dimentional analysis of velocity overshoot effects in ultrashort-channel Si MOSFET's", *IEEE Trans. Electron Devices*, vol. 32, no. 4, pp. 788–792, 1985.

[9.51] T. W. Tang, H. H. Ou, and D. H. Navon, "Prediction of velocity overshoot by a nonlocal hot-carrier transport model", 1985.

[9.52] B. Meinerzhagen and W. L. Engl, "The influence of the thermal equilibrium approximation on the accuracy of classical two-dimensional numerical modeling of silicon submicrometer MOS transistors", *IEEE Trans. Electron Devices*, vol. 35, no. 5, pp. 689–697, 1988.

[9.53] M. Tomizawa, K. Yokoyama, and A. Yoshii, "Nonstationary carrier dynamics in quarter-micron Si MOSFETs", *IEEE Trans. Computer-Aided Des.*, vol. CAD-7, pp. 254–258, 1988.

[9.54] P. J. H. Elias, T. G. Roer, and F. M. Klaassen, "Electron velocity overshoot in sub-micron silicon CMOS transistors", in *ESSDERC*, Notthingham, sept 1990.

[9.55] S. E. Laux and M. V. Fischetti, "Monte-Carlo Simulation of Submicrometer Si n-MOSFET's at 77 and 300 K", *IEEE Electron Device Lett.*, vol. 9, no. 9, pp. 467–469, 1988.

[9.56] S. E. Laux and W. Lee, "Collector signal delay in the presence of velocity overshoot", *IEEE Electron Device Lett.*, vol. 11, no. 4, pp. 174–176, 1990.

[9.57] S. E. Laux and M. V. Fischetti, "Monte Carlo study of velocity overshoot in switching a 0.1-micron CMOS inverter", in *IEDM Tech. Dig.*, 1997, pp. 877–880.

[9.58] C. Jungemann and B. Meinerzhagen, "Impact of the velocity overshoot on the performance of NMOSFETs with gate lengths from 80 to 250nm", in *Proc. ESSDERC*, Leuven (Belgium), 1999, pp. 236–239.

[9.59] J.D. Bude, "MOSFET modeling into the ballistic regime", in *Proc. SISPAD*, 2000, pp. 23–26.

[9.60] M. Rodder, M. Hanratty, D. Rogers, T. Laaksonen, J. C. Hu, S. Murtaza, C.-P. Chao, S. Hantangady, S. Aur, A. Amerasekera, and I.-C. Chen, "A 0.1μm gate length CMOS technology with 30Å gate dielectric for 1.0V-1.5V applications", in *IEDM Tech. Dig.*, 1997, pp. 223–226.

[9.61] L. T. Su, H. Hu, J. B. Jacobs, M. J. Sherony, A. Wei, and D. A. Antoniadis, "Tradeoffs of current drive vs. short-channel effect in deep-submicrometer bulk and SOI MOSFETs", in *IEDM Tech. Dig.*, 1994, pp. 649–652.

[9.62] F. Assad, Z. Ren, D. Vasileska, S. Datta, and M. Lundstrom, "On the performance limits for Si MOSFET's: A theoretical study", *IEEE Trans. Electron Devices*, vol. 47, no. 1, pp. 232–240, 2000.

[9.63] R. Thoma, A. Emunds, B. Meinerzhagen, H. J. Peifer, and W. L. Engl, "Hydrodynamic equations for semiconductors with nonparabolic bandstructures", *IEEE Trans. Electron Devices*, vol. 38, no. 6, pp. 1343–1352, 1991.

[9.64] S. A. Mujtaba, R. W. Dutton, and D. L. Scharfetter, "Semi-empirical local NMOS mobility model for 2-D device simulation incorporating screened minority impurity scattering", in *NUPAD Tech. Dig.*, Honolulu, 1994, vol. 5, pp. 3–6.

[9.65] W. Hänsch and M. Miura-Mattausch, "A new current relation for hot electron transport", in *Proc. NASECODE IV*, 1985, pp. 311–314.

[9.66] S. Takagi, A. Toriumi, M. Iwase, and H. Tango, "On the universality of inversion layer mobility in Si MOSFET's: Part I–Effects of substrate impurity concentration", *IEEE Trans. Electron Devices*, vol. 41, pp. 2357–2362, 1994.

[9.67] J.A. Cooper and D. F.Nelson, "High-field drift velocity of electrons at the Si-SiO_2 interface as determined by a time-of-flight technique", *J. Appl. Phys.*, vol. 54, pp. 1445–1456, 1983.
[9.68] D. F. Nelson and J. A. Cooper, "High Field Surface Drift Velocity in Silicon", in *The Physics of Submicron Structures.* Plenum Press, New York, 1984.
[9.69] H. K. Gummel, "On the definition of the cutoff frequency f_T", *Proc. IEEE*, p. 2159, 1969.
[9.70] S. M. Sze, *Physics of Semiconductors Devices*, Wiley, New York, 1981.
[9.71] C. Jungemann, "Methoden zur Simulation hochenergetischer Elektronen in Ultrakurzkanaltransistoren", Doctor thesis, RWTH Aachen, Aachen, 1995, Shaker.
[9.72] T. Vogelsang and R. K. Hofmann, "Electron transport in strained Si layers on $Si_{1-x}Ge_x$ substrates", *Appl. Phys. Lett.*, vol. 63, pp. 186–188, 1993.
[9.73] F. M. Bufler, P. Graf, S. Keith, and B. Meinerzhagen, "Full band Monte Carlo investigation of electron transport in strained Si grown on $\mathrm{Si}_{1-x}\mathrm{Ge}_x$ substrates", *Appl. Phys. Lett.*, vol. 70, pp. 2144–2146, 1997.
[9.74] S. Keith, F. M. Bufler, and B. Meinerzhagen, "Full band Monte–Carlo device simulation of an 0.1 μm n-channel MOSFET in strained silicon material", in *Proc. ESSDERC*, Stuttgart, Sept. 1997, pp. 200–203.
[9.75] M. V. Fischetti, F. Gamiz, and W. Hänsch, "On the enhanced electron mobility in strained-silicon inversion layers", *J. Appl. Phys.*, vol. 92, pp. 7320–7324, 2002.
[9.76] Y. Taur, C. H. Wann, and D. J. Frank, "25 nm CMOS design considerations", *IEDM Tech. Dig.*, pp. 789–792, 1998.
[9.77] S. Keith, C. Jungemann, and B. Meinerzhagen, "Full band Monte Carlo device simulation of 0.1 – 0.5 μm strained–Si P–MOSFETs", in *Proc. ESSDERC*, Bordeaux, 1998, pp. 312–315.
[9.78] F. M. Bufler, P. Graf, and B. Meinerzhagen, "High-field hole transport in strained Si and SiGe by Monte Carlo simulation: Full band versus analytic band models", *VLSI Design*, vol. 8, pp. 41–45, 1998.
[9.79] F. M. Bufler, "Full–band Monte Carlo simulation of electrons and holes in strained Si and SiGe", Dissertation, Universität Bremen, Bremen, 1997, (H. Utz Verlag Wissenschaft, München: 1998).
[9.80] C. Jungemann, S. Keith, and B. Meinerzhagen, "Full-band Monte Carlo simulation of a 0.12μm-Si-PMOSFET with and without a strained SiGe-channel", in *IEDM Tech. Dig.*, San Francisco (USA), 1998, pp. 897–900.
[9.81] H. Fischer, "Transporteigenschaften von $Si_{1-x}Ge_x$–p–MOSFETs", Dissertation, Universität der Bundeswehr München, München, 1999.
[9.82] T. H. Ning, P. W. Cook, R. H. Dennard, C. M. Osburn, S. E. Schuster, and H.-N. Yu, "1 μm MOSFET VLSI technology: Part iv—Hot-electron design constrains", *IEEE Trans. Electron Devices*, vol. 26, no. 4, pp. 346–353, 1979.
[9.83] E. Takeda and N. Suzuki, "An empirical model for device degradation due to hot-carrier injection", *IEEE Electron Device Lett.*, vol. 4, no. 4, pp. 111–113, 1983.
[9.84] C. Hu, S.C. Tam, F.-C. Hsu, P.-K. Ko, T.-Y. Chan, and K. W. Terrill, "Hot-electron induced MOSFET degradation - model, monitor, and improvement", *IEEE Trans. Electron Devices*, vol. 32, no. 2, pp. 375–385, 1985.
[9.85] J. M. Higman, K. Hess, C. G. Hwang, and R. W. Dutton, "Coupled Monte Carlo-drift diffusion analysis of hot-electron effects in MOSFET's", *IEEE Trans.*

Electron Devices, vol. 36, no. 5, pp. 930–937, 1989.

[9.86] M. V. Fischetti, S. E. Laux, and E. Crabbe, "Understanding hot-electron transport in silicon devices: Is there a shortcut?", *J. Appl. Phys.*, vol. 78, pp. 1058–1087, 1995.

[9.87] C. Jungemann, S. Yamaguchi, and H. Goto, "On the accuracy and efficiency of substrate current calculations for sub-μm n-MOSFET's", *IEEE Electron Device Lett.*, vol. 17, no. 10, pp. 464–466, 1996.

[9.88] I. Izawa, M. Katsube, Y. Yohoyama, K. Hashimoto, E. Kawamura, A. Shimizi, H. Takagi, F. Inoue, H. Shimizu, K. Furumochi, H. Goto, S. Kawamura, K. Watanabe, and K. Aoyama, "A novel embedded SRAM technology with 10 μm^2 full–CMOS cells for 0.25 μm logic devices", in *IEDM Tech. Dig.*, 1994, pp. 941–943.

[9.89] N. Khalil, J. Faricelli, D. Bell, and S. Selberherr, "The extraction of two-dimensional MOS transistor doping via inverse modeling", *IEEE Electron Device Lett.*, vol. 16, no. 1, pp. 17–19, 1995.

[9.90] C. Jungemann, S. Yamaguchi, and H. Goto, "Accurate prediction of hot-carrier effects for a deep sub-μm CMOS technology based on inverse modeling and full band Monte Carlo device simulation", in *Proc. SISPAD*, 1996, vol. 1, pp. 59–60.

[9.91] H. Goto, S. Yamaguchi, and C. Jungemann, "Inverse modeling as a basis for predictive device simulation of deep submicron metal-oxide-semiconductor field effect transistors", *Jpn. J. Appl. Phys.*, vol. 37, pp. 5437–5443, 1998.

[9.92] A. Pacelli and U. Ravaioli, "Analysis of variance-reduction schemes for ensemble Monte-Carlo simulation of semiconductor devices", *Solid–State Electron.*, vol. 41, pp. 599–605, 1997.

[9.93] J. D. Bude and M. Mastrapasqua, "Impact ionization and distribution functions in sub-micron nMOSFET technologies", *IEEE Electron Device Lett.*, vol. 16, no. 10, pp. 439–441, 1995.

[9.94] C. Jungemann, S. Yamaguchi, and H. Goto, "Efficient full band Monte Carlo hot carrier simulation for silicon devices", in *Proc. ESSDERC*, Bologna, 1996, vol. 26, pp. 821–824.

[9.95] C. Jungemann, S. Yamaguchi, and H. Goto, "Is there experimental evidence for a difference between surface and bulk impact ionization in silicon", in *IEDM Tech. Dig.*, 1996, pp. 383–386.

[9.96] C. Jungemann, B. Meinerzhagen, S. Decker, S. Keith, S. Yamaguchi, and H. Goto, "Is physically sound and predictive modeling of NMOS substrate currents possible?", *Solid–State Electron.*, vol. 42, pp. 647–655, 1998.

[9.97] W. Jacobs, "The SATURN Technology CAD system", in *Technology CAD Systems*, F. Fasching, S. Halama, and S. Selberherr, Eds. Springer, Wien, New York, 1993.

[9.98] A. Schütz, S. Selberherr, and H. W. Pötzl, "A two-dimensional model of the avalanche effect in MOS transistors", *Solid–State Electron.*, vol. 25, pp. 177–183, 1982.

[9.99] J. W. Slotboom, G. Streutker, G. J. T. Davids, and P. B. Hartog, "Surface impact ionization in Silicon devices", in *IEDM*, 1987, pp. 494–497.

[9.100] M. Fukuma and W. W. Lui, "MOSFET substrate current model including energy transport", *IEEE Electron Device Lett.*, vol. 8, no. 5, pp. 214–216, 1987.

[9.101] J. W. Slotboom, G. Streutker, M. J. v. Dort, P. H. Woerlee, A. Pruijmboom, and

D. J. Gravesteijn, "Non-local impact ionization in silicon devices", in *IEDM*, 1991, pp. 127–130.

[9.102] M. Takeda, J. Frey, Z. Peng, and N. Goldsman, "Simulation of substrate current characteristics of submicron MOSFETs", *Electronics Letters*, vol. 27, pp. 144–146, 1991.

[9.103] V. M. Agostinelli, Jr., T. J. Bordelon, X. Wang, K. Hasnat, C.-F. Yeap, D. B. Lemersal, Jr., A. F. Tasch, and C. M. Mazier, "Two-dimensional energy-dependent models for the simulation of substrate current in submicron MOSFETs", *IEEE Trans. Electron Devices*, vol. 41, no. 10, pp. 1784–1795, 1994.

[9.104] T. Thurgate and N. Chan, "An impact ionization model for two-dimensional device simulation", *IEEE Trans. Electron Devices*, vol. 32, no. 2, pp. 400–404, 1985.

[9.105] R. Kuhnert, C. Werner, and A. Schütz, "A novel impact-ionization model for 1-μm-MOSFET simulation", *IEEE Trans. Electron Devices*, vol. 32, no. 6, pp. 1057–1063, 1985.

[9.106] B. Meinerzhagen, "Consistent gate and substrate current modeling based on energy transport and the Lucky Electron concept", in *IEDM Tech. Dig.*, 1988, pp. 504–507.

[9.107] C. Jungemann, R. Thoma, and W. L. Engl, "A soft threshold lucky electron model for efficient and accurate numerical device simulation", *Solid-State Electron.*, vol. 39, pp. 1079–1086, 1996.

[9.108] A. G. Chynoweth, "Uniform silicon $p-n$ junctions. II. Ionization rates for electrons", *J. Appl. Phys.*, vol. 31, pp. 1161–1165, 1960.

[9.109] R. van Overstraeten and H. de Man, "Measurement of the ionization rates in diffused silicon p-n junctions", *Solid-State Electron.*, vol. 13, pp. 583–608, 1970.

[9.110] W. Maes, K. de Meyer, and R. Van Overstraeten, "Impact ionization in silicon: A review and update", *Solid-State Electron.*, vol. 33, pp. 705–718, 1990.

[9.111] W. Shockley, "Problems related to $p-n$ junctions in silicon", *Solid-State Electron.*, vol. 2, pp. 35–67, 1961.

[9.112] J. M. Higman, I. C. Kizilyalli, and K. Hess, "Nonlocality of the electron ionization coefficient in n-MOSFET's: An analytic approach", *IEEE Electron Device Lett.*, vol. 9, no. 8, pp. 399–401, 1988.

[9.113] K. Taniguchi, M. Yamaji, K. Sonoda, T. Kunikiyo, and C. Hamaguchi, "Monte Carlo study of impact ionization phenomena in small geometry MOSFET's", in *IEDM Tech. Dig.*, 1994, pp. 355–358.

[9.114] E. Sangiorgi, B. Riccò, and F. Venturi, "MOS^2: An efficient Monte Carlo simulator for MOS devices", *IEEE Trans. Computer-Aided Des.*, vol. 7, no. 2, pp. 259–271, 1988.

[9.115] F. M. Bufler, A. Schenk, and W. Fichtner, "Efficient Monte Carlo device modeling", *IEEE Trans. Electron Devices*, vol. 47, no. 10, pp. 1891–1897, 2000.

[9.116] N. Sano, "Increasing importance of electronic thermal noise in sub-0.1μm Si-MOSFETs", *IEICE Trans. on Electronics*, vol. E83-C, no. 8, pp. 1203–1211, 2000.

[9.117] F. Venturi, R. K. Smith, E. C. Sangiorgi, M. R. Pinto, and B. Riccò, "A general purpose device simulation coupling Poisson and Monte Carlo transport with application to deep submicron MOSFET's", *IEEE Trans. Computer-Aided Des.*, vol. 8, no. 4, pp. 360–369, 1989.

[9.118] C. Jungemann and B. Meinerzhagen, "On the applicability of nonself-consistent Monte Carlo device simulations", *IEEE Trans. Electron Devices*, vol. 49, no. 6, pp. 1072–1074, 2002.

[9.119] B. Neinhüs, S. Decker, P. Graf, F. M. Bufler, and B. Meinerzhagen, "Consistent hydrodynamic and Monte–Carlo simulation of SiGe HBTs based on table models for the relaxation times", *VLSI Design*, vol. 8, pp. 387–391, 1998.

[9.120] H.-J. Peifer, "Monte–Carlo Simulation des Hochenergietransports von Elektronen in submikron MOS–Strukturen", Doctor thesis, RWTH Aachen, Aachen, 1992, Augustinus Buchhandlung.

[9.121] B. Meinerzhagen, "Two-dimensional numerical substrate current modeling for n-channel MOS transistors", in *Proc. NASECODE V*, 1987, pp. 42–59.

[9.122] J. D. Bude, M. R. Pinto, and S. K. Smith, "Monte Carlo simulation of the CHISEL flash memory cell", *IEEE Trans. Electron Devices*, vol. 47, no. 11, pp. 1873–1881, 2000.

[9.123] N. Sano, K. Matsuzawa, M. Mukai, and N. Nakayama, "Influence of thermal noise on drain current in very small Si-MOSFETs", *Jpn. J. Appl. Phys.*, vol. 39, pp. 1974–1978, 2000.

[9.124] H. K. Gummel, "A self-consistent iterative scheme for one-dimensional steady state transistor calculations", *IEEE Trans. Electron Devices*, pp. 455–465, October 1964.

[9.125] Y.-J. Park, D. H. Navon, and T.-W. Tang, "Monte Carlo simulation of bipolar transistors", *IEEE Trans. Electron Devices*, vol. 31, no. 12, pp. 1724–1730, 1984.

[9.126] C. Jungemann, B. Neinhüs, and B. Meinerzhagen, "Comparative study of electron transit times evaluated by DD, HD, and MC device simulation for a SiGe HBT", *IEEE Trans. Electron Devices*, vol. 48, no. 10, pp. 2216–2220, 2001.

[9.127] P. W. Rambo and J. Denavit, "Time stability of Monte Carlo device simulation", *IEEE Trans. Computer–Aided Des.*, vol. 12, pp. 1734–1741, 1993.

[9.128] I. Bork, C. Jungemann, B. Meinerzhagen, and W. L. Engl, "Influence of heat flux on the accuracy of hydrodynamic models for ultrashort Si MOSFETs", in *NUPAD Tech. Dig.*, Honolulu, 1994, vol. 5.

[9.129] C. Jungemann, B. Neinhüs, and B. Meinerzhagen, "Spatial analysis of the electron transit time in a silicon/germanium heterojunction bipolar transistor by drift-diffusion, hydrodynamic, and full-band Monte Carlo device simulation", in *Proc. SISPAD*, Seattle (USA), Sept. 2000, pp. 42–45.

[9.130] J. J. H. van den Biesen, "A simple regional analysis of transient times in bipolar transistors", *Solid–State Electron.*, vol. 29, pp. 529–534, 1986.

[9.131] C. Jungemann, B. Neinhüs, S. Decker, and B. Meinerzhagen, "Hierarchical 2–D DD and HD noise simulations of Si and SiGe devices: Part II—Results", *IEEE Trans. Electron Devices*, vol. 49, no. 7, pp. 1258–1264, 2002.

[9.132] D. Knoll, B. Heinemann, H. J. Osten, K. E. Ehwald, B. Tillack, P. Schley, R. Barth, M. Matthes, K. S. Park, Y. Kim, and W. Winkler, "Si/SiGe:C heterojunction bipolar transistors in an Epi-Free well, single-polysilicon technology", in *IEDM Tech. Dig.*, 1998, pp. 703–706.

[9.133] D. L. Scharfetter, "Measured dependence of lifetime upon defect density and temperature in depletion layers of epitaxial silicon diodes", in *Solid-State Dev. Res. Conf.*, Santa Barbara, USA, 1967.

[9.134] S. Decker, "Numerische Simulation von Si/SiGe–Hochfrequenztransistoren unter besonderer Berücksichtigung des elektronischen Rauschens", Dissertation, Universität Bremen, Aachen, 2002.

[9.135] F. Bonani and G. Ghione, "Generation–recombination noise modelling in semiconductor devices through population or approximate equivalent current density fluctuations", *Solid–State Electron.*, vol. 43, pp. 285–295, 1999.

[9.136] S. P. Voinigescu, M. C. Maliepaard, J. L. Showell, G. E. Babcock, D. Marchesan, M. Schroter, P. Schvan, and D. L. Harame, "A scalable high-frequency noise model for bipolar transistors with application to optimal transistor sizing for low-noise amplifier design", *IEEE Solid-State Circuits*, vol. 32, no. 9, pp. 1430–1439, 1997.

[9.137] F. Herzel and B. Heinemann, "High-frequency noise analysis of Si/SiGe heterojunction bipolar transistors", *Int. J. Electronics*, vol. 81, pp. 37–48, 1996.

[9.138] K. Aufinger and M. Reisch, "RF noise for bipolar transistors-a critical comparison", in *Proc. BCTM*, San Francisco, USA, 2001, pp. 110–113.

[9.139] M. J. Martín-Martínez, S. Pérez, D. Pardo, and T. González, "High injection effects on noise characteristics of Si BJTs and SiGe HBTs", *Microelectronics Reliability*, vol. 41, pp. 847–854, 2001.

[9.140] M. J. Martín, S. Pérez, D. Pardo, and T. González, "Monte Carlo analysis of the noise behavior in Si BJTs and SiGe HBTs at RF frequencies", *J. Appl. Phys.*, vol. 90, no. 3, pp. 1582–1588, 2001.

[9.141] M. J. Martín-Martínez, S. Pérez, D. Pardo, and T. González, "Influence of Ge profile on the noise behavior of SiGe HBTs under high injection conditions", *Physica B*, vol. 314, pp. 381–385, 2002.

[9.142] H. Beneking, *High Speed Semiconductor Devices*, Chapman & Hall, London, 1994.

[9.143] S. E. Laux and K. Hess, "Revisiting the analytic theory of p-n junction impedance: Improvements guided by computer simulation leading to a new equivalent circuit", *IEEE Trans. Electron Devices*, vol. 46, no. 2, pp. 396–412, 1999.

[9.144] G. Niu, J. D. Cressler, S. Zhang, W. E. Ansley, C. S. Webster, and D. L. Harame, "A unified approach to RF and microwave noise parameter modeling in bipolar transistors", *IEEE Trans. Electron Devices*, vol. 48, no. 11, pp. 2568–2574, 2001.

Subject Index

SpringerEngineering

Andreas Schenk

Advanced Physical Models for Silicon Device Simulation

1998. XVIII, 349 pages. 125 figures.
Hardcover **EUR 133,–**
(Recommended retail price)
Net-price subject to local VAT.
ISBN 3-211-83052-9
Computational Microelectronics

The book contains an extensive review on physical models for silicon device simulators. Referencing fundamental experimental and theoretical work, the models are discussed in terms of physical accuracy and application results. The derivation of device models from "first principles" and the fundamental problems therein are then described in detail for the most important examples using some new approaches of "taylored quantum mechanics".

"... this is a well produced book, written in a easy to read style, and will also be a very useful primer for someone starting out in the field who wants to know what can and cannot be done, and a useful source of reference for experienced users ..."

Microelectronics Journal

Springer-Verlag Wien GmbH

A-1201 Wien, Sachsenplatz 4–6, P.O. Box 89, Fax +43.1.330 24 26, e-mail: books@springer.at, Internet: **www.springer.at**
D-69126 Heidelberg, Haberstraße 7, Fax +49.6221.345-4229, e-mail: orders@springer.de
USA, Secaucus, NJ 07096-2485, P.O. Box 2485, Fax +1.201.348-4505, e-mail: orders@springer-ny.com
Eastern Book Service, Japan, Tokyo 113, 3–13, Hongo 3-chome, Bunkyo-ku, Fax +81.3.38 18 08 64, e-mail: orders@svt-ebs.co.jp

Springer-Verlag and the Environment

WE AT SPRINGER-VERLAG FIRMLY BELIEVE THAT AN international science publisher has a special obligation to the environment, and our corporate policies consistently reflect this conviction.

WE ALSO EXPECT OUR BUSINESS PARTNERS – PRINTERS, paper mills, packaging manufacturers, etc. – to commit themselves to using environmentally friendly materials and production processes.

THE PAPER IN THIS BOOK IS MADE FROM NO-CHLORINE pulp and is acid free, in conformance with international standards for paper permanency.

MIX
Papier aus verantwortungsvollen Quellen
Paper from responsible sources
FSC® C105338

If you have any concerns about our products,
you can contact us on
ProductSafety@springernature.com

In case Publisher is established outside the EU,
the EU authorized representative is:
Springer Nature Customer Service Center GmbH
Europaplatz 3, 69115 Heidelberg, Germany

Printed by Libri Plureos GmbH
in Hamburg, Germany